Block Copolymers with Crystallizable Blocks: Synthesis, Self-Assembly and Applications

Block Copolymers with Crystallizable Blocks: Synthesis, Self-Assembly and Applications

Editors

Holger Schmalz
Volker Abetz

MDPI • Basel • Beijing • Wuhan • Barcelona • Belgrade • Manchester • Tokyo • Cluj • Tianjin

Editors
Holger Schmalz
Macromolecular Chemistry II
University of Bayreuth
Bayreuth
Germany

Volker Abetz
Institute of Membrane Research
Helmholtz-Zentrum Hereon
Geesthacht
Germany

Editorial Office
MDPI
St. Alban-Anlage 66
4052 Basel, Switzerland

This is a reprint of articles from the Special Issue published online in the open access journal *Polymers* (ISSN 2073-4360) (available at: www.mdpi.com/journal/polymers/special_issues/blo_copoly_crystal).

For citation purposes, cite each article independently as indicated on the article page online and as indicated below:

LastName, A.A.; LastName, B.B.; LastName, C.C. Article Title. *Journal Name* **Year**, *Volume Number*, Page Range.

ISBN 978-3-0365-3326-1 (Hbk)
ISBN 978-3-0365-3325-4 (PDF)

© 2022 by the authors. Articles in this book are Open Access and distributed under the Creative Commons Attribution (CC BY) license, which allows users to download, copy and build upon published articles, as long as the author and publisher are properly credited, which ensures maximum dissemination and a wider impact of our publications.

The book as a whole is distributed by MDPI under the terms and conditions of the Creative Commons license CC BY-NC-ND.

Contents

About the Editors vii

Holger Schmalz and Volker Abetz
Block Copolymers with Crystallizable Blocks: Synthesis, Self-Assembly and Applications
Reprinted from: *Polymers* **2022**, *14*, 696, doi:10.3390/polym14040696 1

Christian Hils, Ian Manners, Judith Schöbel and Holger Schmalz
Patchy Micelles with a Crystalline Core: Self-Assembly Concepts, Properties, and Applications
Reprinted from: *Polymers* **2021**, *13*, 1481, doi:10.3390/polym13091481 7

Naisheng Jiang and Donghui Zhang
Solution Self-Assembly of Coil-Crystalline Diblock Copolypeptoids Bearing Alkyl Side Chains
Reprinted from: *Polymers* **2021**, *13*, 3131, doi:10.3390/polym13183131 33

Brahim Bessif, Thomas Pfohl and Günter Reiter
Self-Seeding Procedure for Obtaining Stacked Block Copolymer Lamellar Crystals in Solution
Reprinted from: *Polymers* **2021**, *13*, 1676, doi:10.3390/polym13111676 57

Gerald Guerin, Paul A. Rupar and Mitchell A. Winnik
In-Depth Analysis of the Effect of Fragmentation on the Crystallization-Driven Self-Assembly Growth Kinetics of 1D Micelles Studied by Seed Trapping
Reprinted from: *Polymers* **2021**, *13*, 3122, doi:10.3390/polym13183122 67

Zehua Li, Amanda K. Pearce, Andrew P. Dove and Rachel K. O'Reilly
Precise Tuning of Polymeric Fiber Dimensions to Enhance the Mechanical Properties of Alginate Hydrogel Matrices
Reprinted from: *Polymers* **2021**, *13*, 2202, doi:10.3390/polym13132202 83

Md. Mushfequr Rahman
Membrane Separation of Gaseous Hydrocarbons by Semicrystalline Multiblock Copolymers: Role of Cohesive Energy Density and Crystallites of the Polyether Block
Reprinted from: *Polymers* **2021**, *13*, 4181, doi:10.3390/polym13234181 95

Eider Matxinandiarena, Agurtzane Múgica, Manuela Zubitur, Viko Ladelta, George Zapsas and Dario Cavallo et al.
Crystallization and Morphology of Triple Crystalline Polyethylene-*b*-poly(ethylene oxide)-*b*-poly(ε-caprolactone) PE-*b*-PEO-*b*-PCL Triblock Terpolymers
Reprinted from: *Polymers* **2021**, *13*, 3133, doi:10.3390/polym13183133 109

Nicolás María, Jon Maiz, Daniel E. Martínez-Tong, Angel Alegria, Fatimah Algarni and George Zapzas et al.
Phase Transitions in Poly(vinylidene fluoride)/Polymethylene-Based Diblock Copolymers and Blends
Reprinted from: *Polymers* **2021**, *13*, 2442, doi:10.3390/polym13152442 135

Claudio De Rosa, Rocco Di Girolamo, Alessandra Cicolella, Giovanni Talarico and Miriam Scoti
Double Crystallization and Phase Separation in Polyethylene—Syndiotactic Polypropylene Di-Block Copolymers
Reprinted from: *Polymers* **2021**, *13*, 2589, doi:10.3390/polym13162589 163

Nicole Janoszka, Suna Azhdari, Christian Hils, Deniz Coban, Holger Schmalz and André H. Gröschel
Morphology and Degradation of Multicompartment Microparticles Based on Semi-Crystalline Polystyrene-*block*-Polybutadiene-*block*-Poly(L-lactide) Triblock Terpolymers
Reprinted from: *Polymers* **2021**, *13*, 4358, doi:10.3390/polym13244358 . . . . **177**

About the Editors

Holger Schmalz

Holger Schmalz studied chemistry at the University of Bayreuth, Germany, and completed his PhD with Volker Abetz in 2002 on thermoplastic elastomers based on semicrystalline block copolymers. He then joined Ticona GmbH in Frankfurt, Germany, working on the cationic polymerisation of trioxane and impact-modified polyoxymethylene. Since 2004, he is a staff scientist (Akademischer Oberrat) in the department of Macromolecular Chemistry II at the University of Bayreuth. His research interests include block copolymer synthesis via living/controlled polymerisation methods, block copolymer self-assembly in bulk and confinement, and the crystallization-driven self-assembly (CDSA) of block copolymers with crystallizable blocks.

Volker Abetz

Volker Abetz received his diploma from University of Freiburg in 1987. He undertook a PhD in spectroscopic polarimetry of polymeric multicomponent systems with Reimund Stadler (Freiburg) and Gerald G. Fuller (Stanford University), graduating in 1990. In the following years, he worked on polymer blends with Erhard W. Fischer (Max Planck Institute for Polymer Research, Mainz), on interpenetrating polymer networks with Guy C. Meyer (Institute Charles Sadron and University of Strasbourg), and on block copolymers with Reimund Stadler and Axel H.E. Müller at the Universities of Mainz and Bayreuth. He received his habilitation in 2000. In 2004, he joined the University of Potsdam as professor of polymer chemistry. Later, he was appointed as the director of the Institute of Polymer Research at Helmholtz-Zentrum Geesthacht (now Institute of Membrane Research at Helmholtz-Zentrum Hereon) in combination with a professorship for multicomponent polymer systems at the University of Kiel. Since 2012, he has been a professor of physical chemistry at the University of Hamburg joint with his position at Hereon. His research interests include the chemistry and physics of polymers, self-assembly and membranes.

Editorial

Block Copolymers with Crystallizable Blocks: Synthesis, Self-Assembly and Applications

Holger Schmalz [1,*] and Volker Abetz [2,3,*]

1. Macromolecular Chemistry II and Bavarian Polymer Institute, Universität Bayreuth, Universitätsstraße 30, 95440 Bayreuth, Germany
2. Institute of Physical Chemistry, Universität Hamburg, Grindelallee 117, 20146 Hamburg, Germany
3. Institute of Membrane Research, Helmholtz-Zentrum Hereon, Max-Planck-Straße 1, 21502 Geesthacht, Germany
* Correspondence: holger.schmalz@uni-bayreuth.de (H.S.); volker.abetz@hereon.de (V.A.)

Abstract: Block copolymers with crystallizable blocks are a highly interesting class of materials owing to their unique self-assembly behaviour both in bulk and solution. This Special Issue brings together new developments in the synthesis and self-assembly of semicrystalline block copolymers and also addresses potential applications of these exciting materials.

Citation: Schmalz, H.; Abetz, V. Block Copolymers with Crystallizable Blocks: Synthesis, Self-Assembly and Applications. *Polymers* 2022, *14*, 696. https://doi.org/10.3390/polym14040696

Received: 31 January 2022
Accepted: 8 February 2022
Published: 11 February 2022

Publisher's Note: MDPI stays neutral with regard to jurisdictional claims in published maps and institutional affiliations.

Copyright: © 2022 by the authors. Licensee MDPI, Basel, Switzerland. This article is an open access article distributed under the terms and conditions of the Creative Commons Attribution (CC BY) license (https://creativecommons.org/licenses/by/4.0/).

Block copolymers bearing one or more crystallizable blocks have moved into the focus of current research owing to their unique self-assembly behaviour both in bulk and in solution. The bulk morphology and, hence, the properties of semicrystalline block copolymers are influenced by a complex interplay between crystallization and micro phase separation. Depending on the segregation strength (confinement) in the melt, crystallization can either be confined in the pre-existing microphase-separated morphology for strongly segregated melts, whereas for weakly segregated systems, a "breakout crystallization" can occur, which overwrites any existing morphology leading exclusively to lamellar structures [1–9]. This opens a broad parameter space for tuning the properties of semicrystalline block copolymers in bulk. First studies on semicrystalline AB diblock, ABA triblock and multiblock copolymers with one crystallizable block based on poly(ethylene oxide) (PEO), polyester blocks like poly(ε-caprolactone) (PCL), or polyethylene (PE, based on hydrogenated poly(1,4-butadiene)) have already been reported in the mid-1970s to 1980s [10–17]. An important milestone in this field was the development of ABC triblock terpolymers with one or two crystallizable blocks based on polystyrene-*block*-poly(1,4-butadiene)-*block*-poly(ε-caprolactone) (SBC) and the corresponding hydrogenated analogues with PE middle blocks (SEC), reported first by the group of *R. Stadler* in 1996 and 1998, respectively, and intensively studied thereafter together with the group of *A. J. Müller* [18–22]. Shortly after, in 1998, *Floudas* et al. reported on the first μ-ABC miktoarm star terpolymer with two crystallizable PEO and PCL blocks [23]. An important and technically highly relevant application of block copolymers with crystallizable blocks are thermoplastic elastomers. Here, ABC triblock terpolymers with a glassy polystyrene and a semicrystalline PE end block were shown to exhibit superior elastic properties compared to conventional amorphous ABA-type thermoplastic elastomers at moderate deformations [24,25]. Additionally, commercially available multiblock copolymers with semicrystalline polyamide or polyester hard blocks and polyether-based soft blocks are well-known thermoplastic elastomers and have inspired the development of more complex multiblock copolymers with improved elasticity employing well-defined ABA triblock copolymers as soft segments [26]. Some of the semicrystalline multiblock copolymers containing polyether segments, especially poly(ethylene oxide) segments, also attracted interest for gas separation membranes [27–29]. New synthetic concepts give access to even more complex block copolymer architectures such as triblock or tetrablock copolymers with three or even four different crystallizable blocks, [30,31] as well as to the implementation of new semicrystalline blocks, e.g., poly(vinylidene fluoride) (PVDF) [32]. Some of these recent developments are addressed in this Special Issue.

Crystallization-driven self-assembly (CDSA) of block copolymers with one core-forming, crystallizable block has developed to an extremely active and innovative field of research, starting from the first observation of defined cylindrical micelles with crystalline poly(ferrocenyl dimethylsilane) (PFS) cores in 1998 [33] and following the development of living CDSA in the groups of I. Manners and M. A. Winnik [34–42]. This paved the way to a myriad of crystalline-core micellar structures and hierarchical super-structures that were not accessible before via the self-assembly of fully amorphous block copolymers, e.g., cylindrical micelles with defined length, length distribution, and corona chemistries (block type or patchy corona), branched micelles, non-centrosym metric cylindrical micelles, and fascinating micellar superstructures (e.g., 2D lenticular platelets, scarf-shaped micelles, multidimensional micellar assemblies, cross and "wind mill"-like supermicelles). Another intriguing material class is based on amphiphilic crystalline-core micelles with poly(L-lactide) (PLLA) or corresponding stereocomplexes (PLLA/PDLA (poly(D-lactide)), showing interesting potential for biomedical applications, such as controlled release and drug delivery [43,44].

This Special Issue brings together new developments in the synthesis and self-assembly (bulk and solution) of block copolymers with crystallizable blocks, including emerging applications of these exciting materials. In a fundamental work, *Rahman* studied the use of semicrystalline multiblock copolymer membranes with polyether soft segments for hydrocarbon separation [45]. The permeability of hydrocarbons was found to decrease with the number of carbons and polytetrahydrofuran (PTHF)-based systems were superior to PEO-based systems in terms of permeability and permselectivity, making these systems interesting for applications in the petrochemical industry. In addition, the lower performance of multiblock copolymers with longer PEO soft segments was attributed to partial PEO crystallization. The combination of homologation (C1 polymerization) with ring-opening (ROP) or iodine transfer polymerization (ITP) is a facile route to block copolymers with polymethylene (structurally identical to PE) blocks. *Hadjichristidis and Müller* et al. utilized this approach to synthesize PE-b-PEO-b-PCL triblock terpolymers, in which all three blocks are able to crystallize. Here, PE crystallizes first upon cooling from the phase-separated melt followed by PCL and PEO [46]. They note that a combination of different characterization techniques (DSC, WAXS, PLOM) is necessary to fully deduce the complex behaviour of triple crystalline triblock terpolymers. In a joint work with *Maiz* et al. phase transitions in PE-b-PVDF diblock copolymers and blends were studied [47]. Due to the polymorphic nature of semicrystalline PVDF control over crystal structure is crucial, as for example the piezoelectric and ferroelectric β-phase is interesting for applications in electronic devices or renewable energies.Compared to PVDF homopolymer the formation of the β-phase was found to be strongly promoted in PE-b-PVDF diblock copolymers at low cooling rates. Living, stereoselective olefin polymerization is an efficient method for the synthesis of double crystalline diblock copolymers with PE and sPP (syndiotactic polypropylene) blocks, as described by *De Rosa* et al. [48]. By using selective crystalline substrates for the epitaxial crystallization of PE (benzoic acid) and sPP (p-terphenyl), well-ordered morphologies with crystalline lamellae of PE and sPP highly oriented along one direction are accessible. A relatively new approach is the evaporation-induced confinement assembly (EICA) of semicrystalline block copolymers in microemulsions that after solvent evaporation gives rise to microparticles with confinement specific morphologies, e.g., helical cylinders or axially stacked rings. In this context, *Gröschel and Schmalz* et al. have studied the confinement assembly of a series of PS-b-PB-b-PLLA triblock terpolymers [49]. It turned out that over a broad composition range, microparticles with predominantly hexagonally packed core–shell cylinders consisting of a PLLA core, a PB shell and a PS matrix were formed, which upon hydrolysis of the PLLA block resulted in highly porous microparticles with pronounced surface corrugations.

Considering the CDSA of block copolymers with crystallizable blocks in solution, this Special Issue includes two reviews focussing on the preparation and application of micelles with a patch-like microphase-separated (patchy) corona [50], as well as on glycine-based diblock copolypeptoids [51], respectively. Patchy micelles can be prepared by CDSA of triblock terpolymers with crystallizable middle blocks and two incompatible amorphous end blocks, or from mixtures of diblock copolymers with one common crystallizable block. Owing to their unique corona structure, patchy micelles can be utilized as highly efficient surfactants and blend compatibilizers, as nanoparticle templates, and in heterogeneous catalysis. Polypeptoids with N-substituted polyglycine backbones are a promising class of materials as, in contrast to natural polypeptides, they provide a good thermal stability, solubility in organic solvents and protease stability. This can be attributed to the absence of hydrogen bonding and stereogenic centres. Crystallinity can be easily tuned by the length of the alkyl substituents, giving rise to bioinspired worm-like 1D nanofibrils, nanorods and nanosheets. In an intriguing study, *Reiter* et al. prepared stacked lamellar crystals from a PS-*b*-PEO diblock copolymer in solution using a self-seeding approach [52]. By varying the diblock copolymer concentration and employed self-seeding temperature control over size, the number of platelet-like crystals and even the number of stacked lamellae in the crystals was achieved. A seed-trapping protocol was developed by *Guerin and Winnik* et al. to study the impact of seed fragmentation on CDSA to cylindrical micelles at elevated temperatures, where both seed dissolution and fragmentation occur [53]. Seed fragmentation was found to increase with annealing time at elevated temperatures, resulting in a decrease in length of the regrown cylindrical micelles. Furthermore, kinetics follow a stretched exponential that might indicate a fractionation upon crystallization as the rate of unimer addition to the seeds depends on the length and fraction of the crystallizable block. Finally, going toward potential applications in tissue engineering, a systematic study on the reinforcement of alginate hydrogel matrices with fibre-like micelles prepared by living CDSA of a PCL-*b*-PMMA-*b*-PDMA triblock terpolymer (PMMA = poly(methyl methacrylate); PDMA = poly(N,N-dimethyl acrylamide)) is presented by *Dove and O'Reilly* et al. [54]. Varying the micelle length and concentration in the hydrogel revealed an optimum fibre micelle length of 500 nm at a loading of 0.1 wt%, resulting in a significantly increased strain at flow of 37%.

In summary, the manuscripts in this Special Issue provide a nice overview of the recent developments in block copolymers with crystallizable blocks, spanning from synthesis to self-assembly approaches and potential applications.

Conflicts of Interest: The authors declare no conflict of interest.

References

1. Loo, Y.L.; Register, R.A.; Ryan, A.J. Modes of crystallization in block copolymer microdomains: Breakout, templated, and confined. *Macromolecules* **2002**, *35*, 2365–2374. [CrossRef]
2. Müller, A.J.; Balsamo, V.; Arnal, M.L.; Jakob, T.; Schmalz, H.; Abetz, V. Homogeneous nucleation and fractionated crystallization in block copolymers. *Macromolecules* **2002**, *35*, 3048–3058. [CrossRef]
3. van Horn, R.M.; Steffen, M.R.; O'Connor, D. Recent progress in block copolymer crystallization. *Polym. Cryst.* **2018**, *1*, 10039. [CrossRef]
4. Sangroniz, L.; Wang, B.; Su, Y.; Liu, G.; Cavallo, D.; Wang, D.; Müller, A.J. Fractionated crystallization in semicrystalline polymers. *Prog. Polym. Sci.* **2021**, *115*, 101376. [CrossRef]
5. Loo, Y.-L.; Register, R.A. Crystallization within block copolymer mesophases. In *Developments in Block Copolymer Science and Technology*; John Wiley & Sons, Ltd.: Chichester, UK, 2004; pp. 213–243.
6. Li, S.; Register, R.A. Crystallization in copolymers. In *Handbook of Polymer Crystallization*; John Wiley & Sons, Inc.: Hoboken, NJ, USA, 2013; pp. 327–346.
7. Huang, S.; Jiang, S. Structures and morphologies of biocompatible and biodegradable block copolymers. *RSC Adv.* **2014**, *4*, 24566–24583. [CrossRef]
8. Castillo, R.V.; Müller, A.J. Crystallization and morphology of biodegradable or biostable single and double crystalline block copolymers. *Prog. Polym. Sci.* **2009**, *34*, 516–560. [CrossRef]
9. He, W.N.; Xu, J.T. Crystallization assisted self-assembly of semicrystalline block copolymers. *Prog. Polym. Sci.* **2012**, *37*, 1350–1400. [CrossRef]

10. Herman, J.-J.; Jérome, R.; Teyssié, P.; Gervais, M.; Gallot, B. Structural study on styrene/ε-caprolactone block copolymers in absence and in presence of a solvent of the polystyrene block. *Die Makromol. Chem.* **1981**, *182*, 997–1008. [CrossRef]
11. Gervais, M.; Gallot, B. Structural study of polybutadiene-poly(ethylene oxide) block copolymers. Influence of the nature of the amorphous block on the refolding of the poly(ethylene oxide) chains. *Die Makromol. Chem.* **1977**, *178*, 1577–1593. [CrossRef]
12. Morton, M.; Lee, N.-C.; Terrill, E.R. Elastomeric polydiene ABA triblock copolymers with crystalline end blocks. In *ACS Symposium Series*; Mark, J.E., Lal, J., Eds.; American Chemical Society: Washington, DC, USA, 1982; pp. 101–118.
13. O'Malley, J.J.; Stauffer, W.J. Morphology and properties of crystalline polyester-siloxane block copolymers. *Polym. Eng. Sci.* **1977**, *17*, 510–514. [CrossRef]
14. Hirata, E.; Ijitsu, T.; Soen, T.; Hashimoto, T.; Kawai, H. Domain structure and crystalline morphology of AB and ABA type block copolymers of ethylene oxide and isoprene cast from solutions. *Polymer* **1975**, *16*, 249–260. [CrossRef]
15. Robitaille, C.; Prud'homme, J. Thermal and mechanical properties of a poly(ethylene oxide-b-isoprene-b-ethylene oxide) block polymer complexed with NaSCN. *Macromolecules* **1983**, *16*, 665–671. [CrossRef]
16. Heuschen, J.; Jérôme, R.; Teyssié, P. Polycaprolactone-based block copolymers. II. Morphology and crystallization of copolymers of styrene or butadiene and ε-caprolactone. *J. Polym. Sci. Part B Polym. Phys.* **1989**, *27*, 523–544. [CrossRef]
17. Donth, E.; Kretzschmar, H.; Schulze, G.; Garg, D.; Höring, S.; Ulbricht, J. Influence of the chain-end mobility on the melt crystallization of the ethylene oxide (B) sequences in systems containing diblock AB and triblock ABA copolymers with methyl methacrylate (A). *Acta Polym.* **1987**, *38*, 260–270. [CrossRef]
18. Balsamo, V.; von Gyldenfeldt, F.; Stadler, R. Synthesis of SBC, SC and BC block copolymers based on polystyrene (S), polybutadiene (B) and a crystallizable poly(ε-caprolactone) (C) block. *Macromol. Chem. Phys.* **1996**, *197*, 1159–1169. [CrossRef]
19. Balsamo, V.; von Gyldenfeldt, F.; Stadler, R. Thermal behavior and spherulitic superstructures of SBC triblock copolymers based on polystyrene (S), polybutadiene (B) and a crystallizable poly(ε-caprolactone) (C) block. *Macromol. Chem. Phys.* **1996**, *197*, 3317–3341. [CrossRef]
20. Balsamo, V.; von Gyldenfeldt, F.; Stadler, R. "Superductile" semicrystalline ABC triblock copolymers with the polystyrene block (A) as the matrix. *Macromolecules* **1999**, *32*, 1226–1232. [CrossRef]
21. Balsamo, V.; Stadler, R. Ellipsoidal core-shell cylindrical microphases in PS-b-PB-b-PCL triblock copolymers with a crystallizable matrix. *Macromol. Symp.* **1997**, *117*, 153–165. [CrossRef]
22. Balsamo, V.; Müller, A.J.; von Gyldenfeldt, F.; Stadler, R. Ternary ABC block copolymers based on one glassy and two crystallizable blocks: Polystyrene-*block*-polyethylene-*block*-poly(ε-caprolactone). *Macromol. Chem. Phys.* **1998**, *199*, 1063–1070. [CrossRef]
23. Floudas, G.; Reiter, G.; Lambert, O.; Dumas, P. Structure and dynamics of structure formation in model triarm star block copolymers of polystyrene, poly(ethylene oxide), and poly(ε-caprolactone). *Macromolecules* **1998**, *31*, 7279–7290. [CrossRef]
24. Schmalz, H.; Böker, A.; Lange, R.; Krausch, G.; Abetz, V. Synthesis and properties of ABA and ABC triblock copolymers with glassy (A), elastomeric (B), and crystalline (C) blocks. *Macromolecules* **2001**, *34*, 8720–8729. [CrossRef]
25. Schmalz, H.; Abetz, V.; Lange, R. Thermoplastic elastomers based on semicrystalline block copolymers. *Compos. Sci. Technol.* **2003**, *63*, 1179–1186. [CrossRef]
26. Schmalz, H.; van Guldener, V.; Gabriëlse, W.; Lange, R.; Abetz, V. Morphology, surface structure, and elastic properties of PBT-based copolyesters with PEO-b-PEB-b-PEO triblock copolymer soft segments. *Macromolecules* **2002**, *35*, 5491–5499. [CrossRef]
27. Rahman, M.M.; Shishatskiy, S.; Abetz, C.; Georgopanos, P.; Neumann, S.; Khan, M.M.; Filiz, V.; Abetz, V. Influence of temperature upon properties of tailor-made PEBAX® MH 1657 nanocomposite membranes for post-combustion CO_2 capture. *J. Memb. Sci.* **2014**, *469*, 344–354. [CrossRef]
28. Rahman, M.M.; Lillepärg, J.; Neumann, S.; Shishatskiy, S.; Abetz, V. A thermodynamic study of CO_2 sorption and thermal transition of PolyActive™ under elevated pressure. *Polymer* **2016**, *93*, 132–141. [CrossRef]
29. Rahman, M.M.; Abetz, C.; Shishatskiy, S.; Martin, J.; Müller, A.J.; Abetz, V. CO_2 selective PolyActive membrane: Thermal transitions and gas permeance as a function of thickness. *ACS Appl. Mater. Interfaces* **2018**, *10*, 26733–26744. [CrossRef]
30. Ladelta, V.; Zapsas, G.; Abou-Hamad, E.; Gnanou, Y.; Hadjichristidis, N. Tetracrystalline tetrablock quarterpolymers: Four different crystallites under the same roof. *Angew. Chem. Int. Ed.* **2019**, *58*, 16267–16274. [CrossRef]
31. Palacios, J.K.; Mugica, A.; Zubitur, M.; Iturrospe, A.; Arbe, A.; Liu, G.; Wang, D.; Zhao, J.; Hadjichristidis, N.; Müller, A.J. Sequential crystallization and morphology of triple crystalline biodegradable PEO-b-PCL-b-PLLA triblock terpolymers. *RSC Adv.* **2016**, *6*, 4739–4750. [CrossRef]
32. Voet, V.S.D.; Tichelaar, M.; Tanase, S.; Mittelmeijer-Hazeleger, M.C.; ten Brinke, G.; Loos, K. Poly(vinylidene fluoride)/nickel nanocomposites from semicrystalline block copolymer precursors. *Nanoscale* **2013**, *5*, 184–192. [CrossRef]
33. Massey, J.; Power, K.N.; Manners, I.; Winnik, M.A. Self-assembly of a novel organometallic-inorganic block copolymer in solution and the solid state: Nonintrusive observation of novel wormlike poly(ferrocenyldimethylsilane)-b-poly(dimethyl siloxane) micelles. *J. Am. Chem. Soc.* **1998**, *120*, 9533–9540. [CrossRef]
34. MacFarlane, L.; Zhao, C.; Cai, J.; Qiu, H.; Manners, I. Emerging applications for living crystallization-driven self-assembly. *Chem. Sci.* **2021**, *12*, 4661–4682. [CrossRef] [PubMed]
35. Ganda, S.; Stenzel, M.H. Concepts, fabrication methods and applications of living crystallization-driven self-assembly of block copolymers. *Prog. Polym. Sci.* **2020**, *101*, 101195. [CrossRef]
36. Li, X.; Gao, Y.; Harniman, R.; Winnik, M.; Manners, I. Hierarchical assembly of cylindrical block comicelles mediated by spatially confined hydrogen-bonding interactions. *J. Am. Chem. Soc.* **2016**, *138*, 12902–12912. [CrossRef] [PubMed]

37. Li, X.; Gao, Y.; Boott, C.E.; Hayward, D.W.; Harniman, R.; Whittell, G.R.; Richardson, R.M.; Winnik, M.A.; Manners, I. "Cross" supermicelles via the hierarchical assembly of amphiphilic cylindrical triblock comicelles. *J. Am. Chem. Soc.* **2016**, *138*, 4087–4095. [CrossRef]
38. Crassous, J.J.; Schurtenberger, P.; Ballauff, M.; Mihut, A.M. Design of block copolymer micelles via crystallization. *Polymer* **2015**, *62*, A1–A13. [CrossRef]
39. Jia, L.; Zhao, G.; Shi, W.; Coombs, N.; Gourevich, I.; Walker, G.C.; Guerin, G.; Manners, I.; Winnik, M.A. A design strategy for the hierarchical fabrication of colloidal hybrid mesostructures. *Nat. Commun.* **2014**, *5*, 3882. [CrossRef]
40. Rupar, P.A.; Chabanne, L.; Winnik, M.A.; Manners, I. Non-centrosymmetric cylindrical micelles by unidirectional growth. *Science* **2012**, *337*, 559–562. [CrossRef]
41. Gilroy, J.B.; Gädt, T.; Whittell, G.R.; Chabanne, L.; Mitchels, J.M.; Richardson, R.M.; Winnik, M.A.; Manners, I. Monodisperse cylindrical micelles by crystallization-driven living self-assembly. *Nat. Chem.* **2010**, *2*, 566–570. [CrossRef]
42. Gädt, T.; Ieong, N.S.; Cambridge, G.; Winnik, M.A.; Manners, I. Complex and hierarchical micelle architectures from diblock copolymers using living, crystallization-driven polymerizations. *Nat. Mater.* **2009**, *8*, 144–150. [CrossRef]
43. Petzetakis, N.; Dove, A.P.; O'Reilly, R.K. Cylindrical micelles from the living crystallization-driven self-assembly of poly(lactide)-containing block copolymers. *Chem. Sci.* **2011**, *2*, 955–960. [CrossRef]
44. Inam, M.; Cambridge, G.; Pitto-Barry, A.; Laker, Z.P.L.; Wilson, N.R.; Mathers, R.T.; Dove, A.P.; O'Reilly, R.K. 1D vs. 2D shape selectivity in the crystallization-driven self-assembly of polylactide block copolymers. *Chem. Sci.* **2017**, *8*, 4223–4230. [CrossRef] [PubMed]
45. Rahman, M.M. Membrane separation of gaseous hydrocarbons by semicrystalline multiblock copolymers: Role of cohesive energy density and crystallites of the polyether block. *Polymers* **2021**, *13*, 4181. [CrossRef] [PubMed]
46. Matxinandiarena, E.; Múgica, A.; Zubitur, M.; Ladelta, V.; Zapsas, G.; Cavallo, D.; Hadjichristidis, N.; Müller, A.J. Crystallization and morphology of triple crystalline polyethylene-*b*-poly(ethylene oxide)-*b*-poly(ε-caprolactone) PE-*b*-PEO-*b*-PCL triblock terpolymers. *Polymers* **2021**, *13*, 3133. [CrossRef] [PubMed]
47. María, N.; Maiz, J.; Martínez-Tong, D.E.; Alegria, A.; Algarni, F.; Zapzas, G.; Hadjichristidis, N.; Müller, A.J. Phase transitions in poly(vinylidene fluoride)/polymethylene-based diblock copolymers and blends. *Polymers* **2021**, *13*, 2442. [CrossRef]
48. De Rosa, C.; Di Girolamo, R.; Cicolella, A.; Talarico, G.; Scoti, M. Double crystallization and phase separation in polyethylene—syndiotactic polypropylene di-block copolymers. *Polymers* **2021**, *13*, 2589. [CrossRef]
49. Janoszka, N.; Azhdari, S.; Hils, C.; Coban, D.; Schmalz, H.; Gröschel, A.H. Morphology and degradation of multicompartment microparticles based on semi-crystalline polystyrene-*block*-polybutadiene-*block*-poly(L-lactide) triblock terpolymers. *Polymers* **2021**, *13*, 4358. [CrossRef]
50. Hils, C.; Manners, I.; Schöbel, J.; Schmalz, H. Patchy micelles with a crystalline core: Self-assembly concepts, properties, and applications. *Polymers* **2021**, *13*, 1481. [CrossRef]
51. Jiang, N.; Zhang, D. Solution self-assembly of coil-crystalline diblock copolypeptoids bearing alkyl side chains. *Polymers* **2021**, *13*, 3131. [CrossRef]
52. Bessif, B.; Pfohl, T.; Reiter, G. Self-seeding procedure for obtaining stacked block copolymer lamellar crystals in solution. *Polymers* **2021**, *13*, 1676. [CrossRef]
53. Guerin, G.; Rupar, P.A.; Winnik, M.A. In-depth analysis of the effect of fragmentation on the crystallization-driven self-assembly growth kinetics of 1D micelles studied by seed trapping. *Polymers* **2021**, *13*, 3122. [CrossRef]
54. Li, Z.; Pearce, A.K.; Dove, A.P.; O'Reilly, R.K. Precise tuning of polymeric fiber dimensions to enhance the mechanical properties of alginate hydrogel matrices. *Polymers* **2021**, *13*, 2202. [CrossRef] [PubMed]

Review

Patchy Micelles with a Crystalline Core: Self-Assembly Concepts, Properties, and Applications

Christian Hils [1], Ian Manners [2], Judith Schöbel [3,*] and Holger Schmalz [1,4,*]

1. Macromolecular Chemistry II, University of Bayreuth, Universitätsstraße 30, 95440 Bayreuth, Germany; christian.hils@uni-bayreuth.de
2. Department of Chemistry, University of Victoria, 3800 Finnerty Road, Victoria, BC V8P 5C2, Canada; manners@uvic.ca
3. Fraunhofer Institute for Applied Polymer Research IAP, Geiselbergstraße 69, 14476 Potsdam-Golm, Germany
4. Bavarian Polymer Institute (BPI), University of Bayreuth, Universitätsstraße 30, 95440 Bayreuth, Germany
* Correspondence: judith.schoebel@iap.fraunhofer.de (J.S.); holger.schmalz@uni-bayreuth.de (H.S.)

Abstract: Crystallization-driven self-assembly (CDSA) of block copolymers bearing one crystallizable block has emerged to be a powerful and highly relevant method for the production of one- and two-dimensional micellar assemblies with controlled length, shape, and corona chemistries. This gives access to a multitude of potential applications, from hierarchical self-assembly to complex superstructures, catalysis, sensing, nanomedicine, nanoelectronics, and surface functionalization. Related to these applications, patchy crystalline-core micelles, with their unique, nanometer-sized, alternating corona segmentation, are highly interesting, as this feature provides striking advantages concerning interfacial activity, functionalization, and confinement effects. Hence, this review aims to provide an overview of the current state of the art with respect to self-assembly concepts, properties, and applications of patchy micelles with crystalline cores formed by CDSA. We have also included a more general discussion on the CDSA process and highlight block-type co-micelles as a special type of patchy micelle, due to similarities of the corona structure if the size of the blocks is well below 100 nm.

Keywords: crystallization-driven self-assembly (CDSA); crystalline-core micelles; patchy micelles; block copolymers

1. Introduction

The solution self-assembly of block copolymers (BCPs) has paved the way to a vast number of micellar assemblies of various shapes (e.g. spheres, cylinders, vesicles, platelets, core-shell, core-shell-corona, and compartmentalized (core or corona) structures) and hierarchical superstructures, as well as hybrids with fascinating applications in drug delivery and release, as emulsifiers/blend compatibilizers, in nanoelectronics, as responsive materials (temperature, pH, light), templates for nanoparticles, in heterogeneous catalysis, etc. [1–6]. A key prerequisite for controlling/programming the solution self-assembly is the synthesis of well-defined diblock and triblock (linear, star-shaped, ABA- or ABC-type) copolymers via controlled or living polymerization techniques, such as living anionic polymerization, reversible addition−fragmentation chain transfer, nitroxide-mediated, and atom transfer radical polymerization [5–9]. In general, anisotropic polymer micelles can be divided into three main categories: multicompartment core micelles (MCMs), surface-compartmentalized micelles, and a combination of both [2]. MCMs are generally defined as micellar assemblies with a solvophilic corona and a microphase-separated solvophobic core. According to the suggestion of Laschewsky et al., a key feature of multicompartment micelles is that the various sub-domains in the micellar core feature substantially different properties to behave as separate compartments [10,11]. MCMs are commonly prepared via hierarchical self-assembly of suitable building blocks, which provide "sticky patches" [12–15]. Depending on the number and geometrical arrangement (linear, triangular, tetrahedral, etc.) of the "sticky patches", as well as the volume fraction of the solvophilic block, various spherical, cylindrical, sheet-like, and vesicular MCMs are accessible [16–25]. For a deeper insight into this highly relevant topic, the reader is referred to recent extensive reviews on MCMs [26–31]. Surface-compartmentalized micelles are subdivided into micelles with a Janus-type (two opposing faces with different chemistry or polarity) or patch-like, microphase-separated corona, featuring several compartments of different chemistry or polarity (denoted as patchy micelles), as illustrated in Figure 1 for cylindrical micelles. Here, block co-micelles with a block-like arrangement of several (>2) surface compartments along the cylindrical long axis can be regarded as a special case of patchy micelles. It is noted that AB-type diblock co-micelles also represent Janus-type micelles, where the two opposing faces are arranged perpendicular to the cylindrical long axis. The broken symmetry of Janus particles offers efficient and distinctive means of targeting complex materials by hierarchical self-assembly and realize unique properties and applications, like particulate surfactants, optical nanoprobes, biosensors, self-propulsion, and many more [32–41].

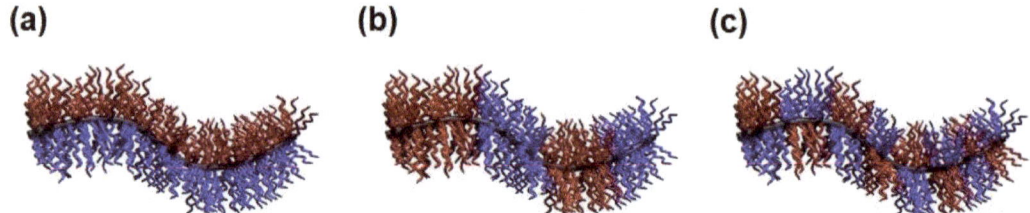

Figure 1. Schematic depiction of a cylindrical (**a**) Janus micelle, (**b**) block co-micelle, and (**c**) patchy micelle.

For the preparation of patchy micelles and polymersomes from amorphous BCPs, three main strategies can be applied: (i) self-assembly of ABC triblock terpolymers in selective solvents for the incompatible A and C blocks [42–48]; (ii) co-assembly of AB and CD diblock copolymers with selective interactions between the B and C blocks (e.g. hydrogen bonding, ionic interactions, solvophobic interactions) [49–52], resulting in patchy micelles with an insoluble mixed B/C core; and (iii) co-assembly of AB and BC diblock copolymers [53–56] where the B block forms the insoluble core. However, mostly spherical micelles or polymersomes with a patchy corona have been reported and only a few reports describe the preparation of one-dimensional (worm-like, cylindrical) assemblies with a patch-like compartmentalized corona, even though theoretical work on mixed polymer brushes predict their existence [57–61]. One of the rare but highly intriguing examples are P*t*BA–*b*–PCEMA–*b*–PGMA (poly(*tert*-butyl acrylate)–*block*–poly(2-cinnamoyloxyethyl methacrylate)–*block*–poly(glyceryl monomethacrylate)) and P*n*BA–*b*–PCEMA–*b*–P*t*BA (P*n*BA: poly(*n*-butyl acrylate)) triblock terpolymers [42,43,45]. For self-assembly, the triblock terpolymers were first dissolved in a good solvent for all blocks (CH_2Cl_2, $CHCl_3$, or THF), followed by the addition of methanol (non-solvent for the middle block) to induce micelle formation. As an intermediate, cylindrical micelles with a patchy corona were formed first, with the P*t*BA blocks forming small circular patches in a corona mainly consisting of PGMA or P*n*BA. Upon further decreasing the solvent quality for the P*t*BA block (addition of MeOH), these cylinders can form double and triple helices via hierarchical self-assembly. This concept has also been applied to triblock terpolymers with a poly(2-hydroxyethyl methacrylate) middle block, having the potential for further modification by esterification of the pendant hydroxy functions [42]. Besides, crystallization-driven self-assembly (CDSA) is a highly versatile tool for the preparation of well-defined cylindrical micelles of controlled length and length distribution, and has proven as a valuable method for the preparation of patchy cylindrical micelles.

This review will focus on the recent developments concerning self-assembly strategies for the production of crystalline-core micelles (CCMs) bearing a patchy corona, and will also address their unique properties and potential applications. As stated above, block co-micelles represent a special case of patchy micelles and thus, will be discussed only briefly. This is not only due to the usually larger size and sequential arrangement of surface compartments in the corona, in contrast to the more alternating arrangement in patchy cylindrical micelles (Figure 1b,c), but is also attributed to the substantially different self-assembly procedure. Block co-micelles are commonly prepared by sequential living CDSA of different diblock copolymers, whereas patchy micelles are formed by simultaneous CDSA of diblock copolymer mixtures or CDSA of ABC triblock terpolymers with crystallizable middle blocks. Hence, this review will be divided into four main sections, starting with a general consideration of CDSA. The second part gives a compact overview over self-assembly strategies used to form cylindrical and platelet-like block co-micelles. The different self-assembly concepts for patchy micelles with crystalline cores will be reviewed in the third section, followed by a discussion on properties and applications of these interesting compartmentalized nanostructures.

2. Crystallization-Driven Self-Assembly (CDSA)

As pointed out in the introduction, the preparation of one-dimensional (1D) cylindrical (or worm-like) micelles with controlled dimensions, low-length dispersities, and tailored corona structures and functionalities still remains a challenge in the self-assembly of fully amorphous BCPs. Besides, the introduction of a crystallizable block, which adds an additional and strong driving force for micelle formation, has turned out to be a highly efficient route to solve these issues. Consequently, the self-assembly of such BCPs, bearing crystallizable blocks, is termed crystallization-driven self-assembly (CDSA) [1,62,63]. This field was pioneered by studies on poly(ferrocenyl dimethylsilane) (PFS)-containing BCPs and is gaining increasing importance for the preparation of well-defined 1D and two-dimensional (2D) assemblies, especially since the discovery of living CDSA (Figure 2) [63–67]. Analogous to the living polymerization of monomers, CDSA can proceed in a living manner, employing small micellar fragments as seeds (Figure 2a: seeded growth) for the addition of unimers (molecularly dissolved BCPs with a crystalizable block). In this approach, the micellar seeds, also termed "stub-like" micelles, are produced by vigorous sonication of long, polydisperse cylindrical micelles prepared by conventional CDSA. Owing to its living nature, the length of the produced cylindrical micelles shows a linear dependence on the unimer/seed ratio employed, and length dispersities are very low (L_w/L_n typically well below 1.1; where L_n is the number average and L_w the weight average micelle length).

Living CDSA can also be realized by using spherical CCMs as seeds [68], by self-seeding [69–71] (Figure 2a), and even directly by polymerization-induced CDSA (Figure 2b) [72–74], i.e., via polymerization in the presence of seed micelles. The self-seeding approach also uses small micellar fragments that are heated in dispersion to a specific annealing temperature (T_a), where most of the crystalline core is molten/dissolved and only a very minor fraction of crystallites survive. These act as seeds in the subsequent CDSA upon cooling (Figure 2a: self-seeding), and the length of the micelles can be controlled by a proper choice of T_a. If T_a is too low, the crystalline cores will not melt/dissolve, and the length distribution of the employed micellar fragments remains unchanged. On the other hand, if T_a is too high, the crystalline cores will melt/dissolve completely, and no crystallites will survive that could act as seeds. As a result, in between these two limiting cases, an increase in micelle length with increasing T_a is observed, as the fraction of surviving crystallites (seeds) decreases with T_a. This range of self-seeding temperatures can be very restricted, making length control difficult. Another drawback of these seed-based protocols is the low amount of cylindrical micelles that can be produced, as commonly rather dilute solutions have to be used. This can be overcome by the living polymerization-induced CDSA approach, enabling the production of uniform cylindrical micelles with concentrations up to ca. 10–20% (w/w solids) within a few hours. In a recent report, it was shown that living CDSA can even be stimulated by light, utilizing the photo-induced cis-trans isomerization in oligo(*p*-phenylenevinylene) (OPV)-based BCPs [75].

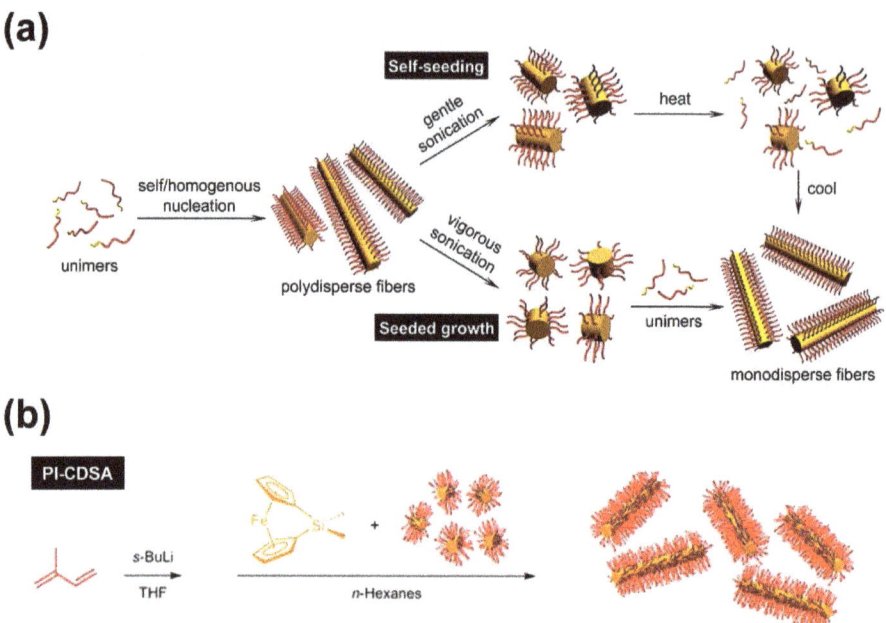

Figure 2. (**a**) Concepts for living CDSA, enabling the production of cylindrical micelles with defined length and narrow length distribution. Self-seeding employing seeds produced by thermal treatment of micelle fragments (top) and seeded growth using small micellar fragments ("stub"-like micelles) as seeds (bottom). (**b**) Living polymerization-induced CDSA (PI-CDSA) utilizing micellar seeds during anionic polymerization of the PFS block. After complete conversion, the reaction was quenched with 4-*tert*-butylphenol. (**a**) Reproduced from [76] with permission of the American Chemical Society (ACS).

Living CDSA has paved the way to a myriad of 1D and 2D micellar assemblies of controlled dimensions, including patchy and block co-micelles (both will be addressed in the next sections) [65,68,77–80], branched micelles [76], platelet-like micelles and co-micelles [81–86], and hierarchical assemblies [81,87–91]. Next to BCPs with a PFS block, a variety of other crystallizable polymer blocks were employed in CDSA, e.g. polyethylene (PE) [68,92–94], poly(ethylene oxide) [95], polyesters (poly(ε-caprolactone) (PCL) or poly(*L*-lactide) (PLLA)) [86,96–101], polycarbonate [102], poly(2-*iso*-propyl-2-oxazoline) (P*i*PrOx) [103,104], liquid crystalline polymers [71,105], poly(vinylidene fluoride) [106], polypeptoids [107,108], and various conjugated polymers (e.g., poly(3-hexyl thiophene) (P3HT) and OPV) [75,109–113].

3. Short Excursion on Block Co-Micelles

Block co-micelles represent a special type of patchy CCM, because of the sequential arrangement of surface compartments and the precisely adjustable size of the blocks, usually leading to larger corona segments than commonly observed for patchy CCMs. Analogous to the synthesis of BCPs, block co-micelles are produced by sequential living CDSA. The characteristic of this process is that the micelles' termini remain "active" after unimer addition is completed. Consequently, addition of a different type of unimer leads to the formation of a blocky structure (Figure 3a) [65,114]. This feature allows for precise control over the block length by adjusting the amount of added unimer.

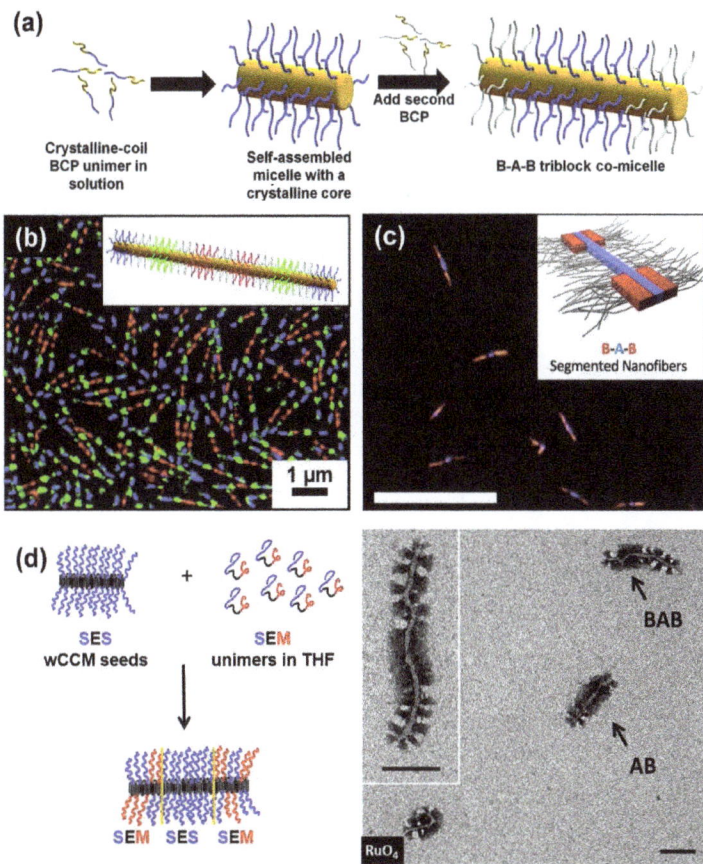

Figure 3. (a) Formation of B–A–B triblock co-micelles via sequential living CDSA in selective solvents for the corona blocks. (b) Structured illumination microscopy image of symmetrical 11-block co-micelles with red, green, and blue fluorescent corona blocks separated by non-fluorescent PDMS spacer blocks. (c) Laser-scanning confocal microscopy image of solid-state, donor–acceptor, coaxial heterojunction nanowires based on B–A–B segmented nanofibers with a semi-crystalline PDHF core (depicted in blue) and a semi-crystalline P3EHT shell (depicted in red) in the outer corona blocks, taken with both blue (PDHF) and red (P3EHT) channels (scale bar: 10 μm). Blue emission from the central PDHF core, as well as red/orange emission from the outer P3EHT segments, due to Förster resonance energy transfer (FRET) were observed. (d) Schematic depiction of the formation of B–A–B triblock co-micelles with patchy outer corona blocks, starting from SES wCCMs as seed micelles and subsequent living CDSA of SEM unimers in THF (left) and corresponding TEM image of patchy block co-micelles (scale bar: 100 nm). (a) Reproduced from [79] with permission of the American Association for the Advancement of Science (AAAS), (b) reproduced from [67] with permission of the Royal Society of Chemistry (RSC), (c) reproduced from [105], and (d) reproduced from [68] with permission of ACS.

Similar to living polymerization techniques, in which the reactivity of the first monomer limits the choice of a second monomer, unimers need to fulfill certain requirements for successful co-crystallization. For example, the micellar cores need to be compatible for epitaxial crystallization, i.e., they should exhibit a similar crystal lattice spacing [115,116]. A common way to fulfill this prerequisite is the use of diblock copolymers bearing the same crystallizable block that induces homoepitaxial growth, as shown first for PFS-containing diblock copolymers to produce B–A–B triblock co-micelles [65]. Within this work, PFS–*b*–polyisoprene (PFS–*b*–PI) cylindrical micelles served as seeds for the nucleation of PFS–*b*–polymethylvinylsilane (PFS–*b*–PMVS) and PFS–*b*–polydimethylsiloxane (PFS–*b*–PDMS) unimers, respectively. For heteroepitaxial growth, different PFS-containing seed micelles were applied to induce CDSA of polyferrocenylgermane (PFG)-containing diblock copolymers [73,83,89]. The crystal lattice spacing of the two core-forming blocks only differs by about 6%, enabling the formation of tri- and pentablock co-micelles as well as 2D co-assemblies.

Living CDSA has opened the door to a huge variety of one dimensional, PFS-containing block co-micelles with tailored numbers, lengths, and composition of corona blocks [114,117–121]. Centrosymmetric and non-centrosymmetric block co-micelles are accessible, and give rise to broad structural complexity [79]. In particular, the introduction of fluorescent corona blocks marks an important step in the development of block co-micelles, since this enables the formation of barcode and RGB micelles (Figure 3b) [67,77,122]. Up to that point, the fabrication of cylindrical nanomaterials with precise, color-tunable compartments of predictable length and number was challenging. Moreover, it is possible to induce fluorescence in the semicrystalline core-forming block by replacing the PFS block by a poly(di-*n*-hexylfluorene) (PDHF) block [78]. B–A–B triblock co-micelles with PDHF core and P3HT outer corona blocks were found to show long-range exciton transport (>200 nm). Inducing secondary crystallization of a poly(3-(2'-hexylethyl)thiophene) (P3EHT) corona block even rendered solid-state donor–acceptor heterojunctions possible (Figure 3c) [105]. These materials bear a high potential for applications in optoelectronics, device fabrication, and sensing [123].

Several other semicrystalline, core-forming blocks—for example, PFG [73,83,89], polycarbonate [102,124], poly(3-heptylselenophene) [109], P3HT [125], OPV [75,126,127], PLLA [128], and PE [68]—were used for the production of block co-micelles. As an example, sequential living CDSA of a polystyrene-*block*-polyethylene-*block*-polystyrene (PS–*b*–PE–*b*–PS; SES) triblock copolymer with a PS–*b*–PE–*b*–PMMA (SEM; PMMA: poly(methyl methacrylate)) triblock terpolymer yielded B–A–B- or A–B–A-type triblock co-micelles with patchy outer or inner B blocks, respectively (Figure 3d) [68]. Interestingly, the choice of seed micelles was crucial for the successful formation of triblock co-micelles, as worm-like SES micelles are accessible on both micelle ends for epitaxial growth, whereas patchy, worm-like SEM micelles show diverse growth behavior, which is predefined by the arrangement of the corona chains at the micelles' ends.

The scope of complex micellar assemblies is further extended by hierarchical self-assembly, using block co-micelles as building blocks for the formation of 2D and three-dimensional (3D) superstructures. There are different strategies to realize hierarchical assemblies—for example, coordination-driven co-assembly [129] or dialysis of amphiphilic block co-micelles against selective solvents, enabling highly efficient side-by-side or end-to-end stacking (Figure 4a,b) [88,130,131], or spatially confined hydrogen-bonding interactions [132,133]. The latter opens access to numerous hierarchical 2D morphologies, such as "I"-shaped, cross, shish-kebab (Figure 4c) or windmill-like (Figure 4d) structures, by precisely tailored interactions between hydrogen donor and hydrogen acceptor units within the block co-micelles. However, not only the attractive interactions by hydrogen-bonding have to be taken into account, but also repulsive interactions caused by steric hindrance of the corona chains. To overcome this problem, tuning the length of the hydrogen acceptor blocks has proven to be a suitable solution, rendering 3D assemblies possible. It is noted that 2D platelet-like hierarchical superstructures, as well as more complex micelle architectures like double- and single-headed, spear-like micelles [90], scarf-like micelles [89], diamond-fiber hybrid structures [81], or platelets with various shapes (rectangular, quasi-hexagonal, and diamond platelet micelles) [82–85] are accessible.

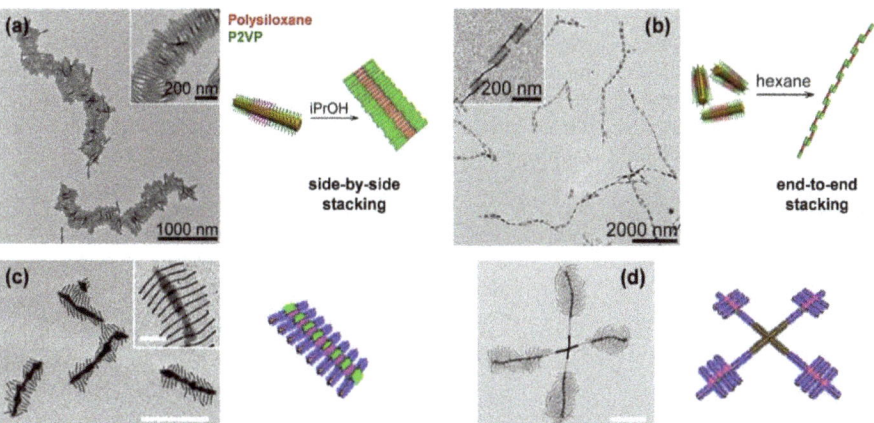

Figure 4. One-dimensional supermicelles by (**a**) side-by-side and (**b**) end-to-end stacking in selective solvents for the outer and middle corona block of B–A–B triblock co-micelles (PFS core), respectively. (**c**) "Shish-kebab" supermicelles (scale bar = 1 µm, inset = 200 nm) formed by hydrogen-bond (H-bond)-mediated co-assembly of an H-bond donor homopolymer (hydroxyl-functionalized poly(vinylmethylsiloxane) (PMVSOH), colored in pink) with B–A–B triblock co-micelles with "neutral" outer corona blocks (poly(*t*-butyl acrylate) (P*t*BA), colored in blue, no H-bond interactions) and an H-bond acceptor middle corona block (P2VP, colored in green). (**d**) "Windmill"-like supermicelles via living CDSA of a PFS-*b*-P*t*BA diblock copolymer from "cross" supermicelles (scale bar = 0.5 µm). The "cross" supermicelles featured an H-bond acceptor corona block (P2VP) at the termini, onto which short CCMs with H-bond donor corona blocks (PMVSOH) that served as seeds for the subsequent living CDSA of the PFS-*b*-P*t*BA diblock copolymer (color code identical to (c)) were immobilized. (**a**,**b**) reproduced from [88] with permission of AAAS, (**c**) reproduced from [133] with permission of ACS, and (**d**) from [132] with permission of Springer Nature.

4. Self-Assembly Concepts for Patchy Micelles with Crystalline Cores

4.1. CDSA of Linear and Star-Shaped Triblock Terpolymers

The most widely used route toward crystalline-core patchy micelles is the CDSA of linear ABC triblock terpolymers with a crystallizable middle block (Table 1) [134]. In contrast to block co-micelles, where the sequential living CDSA of different diblock copolymers results in a block-type segmentation of the corona, the incompatibility of the corona-forming blocks is the driving force for corona segregation in CDSA of triblock terpolymers. This affects the average width of the patches and leads to an alternating, chess-board-like arrangement of the corona patches [135]. Worm-like CCMs (wCCMs) with a patchy corona were first reported in 2008 for triblock terpolymers with a semicrystalline PE middle block and two amorphous outer blocks, namely PS and PMMA (SEM) [93]. Since patchy, worm-like (or cylindrical) CCMs based on these triblock terpolymers have been intensively studied, the self-assembly mechanism will be elucidated in detail on this example.

Initially, the SEM triblock terpolymers are placed in a good solvent for the amorphous blocks and heated above the melting temperature of the semicrystalline PE block in the given solvent (Figure 5a) [94]. Depending on the solvent quality for the PE middle block, different micelle morphologies are formed. In good solvents for the molten PE block (for example, THF or toluene), the triblock terpolymers are molecularly dissolved, i.e., unimers are formed. In bad solvents for PE (for example 1,4-dioxane), the molten PE block collapses, and spherical micelles with an amorphous (molten) PE core are observed. Cooling of the corresponding unimer solution (in good solvents) or dispersion of spherical micelles (bad solvents) results in the nucleation of PE crystallization. In good solvents, the nuclei are stable and able to initiate the bidirectional, 1D epitaxial growth of the remaining unimers to generate wCCMs. However, in bad solvents, the spherical shape of the micelles dictates the final morphology of the CCMs. Consequently, confined crystallization of PE in the respective micellar cores leads to the generation of spherical CCMs. In both cases, the micelle corona exhibits a patch-like, microphase-separated (patchy) structure, whereas for wCCMs the patchy structure of the corona is more pronounced (Figure 5b,c). For wCCMs, an almost alternating arrangement of the PS and PMMA patches in the corona can be deduced from transmission electron microscopy (TEM) [94], and was also confirmed by small-angle neutron scattering studies [135].

Table 1. Overview of self-assembly concepts for patchy micelles with a crystalline core.

Self-Assembly Concept	Employed BCPs	Special Feature	Reference
CDSA of triblock terpolymers			
Linear triblock terpolymers	PS–b–PE–b–PMMA	Control over micelle morphology, length control through seeded growth, co-crystallization with PS–b–PE–b–PS	[93,94,134–136]
	PS–b–PE–b–PDxA [1]	Functional groups for NP incorporation	[137,138]
	PS–b–PFS–b–PMMA	Control over patch size, co-crystallization with diblock co-polymers of varying PS and PMMA block lengths	[139]
	PS–b–PFS–b–PMVS, PI–b–PFS–b–PMMA	Length control through seeded growth	[140]
Star-shaped triblock terpolymers	µ-SIF	Seeded growth, block co-micelles with patchy µ-SIF outer blocks, and middle block based on PFS-b-PDMS	[141]
Non-covalent grafting on carbon nanotubes	PS–b–PE–b–PMMA	Temperature-stable patchy hybrid materials	[142]
Co-assembly of diblock copolymers			
Sterically demanding co-unimers	PFS–b–PMVS, PFS–b–PMVS(C18) [2]	Gradual coassembly of linear and brush-type BCPs	[143]
Strong difference in Flory–Huggins interaction parameters of corona chains	PFS–b–PDMS, PFS–b–PMVS, PFS–b–PI	Different patch arrangements accessible (helical, hemispherical)	[144]
Manipulation of the epitaxial growth rate or the critical dissolution temperature	PFS–b–P2VP, PFS–b–PNiPAM, PFS–b–P2VPQ [3]	Patchy or blocky structures accessible	[80,145]
Addition of crystallizable homopolymer, heating–cooling-aging approach	PFS, PFS–b–PDMS, PFS–b–PI, PFS–b–PMVS, PFS–b–P2VP	PFS crystal fragments serve as seeds, patchy or blocky structures, easy up-scaling	[146]

[1] PDxA: poly(N,N-dialkylaminoethyl methacrylamide). [2] PMVS block alkylated by C18 alkyl chains. [3] Quaternized P2VP.

A facile way to tailor the sizes of the PS and PMMA corona patches is random co-crystallization of an SEM triblock terpolymer with a corresponding SES triblock copolymer, bearing two PS end blocks [136]. A systematic increase of the SES fraction led to a decrease of the PMMA patch size (Figure 7a). Thus, this approach allows to tune the corona structure by a simple co-assembly without the need to synthesize new triblock terpolymers for each desired corona composition. Another efficient way to modify the corona patches is the introduction of functional groups via selective amidation of the PMMA block in SEM triblock terpolymers with different N,N-dialkylethylenediamines [137,138]. CDSA in THF led to patchy wCCMs, for which the patch size and shape could be tuned by varying the block length ratio of the corona blocks (Figure 7b,c) and selective solvent interactions. The functionalized, patchy corona enables an application of these wCCMs as templates for the incorporation of inorganic nanoparticles (NPs), which will be discussed in detail in Section 5.2.

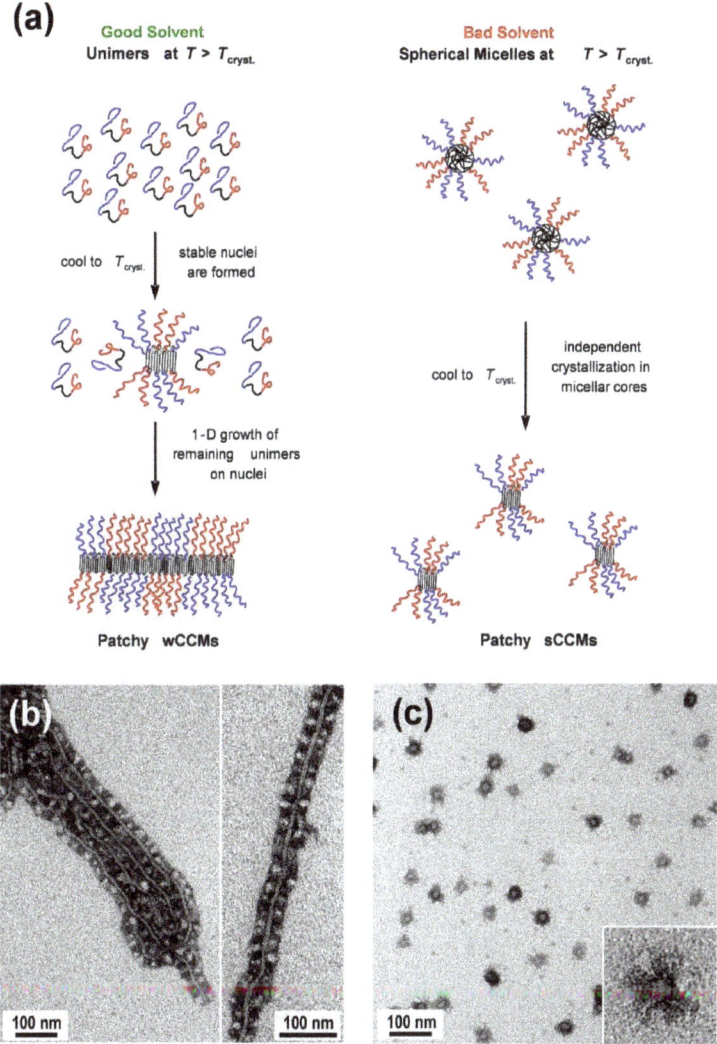

Figure 5. (a) Schematic representation of the proposed mechanism for the formation of patchy worm-like and spherical crystalline-core micelles (wCCMs and sCCMs, respectively) from SEM triblock terpolymers (PS blocks are represented in blue, PE in black, and PMMA in red). TEM images of (b) patchy $S_{340}E_{700}M_{360}$ wCCMs prepared by CDSA in THF and subsequent annealing at 45 °C for 3 h, and (c) patchy $S_{340}E_{700}M_{360}$ sCCMs formed in dimethylacetamide (subscripts denote the respective average degrees of polymerization, PS was selectively stained with RuO_4 vapor and appears dark). Reproduced from [94] with permission of ACS.

The patchy corona structure of SEM wCCMs can also be transferred to multiwalled carbon nanotubes (CNTs) by a non-covalent grafting approach that forms 1D patchy hybrids (Figure 7d) [142]. In contrast to CDSA, which is commonly used to obtain patchy wCCMs, these patchy hybrids were prepared by an ultrasound-assisted process. Here, the PE block selectively adsorbs onto the CNT surface, while the soluble PS and PMMA blocks form the patchy corona. The driving force for CNT functionalization is the high affinity of the PE block to the CNT surface, which was supported by the use of a SEM triblock terpolymer, which is not able to crystallize at room temperature, but successfully generates patchy CNT hybrids.

Different attempts were made to exchange the PE block with another crystallizable block in order to generate patchy wCCMs. Successful examples are triblock terpolymers of PS–b–PFS–b–PMMA, PS–b–PFS–b–PMVS, and PI–b–PFS–b–PMMA [139,140], as well as μ-ABC miktoarm star terpolymers with a crystallizable PFS block (Figure 6a) [141]. The PFS-containing triblock terpolymers were able to undergo a seeded growth protocol for living CDSA in different solvents to form patchy wCCMs of predictable length (Figure 6b). Remarkably, the living CDSA of all triblock terpolymers proceeded rather slowly compared to PFS-containing diblock copolymers, which was attributed to two effects: (i) the comparably high steric hindrance caused by the two corona blocks surrounding the core-forming block, and (ii) the choice of solvent, which did not sufficiently support the crystallization of PFS. For the PS–b–PFS–b–PMMA triblock terpolymers, the corona chain length (core to total corona block ratio) was varied, and co-crystallization of the resulting triblock terpolymers resulted in block co-micelles with a patchy corona. Interestingly, the different micelle blocks were still discernible by TEM analysis because of the different corona thicknesses (Figure 6c).

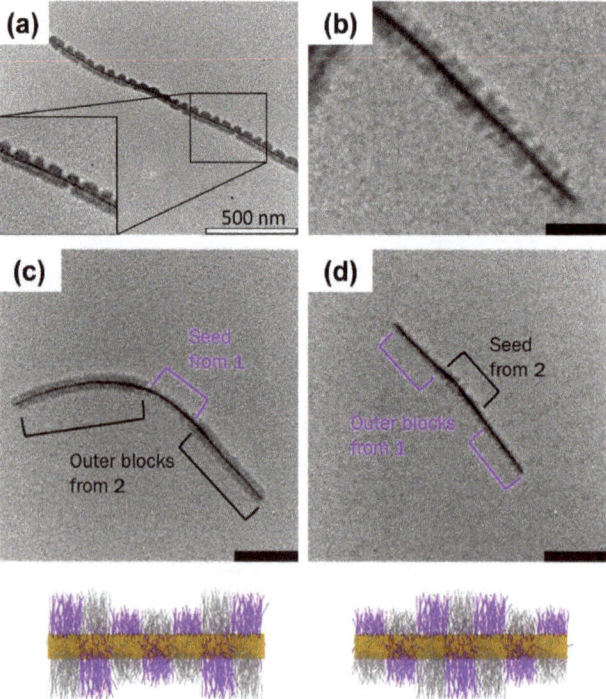

Figure 6. (a) Patchy micelles formed by CDSA of a μ-SIF (polystyrene–*arm*–polyisoprene–*arm*–poly(ferrocenyl dimethylsilane)) miktoarm star terpolymer in ethyl acetate. (b) Patchy cylindrical micelles and (c,d) patchy ABA-type triblock co-micelles with a crystalline PFS core and a patchy PS/PMMA corona prepared in acetone (scale bars = 100 nm). In (c,d), triblock terpolymers with PS and PMMA blocks of different lengths were used to alter the width of the patchy corona in the middle and outer blocks of the triblock co-micelles (in the sketches PS is depicted in light grey and PMMA in purple). (a) Reprinted from [141], and (b–d) from [139] with permission of ACS.

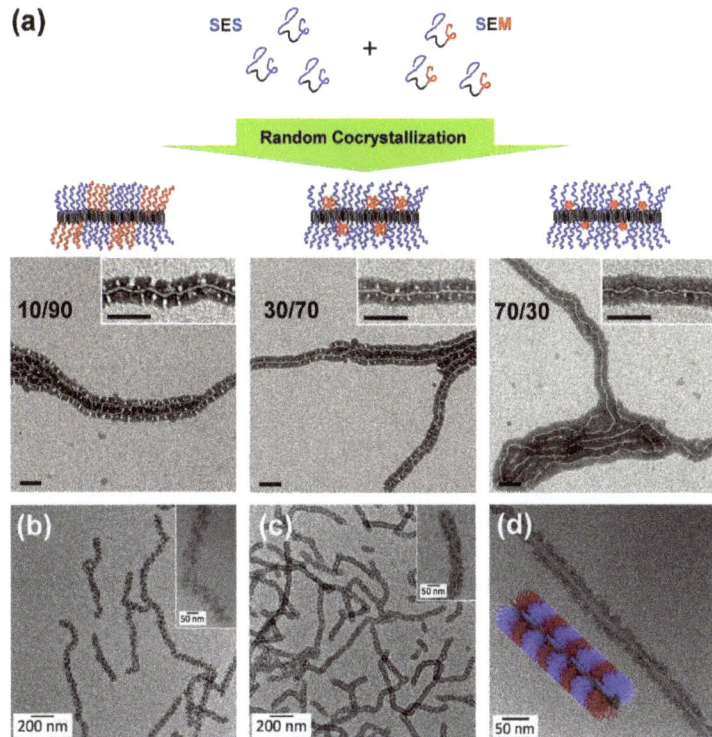

Figure 7. (a) Random co-crystallization of a SES triblock copolymer and a SEM triblock terpolymer, in order to tune the size of the corona patches. TEM images of patchy wCCMs obtained by co-crystallization of $S_{380}E_{880}S_{390}$ with $S_{340}E_{700}M_{360}$ in THF (subscripts denote the respective average degrees of polymerization), revealing a decreasing size of the bright-appearing PMMA corona patches with an increasing amount of $S_{380}E_{880}S_{390}$ (scale bars: 100 nm). TEM images of patchy (b) $S_{415}E_{830}DMA_{420}$ and (c) $S_{660}E_{1350}DMA_{350}$ wCCMs (DMA: *N,N*-dimethylaminoethyl methacrylamide), as well as (d) 1D patchy hybrids with a CNT core and a patchy PS/PMMA corona prepared by ultrasound-assisted, non-covalent grafting of an SEM triblock terpolymer onto CNTs. For all samples, PS was selectively stained with RuO_4 vapor and appears dark. (a) Reproduced from [136] with permission of Elsevier, (b,c) reprinted from [138] with permission of RSC, and (d) reproduced from [142] with permission of ACS.

4.2. Co-Assembly of Diblock Copolymers

The simultaneous co-assembly of PFS-based diblock copolymers represents an alternative way of producing patchy, cylindrical CCMs, next to the use of synthetically more demanding linear or star-shaped triblock terpolymers (Table 1). However, the corona patches of the resulting micelles are usually arranged in a blocky rather than an alternating manner. Consequently, the micelles produced with this approach represent a special case of patchy CCMs. The first example of these patchy block co-micelles was reported in 2014, and is based on the co-crystallization of linear and brush-type BCPs with a crystallizable PFS block [143]. Starting from a linear PFS–*b*–PMVS diblock copolymer, the PMVS corona block was alkylated via thiol–ene functionalization, in order to yield a brush-type BCP with pendant C18 alkyl chains. The brush-type BCPs showed poor crystallization behavior, due to the steric repulsion of the alkyl moieties. However, simultaneous co-crystallization with the linear BCP, applying cylindrical PFS–*b*–PDMS seed micelles, resulted in a gradual integration of the brush-type unimers. Hence, a patchy corona segmentation of the end blocks was observed for the produced B–A–B triblock co-micelles by TEM and atomic force microscopy (AFM) (Figure 8a).

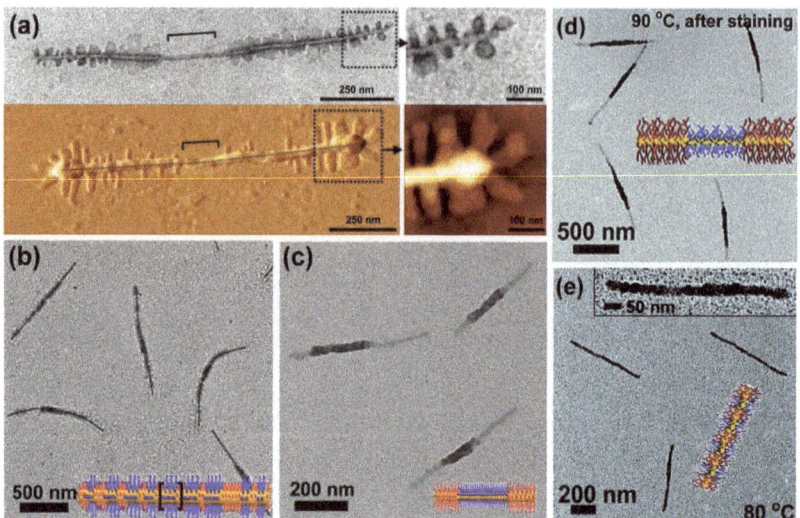

Figure 8. (**a**) TEM (top), as well as AFM topography (bottom left) and phase (bottom right) images of B–A–B triblock co-micelles with patchy end blocks prepared by the co-crystallization of linear and brush-type BCPs with a crystallizable PFS block, employing cylindrical PFS–*b*–PDMS seed micelles. (**b**) Patch-like segmented and (**c**) B–A–B triblock co-micelles produced by controlling the epitaxial growth rate of PFS–*b*–PNiPAM over PFS–*b*–P2VP onto cylindrical PFS–*b*–P2VP seed micelles. Comparable growth rates resulted in patch-like segmentation and dissimilar growth rates in a blocky structure of the corona. (**d**) B–A–B triblock co-micelles and (**e**) patch-like, segmented co-micelles prepared by synergistic self-seeding of a mixture of short PFS–*b*–PNiPAM and PFS–*b*–P2VP cylindrical micelles. In (**d**), the P2VP middle block corona was selectively stained with platin NPs. (**b**–**e**) In the respective sketches, PFS is colored in light orange, P2VP in blue, and PNiPAM in red. (**a**) Reprinted from [143], (**b**,**c**) reprinted from [145] with permission of ACS, and (**d**,**e**) reproduced from [80] with permission of RSC.

The preparation of patchy block co-micelles is not limited to sterically demanding co-blocks, but can be induced by a strong difference in the Flory–Huggins interaction parameter between the corona-forming blocks [144]. Blends of PFS–*b*–PDMS with PFS–*b*–PMVS and PFS–*b*–PI, respectively, were co-crystallized, resulting in a blocky corona segmentation. Staining with Karstedt's catalyst (selective for PI and PMVS) revealed the small corona patches and made two different patch arrangements visible (helical pattern and hemispherical shape). In a subsequent study, the competitive seeded-growth kinetics of the simultaneous co-crystallization of diblock copolymers bearing different corona blocks was investigated [145]. To this end, PFS–*b*–poly(2–vinylpyridine) (PFS–*b*–P2VP) was co-crystallized with two different PFS–*b*–poly(*N*-isopropyl acrylamide) (PFS–*b*–PNiPAM) diblock copolymers using short PFS–*b*–P2VP seed micelles. The length of the PFS block was similar in all used diblock copolymers, but the corona block length of the PFS–*b*–PNiPAM diblock copolymers differed, which affected the epitaxial growth rate of the PFS–*b*–PNiPAM unimers on the seed micelles. If this growth rate was comparable to that of the competing PFS–*b*–P2VP unimers, patchy micelles were observed (Figure 8b). If, on the other hand, the growth rates of the two competing diblock copolymers differed significantly, the formation of block co-micelles was preferred (Figure 8c). Additionally, the epitaxial growth rate of the PFS–*b*–P2VP diblock copolymers was manipulated by quaternization of the P2VP block, which generated a permanent positive charge within the corona chains. Co-crystallization with a PFS–b–PNiPAM diblock copolymer, which yielded a patchy structure with the non-quaternized PFS–b–P2VP, then led to a blocky arrangement of the patches, which could again be attributed to the differing epitaxial growth rates.

Beyond changes in the epitaxial growth rate by manipulation of the corona chains, the crystallization behavior of the PFS core block can also be altered [80]. A variation in the PFS block length affects the so-called critical dissolution temperature (T_c). This temperature describes the point at which the initial average micelle length doubles upon cooling. Heating a mixture of two different micelle fragments with similar T_c values to an annealing temperature (T_a), and $T_a < T_c$ results in separate micelle fragments. If T_a is in the range of the T_c of both micellar fragments, the micellar fragments dissolve partly, and tadpole-shaped fragments are observable. If $T_a > T_c$, self-seeding is taking place and the growth kinetics are dictated by the epitaxial growth rates of the two competing unimer types, i.e., a patchy morphology is observed for similar growth rates and a blocky arrangement of the patches results from dissimilar growth rates (Figure 8d). The self-seeding behavior changes if two diblock copolymers with different T_c values are employed. If T_a is raised above the T_c of one of the diblock copolymers, but is still lower than the T_c of the other diblock copolymer, the diblock copolymer with the lower T_c will partly or almost fully dissolve and epitaxially grow from the remaining micelle seed fragments of both diblock copolymers. This results in either match stick-like micelles or block co-micelles. If T_a is increased well above the T_c of both diblock copolymers, again the growth kinetics determine the final observable corona arrangement—i.e., for similar growth rates, a patchy segmentation is generated (Figure 8e). This concept can also be transferred to mixtures of PFS homopolymers and PFS-based BCPs [146]. Due to the higher T_c of the PFS homopolymer, a certain fraction of PFS homopolymer crystal fragments will survive upon proper choice of T_a; these fragments then act as seeds upon subsequent cooling and annealing. This not only allows the production of cylindrical micelles of uniform length, but also of well-defined block co-micelles or patchy micelles employing a mixture of PFS with different PFS-based BCPs. An important feature of this approach with respect to applications is the comparably easy scale-up, enabling the production of uniform cylindrical micelles of controlled architecture up to concentrations of 10% (w/w solids) or more.

5. Properties and Applications

5.1. Interfacial Activity and Blend Compatibilization

The alternating, patch-like arrangement in the corona of worm-like (or cylindrical) patchy CCMs offers a high potential for a variety of applications. As was shown for amorphous Janus micelles, polymer particles exhibiting two opposing faces made of PS and PMMA (or poly(tert-butyl methacrylate)) serve as excellent particulate surfactants and compatibilizers in polymer blends [147–158]. This originates from the unique interfacial activity of these materials [38]. Patchy wCCMs were proven to show not only a superior interfacial activity compared to cylindrical micelles with a homogeneous PS corona, but also an identical interfacial activity compared to that of Janus micelles at a water–toluene interface (Figure 9a) [159]. Although Janus particles consist of only two clearly separated compartments (or faces), which facilitates the orientation at interfaces, the unique corona structure of patchy micelles is able to adapt to the requirements of the interface, i.e., the respective insoluble block will collapse and the soluble block will expand. Depending on the molecular weight of the corona-forming blocks and thus, the thickness of the corona, the interfacial activity could be tuned, and was shown to increase with thickness (at constant micelle length), which is in good agreement with theoretical predictions [160]. Interestingly, patchy SEM wCCMs can also be hierarchically assembled by a confinement process through emulsification in a toluene-in-water emulsion and subsequent evaporation of the solvents. This leads to microparticles with a highly ordered hexagonal close-packed lattice structure [161].

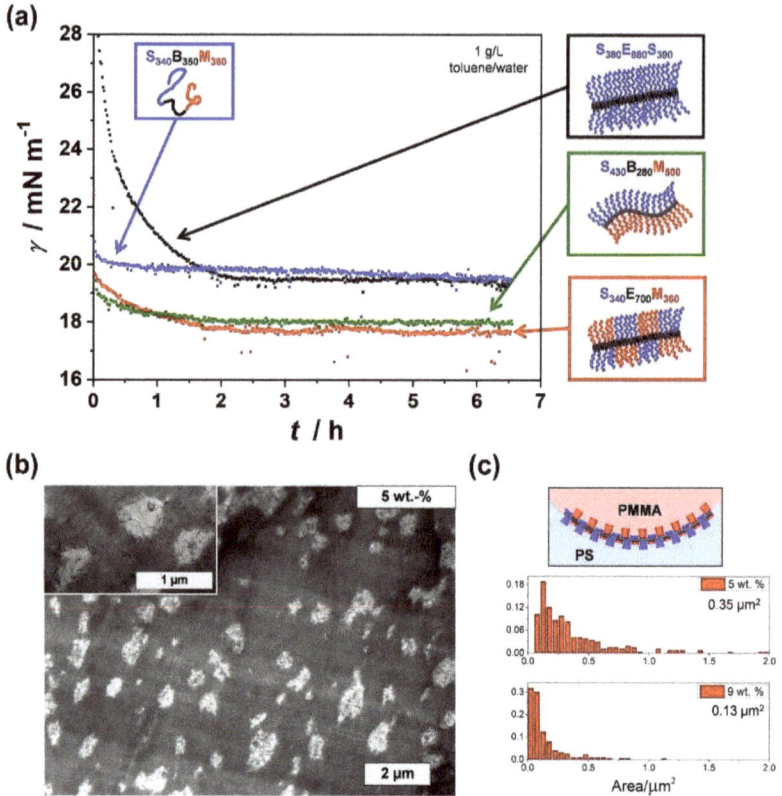

Figure 9. (a) Comparison of interfacial tension isotherms of 1 g·L^{-1} solutions containing SBM unimers, SES wCCMs with a homogeneous PS corona, SEM wCCMs with a patchy PS/PMMA corona, and SBM-based Janus cylinders with opposing PS and PMMA faces (given subscripts correspond to average degrees of polymerization of the respective blocks). (b) TEM image of a solvent-cast PS/PMMA blend (80/20 w/w) compatibilized with 5 wt.%. patchy CNTs (PS/PMMA corona). (c) Schematic representation of the adaption of the patchy PS/PMMA corona to the PS/PMMA blend interface by selective collapse/expansion of the incompatible/compatible corona blocks (top) and histograms of PMMA domain areas for blends with 5 wt.%. and 9 wt.%. patchy CNTs (bottom). (a) Reproduced from [159] with permission of Elsevier and (b,c) reproduced from [162] with permission of ACS.

The excellent interfacial activity of patchy wCCMs can be harnessed for the efficient compatibilization of polymer blends, as reported for solvent-cast PS/PMMA (80/20 w/w) blends [162]. In this work, SEM triblock terpolymers were non-covalently grafted onto the surface of multiwalled CNTs, in order to obtain temperature-stable hybrid compatibilizers with a patchy PS/PMMA corona (patchy CNTs, Figure 7d). The performance of these hybrid compatibilizers was studied depending on their weight fraction, revealing that an increasing filler content considerably reduced the size of the PMMA droplets (minority component) in the blends down to 0.13 µm^2 for the blend with 9 wt.% patchy CNTs (Figure 9b,c). Remarkably, the obtained PMMA domain areas were significantly lower compared to that achieved by using Janus cylinders (L = 2.3 µm, biphasic PS/PMMA corona) as compatibilizers, resulting in domain areas of 10.2 µm^2 and 1.77 µm^2 for 5 wt.% and 10 wt.% Janus cylinders (PS/PMMA = 80/20 w/w), respectively [163]. In addition, the TEM image taken at higher magnification (inset of Figure 9b) shows that well-dispersed patchy CNTs are not only located at the PS/PMMA interface, but are also homogeneously distributed in the PS and PMMA phase. The homogeneous distribution of the patchy CNTs, together with their superior compatibilizing efficiency, can be again attributed to the unique feature of the patchy corona, being able to adapt to their surroundings (PS/PMMA interface, or neat PS and PMMA phases) by selective collapse/expansion of the corona blocks.

5.2. Nanoparticle Templates/Hybrids

Metal and metal oxide NPs are highly attractive materials for a multitude of applications, such as optics, medicine, electronics, or catalysis, originating from their unique optical properties and high surface-to-volume ratio [164–171]. However, the high surface area is an ambivalent feature, as it is useful, for example, in catalysis, but considerably limits the overall stability of NP dispersions, due to agglomeration and Ostwald ripening. Here, the stabilization of NPs with ligands has proven to be a convenient solution to overcome this substantial drawback [172–175]. Another highly efficient method is the use of micellar nanostructures to selectively embed the NPs within functional surface compartments, which not only act as ligands for the NPs, but also keep the NPs' surface accessible and inhibits agglomeration due to spatial separation [38,117,118,127,137,176,177].

In particular, patchy wCCMs, with their well-defined, alternating segmented coronas, have been shown to be versatile NP templates, and even allow the regio-selective incorporation of two different NP types, since the chemistry of the two corona-forming blocks can be tailored to the specific needs of the respective NP [137,138]. In order to obtain these binary-loaded hybrid materials, based on patchy PS–b–PE–b–poly(dimethylaminoethyl methacrylamide) (SEDMA) wCCMs, a two-step procedure for the selective decoration of the patches with NPs was developed (Figure 10a). In the first step, preformed, PS-stabilized gold NPs were mixed with a dispersion of the functional patchy wCCMs, followed by the addition of acetone as a selective solvent for the PDMA block, resulting in a collapse of the PS chains. Due to selective interactions of the PS corona block and the PS-stabilized gold NPs, the NPs were enclosed within the PS patches upon collapse of the PS chains. In the following step, preformed, acetate-stabilized zinc oxide NPs were incorporated in the functional patches by a ligand exchange method. Intrinsic staining provided by the inorganic NPs facilitated an examination of the resulting structures via TEM (Figure 10b,c). The different types of NPs are clearly discernible by their different diameters (D; $D_{\text{gold NP}}$ = 7.9 nm, $D_{\text{zinc oxide NP}}$ = 2.7 nm) and the contrast (heavy metals generate a higher contrast in TEM compared to transition metal oxides). Interestingly, despite the small size of the corona patches (<20 nm), it seems that more than one NP per patch is observable. This might be attributed to the extremely small size of the chosen inorganic NPs (<10 nm).

The selective functionalization of surface-compartmentalized polymeric micelles with inorganic NPs was also shown for PFS-containing triblock co-micelles, featuring a quaternized P2VP corona in the middle [117,118]. Through electrostatic interactions, the middle block was selectively loaded with mercaptoacetic acid-stabilized gold NPs, PbS quantum dots and dextran–magnetite NPs, demonstrating the versatility of block co-micelles as NP templates. Furthermore, NP hybrid materials with block co-micelles derived from co-assembly of diblock copolymers were reported. Here, the spatially confined incorporation of platinum NPs and CdSe quantum dots was enabled by selective interactions with functional corona patches [146,178].

5.3. Heterogeneous Catalysis

As mentioned in the previous section, an application of noble metal and metal oxide NPs in catalysis is highly desirable, because of the high catalytically active surface area of the employed NPs. However, ligands, which are needed for stabilization of the NPs, might inhibit the superior catalytic activity of the NPs by blocking the surface. Even for tailor-made ligands, this is a distinct drawback, since these materials are usually hard to recover after usage. A separation of the catalytically active species from the reaction medium is challenging and expensive. Immobilizing the catalytically active NPs on suitable supports (e.g., inorganic, polymeric) solves this problem of recoverability, while preserving the activity and accessibility of the NPs' surface [179–185]. Nonetheless, agglomeration of the inorganic NPs on the surface of the heterogeneous supports can occur if the NPs are insufficiently confined, resulting in a significant loss of activity over several consecutive catalysis cycles.

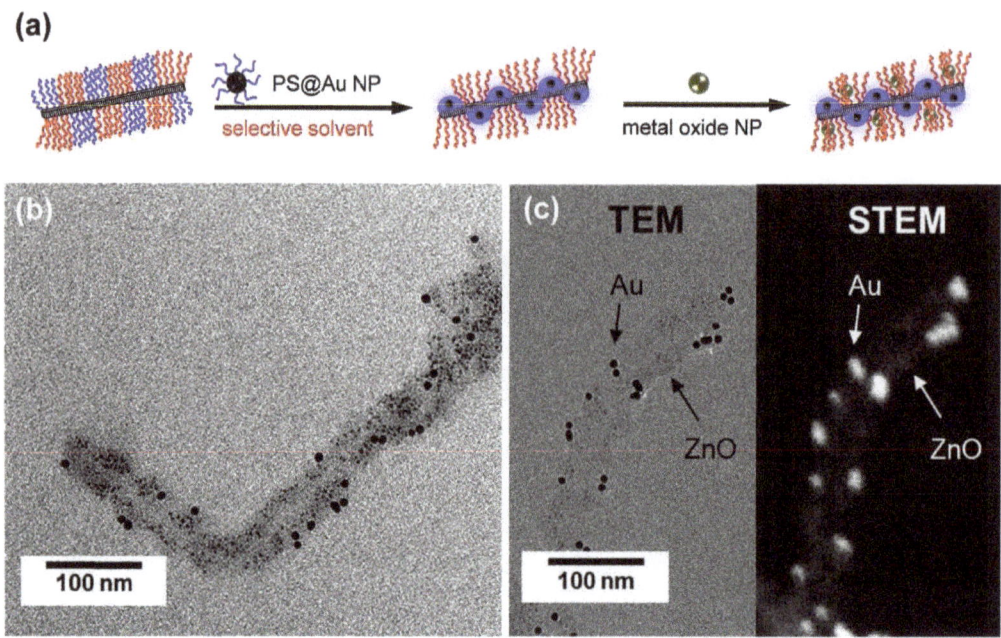

Figure 10. (a) Schematic depiction of the regio-selective, binary loading of patchy SEDMA wCCMs with PS-stabilized gold (Au) NPs and zinc oxide (ZnO) NPs, respectively (PS is displayed in blue, PE in black, and PDMA in red). (b) TEM image of $S_{415}E_{830}DMA_{420}$ wCCMs binary-loaded with Au and ZnO NPs. (c) Bright-field (left) and high-angle annular dark-field scanning transmission electron microscopy (HAADF-STEM, right) images, clearly revealing the binary loading with two different NP types. Reproduced and adapted from [138] with permission of RSC.

The highly regular, alternating arrangement of the corona compartments in patchy wCCMs allows us to efficiently confine inorganic NPs. However, these micellar templates have to be immobilized on a solid support, which provides high accessibility of the reactants to the catalytically active NPs and easy recovery in order to harness these structures for heterogeneous catalysis. This issue was overcome by coating different patchy PS–b–PE–b–poly(dialkylaminoethyl methacrylamide) wCCMs onto the surface of PS nonwovens by means of coaxial electrospinning (Figure 11a,b) [186,187]. The resulting patchy nonwovens were loaded with gold NPs through a simple dip-coating process (Figure 11c), which was driven by a ligand exchange reaction. The hybrid nonwovens were successfully applied as catalysts for the alcoholysis of dimethylphenylsilane (Figure 11d) at room temperature, showing a comparable or even higher catalytic activity than other supports reported before [188–193]. Moreover, the employed patchy hybrid nonwovens were easily recoverable from the reaction medium and reusable in at least 10 consecutive catalysis cycles.

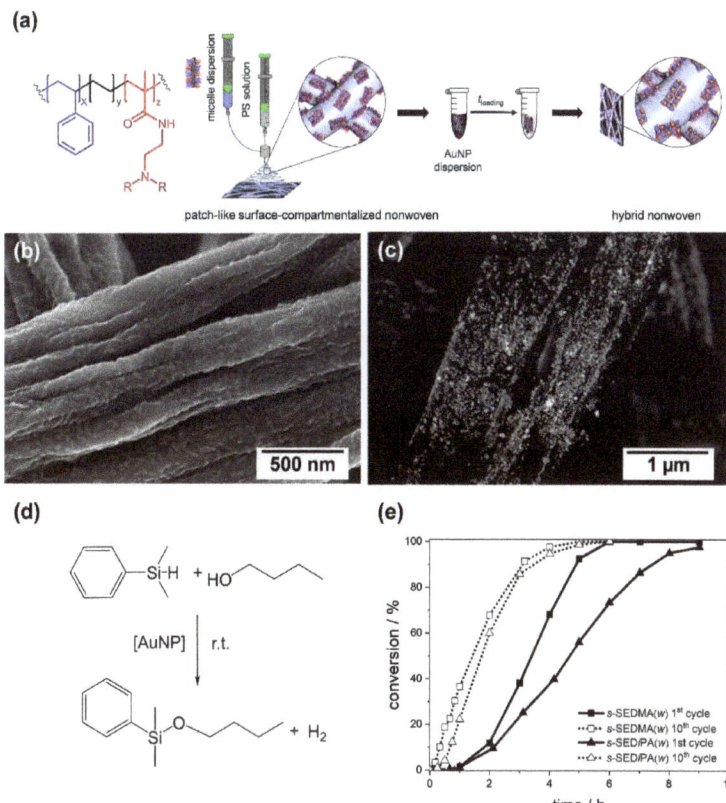

Figure 11. (a) Catalytically active, hybrid nonwovens prepared by a combination of bottom-up (CDSA) and top-down (coaxial electrospinning) approaches. In the first step, patchy nonwovens were prepared by decorating a PS nonwoven with functional, patchy PS–b–PE–b–poly(dialkylaminoethyl methacrylamide) wCCMs by coaxial electrospinning (PS patches are depicted in blue and the functional, tertiary amino group containing patches in red). Subsequently, the patchy nonwovens were loaded with citrate-stabilized Au NPs via a ligand exchange process (citrate against tertiary amino groups in functional patches). (b) Scanning electron microscopy images of a patchy nonwoven (based on $S_{415}E_{830}DMA_{420}$ wCCMs) before and (c) after loading with Au NPs (back-scattered electron detector). (d) Au NP-catalyzed alcoholysis of dimethylphenylsilane in n-butanol. (e) Kinetics of the Au NP-catalyzed alcoholysis of dimethylphenylsilane in n-butanol, employing patchy hybrid nonwovens as catalysts (D*i*PA = poly(di*iso*propylaminoethyl methacrylamide). Reproduced from [186] with permission from RSC.

Since this system offers different possibilities to tune the catalytic activity, an in-depth study on the influence of the patch size and chemistry on the reaction kinetics was conducted. Here, an extended first-order kinetics model was employed, which includes the induction periods observed in the catalytic alcoholysis of dimethylphenylsilane in n-butanol. This study revealed a strong dependence on the accessibility of the reactants to the gold NPs' surface, being mainly controlled by the swellability of the functional patches in n-butanol. The latter depends on both patch chemistry, i.e., poly(N,N-dimethylaminoethyl methacrylamide) (PDMA, more hydrophilic) vs. poly(N,N-di*iso*propylaminoethyl methacrylamide) (PD*i*PA, more hydrophobic) patches, as well as size. As a result, significantly longer induction (t_{ind}) and reaction (t_R) times were observed for the first catalysis cycles compared to the tenth cycles (Figure 11e). Nonwovens with more polar PDMA patches were the most efficient in NP stabilization (prevention of agglomeration), but showed a significantly lower t_R in the first catalysis cycle, due to a strong interaction with the gold NPs' surface. Thus, precise tuning of the patch size and chemistry is needed to optimize the catalysts performance. However, the modular design of the patchy hybrid nonwovens enables a facile adaption to the needs of different catalysis systems—for example, by an exchange of the support material or by varying the type of NPs. Moreover, it is possible to render the functionalized patches thermo-responsive [194], which opens access to catalytic reactions regulated by an inherent temperature control.

6. Conclusions and Outlook

From a conceptual point of view, several strategies exist for the production of patchy micelles with crystalline cores, such as CDSA of triblock terpolymers with crystallizable middle blocks, miktoarm stars, or the co-assembly of diblock copolymers with a common crystallizable block but different corona-forming blocks. However, so far, patchy micelles have been reported only for BCPs with PE or PFS as crystallizable blocks, despite the fact that a large variety of crystallizable polymer blocks has already been utilized in CDSA. Here, ring-opening polymerization of lactones or lactides, in combination with controlled radical polymerization techniques, might be another promising alternative, as BCPs based on PCL or PLLA as crystallizable blocks are readily accessible. Moreover, PFS could be replaced by ruthenocene-based BCPs, which show a higher degree of crystallinity but are less studied for CDSA. Finally, patchy micelles could be derived from the simultaneous heteroepitaxial growth of two crystallizable di- or triblock copolymers bearing different core- and corona-forming blocks, inducing segmentation within the core as well as in the corona.

The alternating arrangement of the corona patches emerges as an excellent feature for the stabilization and confinement of metal and metal oxide nanoparticles, opening applications in heterogeneous catalysis. Yet this has been shown only for the gold nanoparticle-catalyzed alcoholysis of silanes, and it is anticipated that this concept can be transferred to other relevant catalytic processes like heterogeneous hydrogenation. By incorporating different nanoparticle types, even cascade reactions might be realizable. Most interestingly, the interfacial activity of patchy, worm-like (or cylindrical) micelles is equivalent to that of Janus micelles, the latter being, however, more difficult to produce. Thus, patchy micelles might be utilized in interfacial catalysis, as well as in the efficient stabilization of emulsions or compatibilization of polymer blends.

Author Contributions: C.H., I.M., J.S., and H.S. contributed to writing (review and editing). All authors have read and agreed to the published version of the manuscript.

Funding: This research was funded by the German Research Foundation within the collaborative research center SFB 840 (project A2).

Institutional Review Board Statement: Not applicable.

Informed Consent Statement: Not applicable.

Data Availability Statement: Not applicable.

Acknowledgments: C.H. acknowledges support by the Graduate School of the University of Bayreuth.

Conflicts of Interest: The authors declare no conflict of interest.

References

1. Tritschler, U.; Pearce, S.; Gwyther, J.; Whittell, G.R.; Manners, I. 50th Anniversary Perspective: Functional Nanoparticles from the Solution Self-Assembly of Block Copolymers. *Macromolecules* **2017**, *50*, 3439–3463. [CrossRef]
2. Du, J.; O'Reilly, R.K. Anisotropic particles with patchy, multicompartment and Janus architectures: Preparation and application. *Chem. Soc. Rev.* **2011**, *40*, 2402–2416. [CrossRef] [PubMed]
3. Wyman, I.W.; Liu, G. Micellar structures of linear triblock terpolymers: Three blocks but many possibilities. *Polymer* **2013**, *54*, 1950–1978. [CrossRef]

4. Schacher, F.H.; Rupar, P.A.; Manners, I. Functional Block Copolymers: Nanostructured Materials with Emerging Applications. *Angew. Chem. Int. Ed.* **2012**, *51*, 7898–7921. [CrossRef] [PubMed]
5. Matyjaszewski, K.; Möller, M. (Eds.) *Polymer Science: A Comprehensive Reference*; Elsevier Science: Amsterdam, The Netherlands, 2012; ISBN 978-0-08-087862-1.
6. Feng, H.; Lu, X.; Wang, W.; Kang, N.-G.; Mays, J.W. Block Copolymers: Synthesis, Self-Assembly, and Applications. *Polymers* **2017**, *9*, 494. [CrossRef]
7. Müller, A.H.E.; Matyjaszewski, K. (Eds.) *Controlled and Living Polymerizations*; Wiley: Hoboken, NJ, USA, 2009; ISBN 9783527324927.
8. Jennings, J.; He, G.; Howdle, S.M.; Zetterlund, P.B. Block copolymer synthesis by controlled/living radical polymerisation in heterogeneous systems. *Chem. Soc. Rev.* **2016**, *45*, 5055–5084. [CrossRef]
9. Hadjichristidis, N.; Pitsikalis, M.; Iatrou, H. Synthesis of Block Copolymers. In *Block Copolymers I. Advances in Polymer Science*; Abetz, V., Ed.; Springer: Berlin/Heidelberg, Germany, 2005; Volume 189.
10. Lutz, J.; Laschewsky, A. Multicompartment Micelles: Has the Long-Standing Dream Become a Reality? *Macromol. Chem. Phys.* **2005**, *206*, 813–817. [CrossRef]
11. Laschewsky, A. Polymerized micelles with compartments. *Curr. Opin. Colloid Interface Sci.* **2003**, *8*, 274–281. [CrossRef]
12. Gröschel, A.H.; Walther, A.; Löbling, T.I.; Schacher, F.H.; Schmalz, H.; Müller, A.H.E. Guided hierarchical co-assembly of soft patchy nanoparticles. *Nature* **2013**, *503*, 247–251. [CrossRef]
13. Li, W.; Palis, H.; Mérindol, R.; Majimel, J.; Ravaine, S.; Duguet, E. Colloidal molecules and patchy particles: Complementary concepts, synthesis and self-assembly. *Chem. Soc. Rev.* **2020**, *49*, 1955–1976. [CrossRef]
14. Lunn, D.J.; Finnegan, J.R.; Manners, I. Self-assembly of "patchy" nanoparticles: A versatile approach to functional hierarchical materials. *Chem. Sci.* **2015**, *6*, 3663–3673. [CrossRef] [PubMed]
15. Zhang, K.; Jiang, M.; Chen, D. Self-assembly of particles—The regulatory role of particle flexibility. *Prog. Polym. Sci.* **2012**, *37*, 445–486. [CrossRef]
16. Löbling, T.I.; Ikkala, O.; Gröschel, A.H.; Müller, A.H.E. Controlling Multicompartment Morphologies Using Solvent Conditions and Chemical Modification. *ACS Macro Lett.* **2016**, *5*, 1044–1048. [CrossRef]
17. Fang, B.; Walther, A.; Wolf, A.; Xu, Y.; Yuan, J.; Müller, A.H.E. Undulated Multicompartment Cylinders by the Controlled and Directed Stacking of Polymer Micelles with a Compartmentalized Corona. *Angew. Chem. Int. Ed.* **2009**, *48*, 2877–2880. [CrossRef]
18. Lee, S.; Jang, S.; Kim, K.; Jeon, J.; Kim, S.-S.; Sohn, B.-H. Branched and crosslinked supracolloidal chains with diblock copolymer micelles having three well-defined patches. *Chem. Commun.* **2016**, *52*, 9430–9433. [CrossRef]
19. Walther, A.; Müller, A.H.E. Formation of hydrophobic bridges between multicompartment micelles of miktoarm star terpolymers in water. *Chem. Commun.* **2009**, *7*, 1127–1129. [CrossRef] [PubMed]
20. Kong, W.; Jiang, W.; Zhu, Y.; Li, B. Highly Symmetric Patchy Multicompartment Nanoparticles from the Self-Assembly of ABC Linear Terpolymers in C-Selective Solvents. *Langmuir* **2012**, *28*, 11714–11724. [CrossRef]
21. Kim, K.; Jang, S.; Jeon, J.; Kang, D.; Sohn, B.-H. Fluorescent Supracolloidal Chains of Patchy Micelles of Diblock Copolymers Functionalized with Fluorophores. *Langmuir* **2018**, *34*, 4634–4639. [CrossRef]
22. Nghiem, T.; Chakroun, R.; Janoszka, N.; Chen, C.; Klein, K.; Wong, C.K.; Gröschel, A.H. pH-Controlled Hierarchical Assembly/Disassembly of Multicompartment Micelles in Water. *Macromol. Rapid Commun.* **2020**, *41*, 2000301. [CrossRef]
23. Skrabania, K.; Berlepsch, H.V.; Böttcher, C.; Laschewsky, A. Synthesis of Ternary, Hydrophilic−Lipophilic−Fluorophilic Block Copolymers by Consecutive RAFT Polymerizations and Their Self-Assembly into Multicompartment Micelles. *Macromolecules* **2010**, *43*, 271–281. [CrossRef]
24. Löbling, T.I.; Borisov, O.; Haataja, J.S.; Ikkala, O.; Gröschel, A.H.; Müller, A.H.E. Rational design of ABC triblock terpolymer solution nanostructures with controlled patch morphology. *Nat. Commun.* **2016**, *7*, 12097. [CrossRef]
25. Nghiem, T.-L.; Löbling, T.I.; Gröschel, A.H. Supracolloidal chains of patchy micelles in water. *Polym. Chem.* **2017**, *9*, 1583–1592. [CrossRef]
26. Moughton, A.O.; Hillmyer, M.A.; Lodge, T.P. Multicompartment Block Polymer Micelles. *Macromolecules* **2012**, *45*, 2–19. [CrossRef]
27. Wang, L.; Lin, J. Discovering multicore micelles: Insights into the self-assembly of linear ABC terpolymers in midblock-selective solvents. *Soft Matter* **2011**, *7*, 3383–3391. [CrossRef]
28. Gröschel, A.H.; Müller, A.H.E. Self-assembly concepts for multicompartment nanostructures. *Nanoscale* **2015**, *7*, 11841–11876. [CrossRef] [PubMed]
29. Wong, C.K.; Qiang, X.; Müller, A.H.E.; Gröschel, A.H. Self-Assembly of block copolymers into internally ordered microparticles. *Prog. Polym. Sci.* **2020**, *102*, 101211. [CrossRef]
30. Pelras, T.; Mahon, C.S.; Müllner, M. Synthesis and Applications of Compartmentalised Molecular Polymer Brushes. *Angew. Chem. Int. Ed.* **2018**, *57*, 6982–6994. [CrossRef] [PubMed]
31. Nayanathara, U.; Kermaniyan, S.S.; Such, G.K. Multicompartment Polymeric Nanocarriers for Biomedical Applications. *Macromol. Rapid Commun.* **2020**, *41*, 2000298. [CrossRef]
32. Marschelke, C.; Fery, A.; Synytska, A. Janus particles: From concepts to environmentally friendly materials and sustainable applications. *Colloid Polym. Sci.* **2020**, *298*, 841–865. [CrossRef]
33. Fan, X.; Yang, J.; Loh, X.J.; Li, Z. Polymeric Janus Nanoparticles: Recent Advances in Synthetic Strategies, Materials Properties, and Applications. *Macromol. Rapid Commun.* **2019**, *40*, 1800203. [CrossRef]

34. Agrawal, G.; Agrawal, R. Janus Nanoparticles: Recent Advances in Their Interfacial and Biomedical Applications. *ACS Appl. Nano Mater.* **2019**, *2*, 1738–1757. [CrossRef]
35. Zhang, J.; Grzybowski, B.A.; Granick, S. Janus Particle Synthesis, Assembly, and Application. *Langmuir* **2017**, *33*, 6964–6977. [CrossRef]
36. Deng, R.; Liang, F.; Zhu, J.; Yang, Z. Recent advances in the synthesis of Janus nanomaterials of block copolymers. *Mater. Chem. Front.* **2016**, *1*, 431–443. [CrossRef]
37. Pang, X.; Wan, C.; Wang, M.; Lin, Z. Strictly Biphasic Soft and Hard Janus Structures: Synthesis, Properties, and Applications. *Angew. Chem. Int. Ed.* **2014**, *53*, 5524–5538. [CrossRef]
38. Walther, A.; Müller, A.H.E. Janus Particles: Synthesis, Self-Assembly, Physical Properties, and Applications. *Chem. Rev.* **2013**, *113*, 5194–5261. [CrossRef] [PubMed]
39. Hu, J.; Zhou, S.; Sun, Y.; Fang, X.; Wu, L. Fabrication, properties and applications of Janus particles. *Chem. Soc. Rev.* **2012**, *41*, 4356–4378. [CrossRef]
40. Loget, G.; Kuhn, A. Bulk synthesis of Janus objects and asymmetric patchy particles. *J. Mater. Chem.* **2012**, *22*, 15457–15474. [CrossRef]
41. Wurm, F.; Kilbinger, A.F.M. Polymeric Janus Particles. *Angew. Chem. Int. Ed.* **2009**, *48*, 8412–8421. [CrossRef]
42. Dou, H.; Liu, G.; Dupont, J.; Hong, L. Triblock terpolymer helices self-assembled under special solvation conditions. *Soft Matter* **2010**, *6*, 4214–4222. [CrossRef]
43. Dupont, J.; Liu, G.; Niihara, K.-I.; Kimoto, R.; Jinnai, H. Self-Assembled ABC Triblock Copolymer Double and Triple Helices. *Angew. Chem. Int. Ed.* **2009**, *48*, 6144–6147. [CrossRef]
44. Hoppenbrouwers, E.; Li, Z.; Liu, G. Triblock Nanospheres with Amphiphilic Coronal Chains. *Macromolecules* **2003**, *36*, 876–881. [CrossRef]
45. Hu, J.; Njikang, G.; Liu, G. Twisted ABC Triblock Copolymer Cylinders with Segregated A and C Coronal Chains. *Macromolecules* **2008**, *41*, 7993–7999. [CrossRef]
46. Liu, X.; Ding, Y.; Liu, J.; Lin, S.; Zhuang, Q. Evolution in the morphological behaviour of a series of fluorine-containing ABC miktoarm star terpolymers. *Eur. Polym. J.* **2019**, *116*, 342–351. [CrossRef]
47. Njikang, G.; Han, D.; Wang, J.; Liu, G. ABC Triblock Copolymer Micelle-Like Aggregates in Selective Solvents for A and C. *Macromolecules* **2008**, *41*, 9727–9735. [CrossRef]
48. Zhang, W.; He, J.X.; Liu, Q.; Ke, G.Q.; Dong, X. Synthesis of Block Terpolymer PS-PDMAEMA-PMMA via ATRP and its Self-Assembly in Selective Solvents. *Adv. Mater. Res.* **2014**, *1049–1050*, 137–141. [CrossRef]
49. Kuo, S.-W.; Tung, P.-H.; Lai, C.-L.; Jeong, K.-U.; Chang, F.-C. Supramolecular Micellization of Diblock Copolymer Mixtures Mediated by Hydrogen Bonding for the Observation of Separated Coil and Chain Aggregation in Common Solvents. *Macromol. Rapid Commun.* **2007**, *29*, 229–233. [CrossRef]
50. Kuo, S.-W.; Tung, P.-H.; Chang, F.-C. Hydrogen bond mediated supramolecular micellization of diblock copolymer mixture in common solvents. *Eur. Polym. J.* **2009**, *45*, 1924–1935. [CrossRef]
51. Voets, I.K.; De Keizer, A.; Leermakers, F.A.; Debuigne, A.; Jerôme, R.; Detrembleur, C.; Stuart, M.A.C. Electrostatic hierarchical co-assembly in aqueous solutions of two oppositely charged double hydrophilic diblock copolymers. *Eur. Polym. J.* **2009**, *45*, 2913–2925. [CrossRef]
52. Lopresti, C.; Massignani, M.; Fernyhough, C.; Blanazs, A.; Ryan, A.J.; Madsen, J.; Warren, N.J.; Armes, S.P.; Lewis, A.L.; Chirasatitsin, S.; et al. Controlling Polymersome Surface Topology at the Nanoscale by Membrane Confined Polymer/Polymer Phase Separation. *ACS Nano* **2011**, *5*, 1775–1784. [CrossRef] [PubMed]
53. Hu, J.; Liu, G. Chain Mixing and Segregation in B−C and C−D Diblock Copolymer Micelles. *Macromolecules* **2005**, *38*, 8058–8065. [CrossRef]
54. Srinivas, G.; Pitera, J.W. Soft Patchy Nanoparticles from Solution-Phase Self-Assembly of Binary Diblock Copolymers. *Nano Lett.* **2008**, *8*, 611–618. [CrossRef]
55. Zheng, Y.; Liu, G.; Yan, X. Polymer Nano- and Microspheres with Bumpy and Chain-Segregated Surfaces. *J. Am. Chem. Soc.* **2005**, *127*, 15358–15359. [CrossRef] [PubMed]
56. Christian, D.A.; Tian, A.; Ellenbroek, W.G.; Levental, I.; Rajagopal, K.; Janmey, P.A.; Liu, A.J.; Baumgart, T.; Discher, D.E. Spotted vesicles, striped micelles and Janus assemblies induced by ligand binding. *Nat. Mater.* **2009**, *8*, 843–849. [CrossRef]
57. Hsu, H.-P.; Paul, W.; Binder, K. One- and Two-Component Bottle-Brush Polymers: Simulations Compared to Theoretical Predictions. *Macromol. Theory Simul.* **2007**, *16*, 660–689. [CrossRef]
58. De Jong, J.; ten Brinke, G. Conformational Aspects and Intramolecular Phase Separation of Alternating Copolymacromonomers: A Computer Simulation Study. *Macromol. Theory Simul.* **2004**, *13*, 318–327. [CrossRef]
59. Stepanyan, R.; Subbotin, A.; ten Brinke, G. Comb Copolymer Brush with Chemically Different Side Chains. *Macromolecules* **2002**, *35*, 5640–5648. [CrossRef]
60. Theodorakis, P.E.; Paul, W.; Binder, K. Interplay between Chain Collapse and Microphase Separation in Bottle-Brush Polymers with Two Types of Side Chains. *Macromolecules* **2010**, *43*, 5137–5148. [CrossRef]
61. Hsu, H.-P.; Paul, W.; Binder, K. Intramolecular phase separation of copolymer "bottle brushes": No sharp phase transition but a tunable length scale. *EPL Europhys. Lett.* **2006**, *76*, 526–532. [CrossRef]

62. He, W.-N.; Xu, J.-T. Crystallization assisted self-assembly of semicrystalline block copolymers. *Prog. Polym. Sci.* **2012**, *37*, 1350–1400. [CrossRef]
63. Ganda, S.; Stenzel, M.H. Concepts, fabrication methods and applications of living crystallization-driven self-assembly of block copolymers. *Prog. Polym. Sci.* **2020**, *101*, 101195. [CrossRef]
64. Hailes, R.L.N.; Oliver, A.M.; Gwyther, J.; Whittell, G.R.; Manners, I. Polyferrocenylsilanes: Synthesis, properties, and applications. *Chem. Soc. Rev.* **2016**, *45*, 5358–5407. [CrossRef] [PubMed]
65. Wang, X.; Guerin, G.; Wang, H.; Wang, Y.; Manners, I.; Winnik, M.A. Cylindrical Block Copolymer Micelles and Co-Micelles of Controlled Length and Architecture. *Science* **2007**, *317*, 644–647. [CrossRef] [PubMed]
66. Gilroy, J.B.; Gädt, T.; Whittell, G.R.; Chabanne, L.; Mitchels, J.M.; Richardson, R.M.; Winnik, M.A.; Manners, I. Monodisperse cylindrical micelles by crystallization-driven living self-assembly. *Nat. Chem.* **2010**, *2*, 566–570. [CrossRef] [PubMed]
67. MacFarlane, L.; Zhao, C.; Cai, J.; Qiu, H.; Manners, I. Emerging applications for living crystallization-driven self-assembly. *Chem. Sci.* **2021**, *12*, 4661–4682. [CrossRef]
68. Schmelz, J.; Schedl, A.E.; Steinlein, C.; Manners, I.; Schmalz, H. Length Control and Block-Type Architectures in Worm-like Micelles with Polyethylene Cores. *J. Am. Chem. Soc.* **2012**, *134*, 14217–14225. [CrossRef]
69. Qian, J.; Lu, Y.; Chia, A.; Zhang, M.; Rupar, P.A.; Gunari, N.; Walker, G.C.; Cambridge, G.; He, F.; Guerin, G.; et al. Self-Seeding in One Dimension: A Route to Uniform Fiber-like Nanostructures from Block Copolymers with a Crystallizable Core-Forming Block. *ACS Nano* **2013**, *7*, 3754–3766. [CrossRef]
70. Qian, J.; Guerin, G.; Lu, Y.; Cambridge, G.; Manners, I.; Winnik, M.A. Self-Seeding in One Dimension: An Approach To Control the Length of Fiberlike Polyisoprene-Polyferrocenylsilane Block Copolymer Micelles. *Angew. Chem. Int. Ed.* **2011**, *50*, 1622–1625. [CrossRef] [PubMed]
71. Li, X.; Jin, B.; Gao, Y.; Hayward, D.W.; Winnik, M.A.; Luo, Y.; Manners, I. Monodisperse Cylindrical Micelles of Controlled Length with a Liquid-Crystalline Perfluorinated Core by 1D "Self-Seeding". *Angew. Chem. Int. Ed.* **2016**, *55*, 11392–11396. [CrossRef] [PubMed]
72. Boott, C.E.; Gwyther, J.; Harniman, R.L.; Hayward, D.W.; Manners, I. Scalable and uniform 1D nanoparticles by synchronous polymerization, crystallization and self-assembly. *Nat. Chem.* **2017**, *9*, 785–792. [CrossRef] [PubMed]
73. Oliver, A.M.; Gwyther, J.; Boott, C.E.; Davis, S.; Pearce, S.; Manners, I. Scalable Fiber-like Micelles and Block Co-micelles by Polymerization-Induced Crystallization-Driven Self-Assembly. *J. Am. Chem. Soc.* **2018**, *140*, 18104–18114. [CrossRef]
74. Sha, Y.; Rahman, A.; Zhu, T.; Cha, Y.; McAlister, C.W.; Tang, C. ROMPI-CDSA: Ring-opening metathesis polymerization-induced crystallization-driven self-assembly of metallo-block copolymers. *Chem. Sci.* **2019**, *10*, 9782–9787. [CrossRef] [PubMed]
75. Shin, S.; Menk, F.; Kim, Y.; Lim, J.; Char, K.; Zentel, R.; Choi, T.-L. Living Light-Induced Crystallization-Driven Self-Assembly for Rapid Preparation of Semiconducting Nanofibers. *J. Am. Chem. Soc.* **2018**, *140*, 6088–6094. [CrossRef] [PubMed]
76. Qiu, H.; Gao, Y.; Du, V.A.; Harniman, R.; Winnik, M.A.; Manners, I. Branched Micelles by Living Crystallization-Driven Block Copolymer Self-Assembly under Kinetic Control. *J. Am. Chem. Soc.* **2015**, *137*, 2375–2385. [CrossRef] [PubMed]
77. Hudson, Z.M.; Lunn, D.J.; Winnik, M.A.; Manners, I. Colour-tunable fluorescent multiblock micelles. *Nat. Commun.* **2014**, *5*, 3372. [CrossRef] [PubMed]
78. Jin, X.-H.; Price, M.B.; Finnegan, J.R.; Boott, C.E.; Richter, J.M.; Rao, A.; Menke, S.M.; Friend, R.H.; Whittell, G.R.; Manners, I. Long-range exciton transport in conjugated polymer nanofibers prepared by seeded growth. *Science* **2018**, *360*, 897–900. [CrossRef] [PubMed]
79. Rupar, P.A.; Chabanne, L.; Winnik, M.A.; Manners, I. Non-Centrosymmetric Cylindrical Micelles by Unidirectional Growth. *Science* **2012**, *337*, 559–562. [CrossRef] [PubMed]
80. Xu, J.; Zhou, H.; Yu, Q.; Guerin, G.; Manners, I.; Winnik, M.A. Synergistic self-seeding in one-dimension: A route to patchy and block comicelles with uniform and controllable length. *Chem. Sci.* **2019**, *10*, 2280–2284. [CrossRef]
81. He, X.; He, Y.; Hsiao, M.-S.; Harniman, R.L.; Pearce, S.; Winnik, M.A.; Manners, I. Complex and Hierarchical 2D Assemblies via Crystallization-Driven Self-Assembly of Poly(L-lactide) Homopolymers with Charged Termini. *J. Am. Chem. Soc.* **2017**, *139*, 9221–9228. [CrossRef]
82. He, X.; Hsiao, M.-S.; Boott, C.E.; Harniman, R.L.; Nazemi, A.; Li, X.; Winnik, M.A.; Manners, I. Two-dimensional assemblies from crystallizable homopolymers with charged termini. *Nat. Mater.* **2017**, *16*, 481–488. [CrossRef]
83. Nazemi, A.; He, X.; Macfarlane, L.R.; Harniman, R.L.; Hsiao, M.-S.; Winnik, M.A.; Faul, C.F.J.; Manners, I. Uniform "Patchy" Platelets by Seeded Heteroepitaxial Growth of Crystallizable Polymer Blends in Two Dimensions. *J. Am. Chem. Soc.* **2017**, *139*, 4409–4417. [CrossRef]
84. Pearce, S.; He, X.; Hsiao, M.-S.; Harniman, R.L.; Macfarlane, L.R.; Manners, I. Uniform, High-Aspect-Ratio, and Patchy 2D Platelets by Living Crystallization-Driven Self-Assembly of Crystallizable Poly(ferrocenyldimethylsilane)-Based Homopolymers with Hydrophilic Charged Termini. *Macromolecules* **2019**, *52*, 6068–6079. [CrossRef]
85. Qiu, H.; Gao, Y.; Boott, C.E.; Gould, O.E.C.; Harniman, R.L.; Miles, M.J.; Webb, S.E.D.; Winnik, M.A.; Manners, I. Uniform patchy and hollow rectangular platelet micelles from crystallizable polymer blends. *Science* **2016**, *352*, 697–701. [CrossRef] [PubMed]
86. Inam, M.; Cambridge, G.; Pitto-Barry, A.; Laker, Z.P.L.; Wilson, N.R.; Mathers, R.T.; Dove, A.P.; O'Reilly, R.K. 1D vs. 2D shape selectivity in the crystallization-driven self-assembly of polylactide block copolymers. *Chem. Sci.* **2017**, *8*, 4223–4230. [CrossRef] [PubMed]

87. Gould, O.E.; Qiu, H.; Lunn, D.J.; Rowden, J.; Harniman, R.L.; Hudson, Z.M.; Winnik, M.A.; Miles, M.J.; Manners, I. Transformation and patterning of supermicelles using dynamic holographic assembly. *Nat. Commun.* **2015**, *6*, 10009. [CrossRef]
88. Qiu, H.; Hudson, Z.M.; Winnik, M.A.; Manners, I. Multidimensional hierarchical self-assembly of amphiphilic cylindrical block comicelles. *Science* **2015**, *347*, 1329–1332. [CrossRef]
89. Gädt, T.; Ieong, N.S.; Cambridge, G.; Winnik, M.A.; Manners, I. Complex and hierarchical micelle architectures from diblock copolymers using living, crystallization-driven polymerizations. *Nat. Mater.* **2009**, *8*, 144–150. [CrossRef]
90. Hudson, Z.M.; Boott, C.E.; Robinson, M.E.; Rupar, P.A.; Winnik, M.A.; Manners, I. Tailored hierarchical micelle architectures using living crystallization-driven self-assembly in two dimensions. *Nat. Chem.* **2014**, *6*, 893–898. [CrossRef]
91. Dou, H.; Li, M.; Qiao, Y.; Harniman, R.; Li, X.; Boott, C.E.; Mann, S.; Manners, I. Higher-order assembly of crystalline cylindrical micelles into membrane-extendable colloidosomes. *Nat. Commun.* **2017**, *8*, 426. [CrossRef]
92. Fan, B.; Liu, L.; Li, J.-H.; Ke, X.-X.; Xu, J.-T.; Du, B.-Y.; Fan, Z.-Q. Crystallization-driven one-dimensional self-assembly of polyethylene-*b*-poly(*tert*-butylacrylate) diblock copolymers in DMF: Effects of crystallization temperature and the corona-forming block. *Soft Matter* **2015**, *12*, 67–76. [CrossRef]
93. Schmalz, H.; Schmelz, J.; Drechsler, M.; Yuan, J.; Walther, A.; Schweimer, K.; Mihut, A.M. Thermo-Reversible Formation of Wormlike Micelles with a Microphase-Separated Corona from a Semicrystalline Triblock Terpolymer. *Macromolecules* **2008**, *41*, 3235–3242. [CrossRef]
94. Schmelz, J.; Karg, M.; Hellweg, T.; Schmalz, H. General Pathway toward Crystalline-Core Micelles with Tunable Morphology and Corona Segregation. *ACS Nano* **2011**, *5*, 9523–9534. [CrossRef] [PubMed]
95. Xiong, H.; Chen, C.-K.; Lee, K.; Van Horn, R.M.; Liu, Z.; Ren, B.; Quirk, R.P.; Thomas, E.L.; Lotz, B.; Ho, R.-M.; et al. Scrolled Polymer Single Crystals Driven by Unbalanced Surface Stresses: Rational Design and Experimental Evidence. *Macromolecules* **2011**, *44*, 7758–7766. [CrossRef]
96. Coe, Z.; Weems, A.; Dove, A.P.; O'Reilly, R.K. Synthesis of Monodisperse Cylindrical Nanoparticles via Crystallization-driven Self-assembly of Biodegradable Block Copolymers. *J. Vis. Exp.* **2019**, *20*, e59772. [CrossRef]
97. Arno, M.C.; Inam, M.; Coe, Z.; Cambridge, G.; MacDougall, L.J.; Keogh, R.; Dove, A.P.; O'Reilly, R.K. Precision Epitaxy for Aqueous 1D and 2D Poly(ε-caprolactone) Assemblies. *J. Am. Chem. Soc.* **2017**, *139*, 16980–16985. [CrossRef] [PubMed]
98. Petzetakis, N.; Dove, A.P.; O'Reilly, R.K. Cylindrical micelles from the living crystallization-driven self-assembly of poly(lactide)-containing block copolymers. *Chem. Sci.* **2011**, *2*, 955–960. [CrossRef]
99. Arno, M.C.; Inam, M.; Weems, A.C.; Li, Z.; Binch, A.L.A.; Platt, C.I.; Richardson, S.M.; Hoyland, J.A.; Dove, A.P.; O'Reilly, R.K. Exploiting the role of nanoparticle shape in enhancing hydrogel adhesive and mechanical properties. *Nat. Commun.* **2020**, *11*, 1–9. [CrossRef] [PubMed]
100. Yu, W.; Inam, M.; Jones, J.R.; Dove, A.P.; O'Reilly, R.K. Understanding the CDSA of poly(lactide) containing triblock copolymers. *Polym. Chem.* **2017**, *8*, 5504–5512. [CrossRef]
101. Tong, Z.; Su, Y.; Jiang, Y.; Xie, Y.; Chen, S.; O'Reilly, R.K. Spatially Restricted Templated Growth of Poly(ε-caprolactone) from Carbon Nanotubes by Crystallization-Driven Self-Assembly. *Macromolecules* **2021**, *54*, 2844–2851. [CrossRef]
102. Finnegan, J.R.; He, X.; Street, S.T.G.; Garcia-Hernandez, J.D.; Hayward, D.W.; Harniman, R.L.; Richardson, R.M.; Whittell, G.R.; Manners, I. Extending the Scope of "Living" Crystallization-Driven Self-Assembly: Well-Defined 1D Micelles and Block Comicelles from Crystallizable Polycarbonate Block Copolymers. *J. Am. Chem. Soc.* **2018**, *140*, 17127–17140. [CrossRef]
103. Rudolph, T.; von der Lühe, M.; Hartlieb, M.; Norsic, S.; Schubert, U.S.; Boisson, C.; D'Agosto, F.; Schacher, F.H. Toward Anisotropic Hybrid Materials: Directional Crystallization of Amphiphilic Polyoxazoline-Based Triblock Terpolymers. *ACS Nano* **2015**, *9*, 10085–10098. [CrossRef] [PubMed]
104. Finnegan, J.; Pilkington, E.; Alt, K.; Rahim, A.; Kent, S.J.; Davis, T.P.; Kempe, K. Stealth Nanorods via the Aqueous Living Crystallisation-Driven Self-Assembly of Poly(2-oxazoline)s. *Chem. Sci.* **2021**. [CrossRef]
105. Shaikh, H.; Jin, X.-H.; Harniman, R.L.; Richardson, R.M.; Whittell, G.R.; Manners, I. Solid-State Donor-Acceptor Coaxial Heterojunction Nanowires via Living Crystallization-Driven Self-Assembly. *J. Am. Chem. Soc.* **2020**, *142*, 13469–13480. [CrossRef]
106. Folgado, E.; Mayor, M.; Cot, D.; Ramonda, M.; Godiard, F.; Ladmiral, V.; Semsarilar, M. Preparation of well-defined 2D-lenticular aggregates by self-assembly of PNIPAM-*b*-PVDF amphiphilic diblock copolymers in solution. *Polym. Chem.* **2021**, *12*, 1465–1475. [CrossRef]
107. Kang, L.; Chao, A.; Zhang, M.; Yu, T.; Wang, J.; Wang, Q.; Yu, H.; Jiang, N.; Zhang, D. Modulating the Molecular Geometry and Solution Self-Assembly of Amphiphilic Polypeptoid Block Copolymers by Side Chain Branching Pattern. *J. Am. Chem. Soc.* **2021**, *143*, 5890–5902. [CrossRef]
108. Wei, Y.; Liu, F.; Li, M.; Li, Z.; Sun, J. Dimension control on self-assembly of a crystalline core-forming polypeptoid block copolymer: 1D nanofibers versus 2D nanosheets. *Polym. Chem.* **2021**, *12*, 1147–1154. [CrossRef]
109. Kynaston, E.L.; Nazemi, A.; MacFarlane, L.R.; Whittell, G.R.; Faul, C.F.J.; Manners, I. Uniform Polyselenophene Block Copolymer Fiberlike Micelles and Block Co-micelles via Living Crystallization-Driven Self-Assembly. *Macromolecules* **2018**, *51*, 1002–1010. [CrossRef]
110. Kim, Y.-J.; Cho, C.-H.; Paek, K.; Jo, M.; Park, M.-K.; Lee, N.-E.; Kim, Y.-J.; Kim, B.J.; Lee, E. Precise Control of Quantum Dot Location within the P3HT-*b*-P2VP/QD Nanowires Formed by Crystallization-Driven 1D Growth of Hybrid Dimeric Seeds. *J. Am. Chem. Soc.* **2014**, *136*, 2767–2774. [CrossRef]

111. Patra, S.K.; Ahmed, R.; Whittell, G.R.; Lunn, D.J.; Dunphy, E.L.; Winnik, M.A.; Manners, I. Cylindrical Micelles of Controlled Length with a π-Conjugated Polythiophene Core via Crystallization-Driven Self-Assembly. *J. Am. Chem. Soc.* **2011**, *133*, 8842–8845. [CrossRef]
112. Li, X.; Wolanin, P.J.; MacFarlane, L.R.; Harniman, R.L.; Qian, J.; Gould, O.E.C.; Dane, T.G.; Rudin, J.; Cryan, M.J.; Schmaltz, T.; et al. Uniform electroactive fibre-like micelle nanowires for organic electronics. *Nat. Commun.* **2017**, *8*, 15909. [CrossRef]
113. MacFarlane, L.R.; Shaikh, H.; Garcia-Hernandez, J.D.; Vespa, M.; Fukui, T.; Manners, I. Functional nanoparticles through π-conjugated polymer self-assembly. *Nat. Rev. Mater.* **2021**, *6*, 7–26. [CrossRef]
114. Boott, C.E.; Laine, R.F.; Mahou, P.; Finnegan, J.R.; Leitao, E.M.; Webb, S.E.D.; Kaminski, C.F.; Manners, I. In Situ Visualization of Block Copolymer Self-Assembly in Organic Media by Super-Resolution Fluorescence Microscopy. *Chem. Eur. J.* **2015**, *21*, 18539–18542. [CrossRef] [PubMed]
115. Wittmann, J.C.; Hodge, A.M.; Lotz, B. Epitaxial crystallization of polymers onto benzoic acid: Polyethylene and paraffins, aliphatic polyesters, and polyamides. *J. Polym. Sci. Polym. Phys. Ed.* **1983**, *21*, 2495–2509. [CrossRef]
116. Wittmann, J.C.; Lotz, B. Epitaxial crystallization of polyethylene on organic substrates: A reappraisal of the mode of action of selected nucleating agents. *J. Polym. Sci. Polym. Phys. Ed.* **1981**, *19*, 1837–1851. [CrossRef]
117. Wang, H.; Lin, W.; Fritz, K.P.; Scholes, G.D.; Winnik, M.A.; Manners, I. Cylindrical Block Co-Micelles with Spatially Selective Functionalization by Nanoparticles. *J. Am. Chem. Soc.* **2007**, *129*, 12924–12925. [CrossRef]
118. Wang, H.; Patil, A.J.; Liu, K.; Petrov, S.; Mann, S.; Winnik, M.A.; Manners, I. Fabrication of Continuous and Segmented Polymer/Metal Oxide Nanowires Using Cylindrical Micelles and Block Comicelles as Templates. *Adv. Mater.* **2009**, *21*, 1805–1808. [CrossRef]
119. Rupar, P.A.; Cambridge, G.; Winnik, M.A.; Manners, I. Reversible Cross-Linking of Polyisoprene Coronas in Micelles, Block Comicelles, and Hierarchical Micelle Architectures Using Pt(0)–Olefin Coordination. *J. Am. Chem. Soc.* **2011**, *133*, 16947–16957. [CrossRef]
120. Nazemi, A.; Boott, C.E.; Lunn, D.J.; Gwyther, J.; Hayward, D.W.; Richardson, R.M.; Winnik, M.A.; Manners, I. Monodisperse Cylindrical Micelles and Block Comicelles of Controlled Length in Aqueous Media. *J. Am. Chem. Soc.* **2016**, *138*, 4484–4493. [CrossRef]
121. Qiu, H.; Cambridge, G.; Winnik, M.A.; Manners, I. Multi-Armed Micelles and Block Co-micelles via Crystallization-Driven Self-Assembly with Homopolymer Nanocrystals as Initiators. *J. Am. Chem. Soc.* **2013**, *135*, 12180–12183. [CrossRef]
122. He, F.; Gädt, T.; Manners, I.; Winnik, M.A. Fluorescent "Barcode" Multiblock Co-Micelles via the Living Self-Assembly of Di- and Triblock Copolymers with a Crystalline Core-Forming Metalloblock. *J. Am. Chem. Soc.* **2011**, *133*, 9095–9103. [CrossRef]
123. Street, S.T.G.; He, Y.; Jin, X.-H.; Hodgson, L.; Verkade, P.; Manners, I. Cellular uptake and targeting of low dispersity, dual emissive, segmented block copolymer nanofibers. *Chem. Sci.* **2020**, *11*, 8394–8408. [CrossRef]
124. He, X.; Finnegan, J.R.; Hayward, D.W.; MacFarlane, L.R.; Harniman, R.L.; Manners, I. Living Crystallization-Driven Self-Assembly of Polymeric Amphiphiles: Low-Dispersity Fiber-like Micelles from Crystallizable Phosphonium-Capped Polycarbonate Homopolymers. *Macromolecules* **2020**, *53*, 10591–10600. [CrossRef]
125. Qian, J.; Li, X.; Lunn, D.J.; Gwyther, J.; Hudson, Z.M.; Kynaston, E.; Rupar, P.A.; Winnik, M.A.; Manners, I. Uniform, High Aspect Ratio Fiber-like Micelles and Block Co-micelles with a Crystalline π-Conjugated Polythiophene Core by Self-Seeding. *J. Am. Chem. Soc.* **2014**, *136*, 4121–4124. [CrossRef] [PubMed]
126. Tao, D.; Feng, C.; Cui, Y.; Yang, X.; Manners, I.; Winnik, M.A.; Huang, X. Monodisperse Fiber-like Micelles of Controlled Length and Composition with an Oligo(p-phenylenevinylene) Core via "Living" Crystallization-Driven Self-Assembly. *J. Am. Chem. Soc.* **2017**, *139*, 7136–7139. [CrossRef] [PubMed]
127. Tao, D.; Wang, Z.; Huang, X.; Tian, M.; Lu, G.; Manners, I.; Winnik, M.A.; Feng, C. Continuous and Segmented Semiconducting Fiber-like Nanostructures with Spatially Selective Functionalization by Living Crystallization-Driven Self-Assembly. *Angew. Chem. Int. Ed.* **2020**, *59*, 8232–8239. [CrossRef] [PubMed]
128. He, Y.; Eloi, J.-C.; Harniman, R.L.; Richardson, R.M.; Whittell, G.R.; Mathers, R.T.; Dove, A.P.; O'Reilly, R.K.; Manners, I. Uniform Biodegradable Fiber-Like Micelles and Block Comicelles via "Living" Crystallization-Driven Self-Assembly of Poly(L-lactide) Block Copolymers: The Importance of Reducing Unimer Self-Nucleation via Hydrogen Bond Disruption. *J. Am. Chem. Soc.* **2019**, *141*, 19088–19098. [CrossRef]
129. Lunn, D.J.; Gould, O.E.C.; Whittell, G.R.; Armstrong, D.P.; Mineart, K.P.; Winnik, M.A.; Spontak, R.J.; Pringle, P.G.; Manners, I. Microfibres and macroscopic films from the coordination-driven hierarchical self-assembly of cylindrical micelles. *Nat. Commun.* **2016**, *7*, 12371. [CrossRef]
130. Li, X.; Gao, Y.; Boott, C.E.; Hayward, D.W.; Harniman, R.; Whittell, G.R.; Richardson, R.M.; Winnik, M.A.; Manners, I. "Cross" Supermicelles via the Hierarchical Assembly of Amphiphilic Cylindrical Triblock Comicelles. *J. Am. Chem. Soc.* **2016**, *138*, 4087–4095. [CrossRef]
131. Qiu, H.; Russo, G.; Rupar, P.A.; Chabanne, L.; Winnik, M.A.; Manners, I. Tunable Supermicelle Architectures from the Hierarchical Self-Assembly of Amphiphilic Cylindrical B-A-B Triblock Co-Micelles. *Angew. Chem. Int. Ed.* **2012**, *51*, 11882–11885. [CrossRef]
132. Li, X.; Gao, Y.; Boott, C.E.; Winnik, M.A.; Manners, I. Non-covalent synthesis of supermicelles with complex architectures using spatially confined hydrogen-bonding interactions. *Nat. Commun.* **2015**, *6*, 8127. [CrossRef]
133. Li, X.; Gao, Y.; Harniman, R.; Winnik, M.; Manners, I. Hierarchical Assembly of Cylindrical Block Comicelles Mediated by Spatially Confined Hydrogen-Bonding Interactions. *J. Am. Chem. Soc.* **2016**, *138*, 12902–12912. [CrossRef]

134. Schmelz, J.; Schacher, F.H.; Schmalz, H. Cylindrical crystalline-core micelles: Pushing the limits of solution self-assembly. *Soft Matter* **2013**, *9*, 2101–2107. [CrossRef]
135. Rosenfeldt, S.; Lüdel, F.; Schulreich, C.; Hellweg, T.; Radulescu, A.; Schmelz, J.; Schmalz, H.; Harnau, L. Patchy worm-like micelles: Solution structure studied by small-angle neutron scattering. *Phys. Chem. Chem. Phys.* **2012**, *14*, 12750–12756. [CrossRef] [PubMed]
136. Schmelz, J.; Schmalz, H. Corona structure on demand: Tailor-made surface compartmentalization in worm-like micelles via random cocrystallization. *Polymer* **2012**, *53*, 4333–4337. [CrossRef]
137. Schöbel, J.; Karg, M.; Rosenbach, D.; Krauss, G.; Greiner, A.; Schmalz, H. Patchy Wormlike Micelles with Tailored Functionality by Crystallization-Driven Self-Assembly: A Versatile Platform for Mesostructured Hybrid Materials. *Macromolecules* **2016**, *49*, 2761–2771. [CrossRef]
138. Schöbel, J.; Hils, C.; Weckwerth, A.; Schlenk, M.; Bojer, C.; Stuart, M.C.A.; Breu, J.; Förster, S.; Greiner, A.; Karg, M.; et al. Strategies for the selective loading of patchy worm-like micelles with functional nanoparticles. *Nanoscale* **2018**, *10*, 18257–18268. [CrossRef]
139. Oliver, A.M.; Gwyther, J.; Winnik, M.A.; Manners, I. Cylindrical Micelles with "Patchy" Coronas from the Crystallization-Driven Self-Assembly of ABC Triblock Terpolymers with a Crystallizable Central Polyferrocenyldimethylsilane Segment. *Macromolecules* **2017**, *51*, 222–231. [CrossRef]
140. Oliver, A.M.; Spontak, R.J.; Manners, I. Solution self-assembly of ABC triblock terpolymers with a central crystallizable poly(ferrocenyldimethylsilane) core-forming segment. *Polym. Chem.* **2019**, *10*, 2559–2569. [CrossRef]
141. Nunns, A.; Whittell, G.R.; Winnik, M.A.; Manners, I. Crystallization-Driven Solution Self-Assembly of µ-ABC Miktoarm Star Terpolymers with Core-Forming Polyferrocenylsilane Blocks. *Macromolecules* **2014**, *47*, 8420–8428. [CrossRef]
142. Gegenhuber, T.; Gröschel, A.H.; Löbling, T.I.; Drechsler, M.; Ehlert, S.; Förster, S.; Schmalz, H. Noncovalent Grafting of Carbon Nanotubes with Triblock Terpolymers: Toward Patchy 1D Hybrids. *Macromolecules* **2015**, *48*, 1767–1776. [CrossRef]
143. Finnegan, J.R.; Lunn, D.J.; Gould, O.E.C.; Hudson, Z.M.; Whittell, G.R.; Winnik, M.A.; Manners, I. Gradient Crystallization-Driven Self-Assembly: Cylindrical Micelles with "Patchy" Segmented Coronas via the Coassembly of Linear and Brush Block Copolymers. *J. Am. Chem. Soc.* **2014**, *136*, 13835–13844. [CrossRef] [PubMed]
144. Cruz, M.; Xu, J.; Yu, Q.; Guerin, G.; Manners, I.; Winnik, M.A. Visualizing Nanoscale Coronal Segregation in Rod-Like Micelles Formed by Co-Assembly of Binary Block Copolymer Blends. *Macromol. Rapid Commun.* **2018**, *39*, 1800397. [CrossRef] [PubMed]
145. Xu, J.; Zhou, H.; Yu, Q.; Manners, I.; Winnik, M.A. Competitive Self-Assembly Kinetics as a Route To Control the Morphology of Core-Crystalline Cylindrical Micelles. *J. Am. Chem. Soc.* **2018**, *140*, 2619–2628. [CrossRef]
146. Song, S.; Liu, X.; Nikbin, E.; Howe, J.Y.; Yu, Q.; Manners, I.; Winnik, M.A. Uniform 1D Micelles and Patchy & Block Comicelles via Scalable, One-Step Crystallization-Driven Block Copolymer Self-Assembly. *J. Am. Chem. Soc.* **2021**, *143*, 6266–6280.
147. Bahrami, R.; Löbling, T.I.; Gröschel, A.H.; Schmalz, H.; Müller, A.H.E.; Altstädt, V. The Impact of Janus Nanoparticles on the Compatibilization of Immiscible Polymer Blends under Technologically Relevant Conditions. *ACS Nano* **2014**, *8*, 10048–10056. [CrossRef] [PubMed]
148. Bärwinkel, S.; Bahrami, R.; Löbling, T.I.; Schmalz, H.; Müller, A.H.E.; Altstädt, V. Polymer Foams Made of Immiscible Polymer Blends Compatibilized by Janus Particles-Effect of Compatibilization on Foam Morphology. *Adv. Eng. Mater.* **2016**, *18*, 814–825. [CrossRef]
149. Yang, Q.; Loos, K. Janus nanoparticles inside polymeric materials: Interfacial arrangement toward functional hybrid materials. *Polym. Chem.* **2017**, *8*, 641–654. [CrossRef]
150. Bahrami, R.; Löbling, T.I.; Schmalz, H.; Müller, A.H.E.; Altstädt, V. Synergistic effects of Janus particles and triblock terpolymers on toughness of immiscible polymer blends. *Polymer* **2017**, *109*, 229–237. [CrossRef]
151. Walther, A.; André, X.; Drechsler, M.; Abetz, V.; Müller, A.H.E. Janus Discs. *J. Am. Chem. Soc.* **2007**, *129*, 6187–6198. [CrossRef] [PubMed]
152. Bryson, K.C.; Löbling, T.I.; Müller, A.H.E.; Russell, T.P.; Hayward, R.C. Using Janus Nanoparticles To Trap Polymer Blend Morphologies during Solvent-Evaporation-Induced Demixing. *Macromolecules* **2015**, *48*, 4220–4227. [CrossRef]
153. Ruhland, T.M.; Gröschel, A.H.; Walther, A.; Müller, A.H.E. Janus Cylinders at Liquid–Liquid Interfaces. *Langmuir* **2011**, *27*, 9807–9814. [CrossRef] [PubMed]
154. Walther, A.; Drechsler, M.; Müller, A.H.E. Structures of amphiphilic Janus discs in aqueous media. *Soft Matter* **2009**, *5*, 385–390. [CrossRef]
155. Walther, A.; Matussek, K.; Müller, A.H.E. Engineering Nanostructured Polymer Blends with Controlled Nanoparticle Location using Janus Particles. *ACS Nano* **2008**, *2*, 1167–1178. [CrossRef] [PubMed]
156. Walther, A.; Hoffmann, M.; Müller, A.H.E. Emulsion Polymerization Using Janus Particles as Stabilizers. *Angew. Chem. Int. Ed.* **2008**, *47*, 711–714. [CrossRef]
157. Ruhland, T.M.; Gröschel, A.H.; Ballard, N.; Skelhon, T.S.; Walther, A.; Müller, A.H.E.; Bon, S.A.F. Influence of Janus Particle Shape on Their Interfacial Behavior at Liquid–Liquid Interfaces. *Langmuir* **2013**, *29*, 1388–1394. [CrossRef] [PubMed]
158. Gröschel, A.H.; Löbling, T.I.; Petrov, P.D.; Müllner, M.; Kuttner, C.; Wieberger, F.; Müller, A.H.E. Janus Micelles as Effective Supracolloidal Dispersants for Carbon Nanotubes. *Angew. Chem. Int. Ed.* **2013**, *52*, 3602–3606. [CrossRef] [PubMed]
159. Schmelz, J.; Pirner, D.; Krekhova, M.; Ruhland, T.M.; Schmalz, H. Interfacial activity of patchy worm-like micelles. *Soft Matter* **2013**, *9*, 11173–11177. [CrossRef]
160. Böker, A.; He, J.; Emrick, T.; Russell, T.P. Self-assembly of nanoparticles at interfaces. *Soft Matter* **2007**, *3*, 1231–1248. [CrossRef]

161. Dai, X.; Qiang, X.; Hils, C.; Schmalz, H.; Gröschel, A.H. Frustrated Microparticle Morphologies of a Semicrystalline Triblock Terpolymer in 3D Soft Confinement. *ACS Nano* **2021**, *15*, 1111–1120. [CrossRef]
162. Gegenhuber, T.; Krekhova, M.; Schöbel, J.; Gröschel, A.H.; Schmalz, H. "Patchy" Carbon Nanotubes as Efficient Compatibilizers for Polymer Blends. *ACS Macro Lett.* **2016**, *5*, 306–310. [CrossRef]
163. Ruhland, T.M. *Janus Cylinders at Interfaces*; University of Bayreuth: Bayreuth, Germany, 2009.
164. Haruta, M. When Gold Is Not Noble: Catalysis by Nanoparticles. *Chem. Rec.* **2003**, *3*, 75–87. [CrossRef]
165. Daniel, M.-C.; Astruc, D. Gold Nanoparticles: Assembly, Supramolecular Chemistry, Quantum-Size-Related Properties, and Applications toward Biology, Catalysis, and Nanotechnology. *Chem. Rev.* **2004**, *104*, 293–346. [CrossRef]
166. Shenhar, R.; Norsten, T.B.; Rotello, V.M. Polymer-Mediated Nanoparticle Assembly: Structural Control and Applications. *Adv. Mater.* **2005**, *17*, 657–669. [CrossRef]
167. Campelo, J.M.; Luna, D.; Luque, R.; Marinas, J.M.; Romero, A.A. Sustainable Preparation of Supported Metal Nanoparticles and Their Applications in Catalysis. *ChemSusChem* **2009**, *2*, 18–45. [CrossRef]
168. Rycenga, M.; Cobley, C.M.; Zeng, J.; Li, W.; Moran, C.H.; Zhang, Q.; Qin, D.; Xia, Y. Controlling the Synthesis and Assembly of Silver Nanostructures for Plasmonic Applications. *Chem. Rev.* **2011**, *111*, 3669–3712. [CrossRef] [PubMed]
169. Giljohann, D.A.; Seferos, D.S.; Daniel, W.L.; Massich, M.D.; Patel, P.C.; Mirkin, C.A. Gold Nanoparticles for Biology and Medicine. *Angew. Chem. Int. Ed.* **2010**, *49*, 3280–3294. [CrossRef]
170. Dykman, L.; Khlebtsov, N. Gold nanoparticles in biomedical applications: Recent advances and perspectives. *Chem. Soc. Rev.* **2011**, *41*, 2256–2282. [CrossRef]
171. Saha, K.; Agasti, S.S.; Kim, C.; Li, X.; Rotello, V.M. Gold Nanoparticles in Chemical and Biological Sensing. *Chem. Rev.* **2012**, *112*, 2739–2779. [CrossRef]
172. Astruc, D.; Lu, F.; Aranzaes, J.R. Nanoparticles as Recyclable Catalysts: The Frontier between Homogeneous and Heterogeneous Catalysis. *Angew. Chem. Int. Ed.* **2005**, *44*, 7852–7872. [CrossRef]
173. Burda, C.; Chen, X.; Narayanan, R.; El-Sayed, M.A. Chemistry and Properties of Nanocrystals of Different Shapes. *Chem. Rev.* **2005**, *105*, 1025–1102. [CrossRef] [PubMed]
174. Murphy, C.J.; Sau, T.K.; Gole, A.M.; Orendorff, C.J.; Gao, J.; Gou, L.; Hunyadi, S.E.; Li, T. Anisotropic Metal Nanoparticles: Synthesis, Assembly, and Optical Applications. *J. Phys. Chem. B* **2005**, *109*, 13857–13870. [CrossRef]
175. Lu, A.-H.; Salabas, E.-L.; Schüth, F. Magnetic Nanoparticles: Synthesis, Protection, Functionalization, and Application. *Angew. Chem. Int. Ed.* **2007**, *46*, 1222–1244. [CrossRef]
176. Wang, J.; Li, W.; Zhu, J. Encapsulation of inorganic nanoparticles into block copolymer micellar aggregates: Strategies and precise localization of nanoparticles. *Polymer* **2014**, *55*, 1079–1096. [CrossRef]
177. Zhou, T.; Dong, B.; Qi, H.; Mei, S.; Li, C.Y. Janus hybrid hairy nanoparticles. *J. Polym. Sci. Part B Polym. Phys.* **2014**, *52*, 1620–1640. [CrossRef]
178. Zhang, Y.; Pearce, S.; Eloi, J.-C.; Harniman, R.L.; Tian, J.; Cordoba, C.; Kang, Y.; Fukui, T.; Qiu, H.; Blackburn, A.; et al. Dendritic Micelles with Controlled Branching and Sensor Applications. *J. Am. Chem. Soc.* **2021**, *143*, 5805–5814. [CrossRef]
179. Liu, L.; Corma, A. Metal Catalysts for Heterogeneous Catalysis: From Single Atoms to Nanoclusters and Nanoparticles. *Chem. Rev.* **2018**, *118*, 4981–5079. [CrossRef] [PubMed]
180. Van Deelen, T.W.; Mejía, C.H.; de Jong, K.P. Control of metal-support interactions in heterogeneous catalysts to enhance activity and selectivity. *Nat. Catal.* **2019**, *2*, 955–970. [CrossRef]
181. Li, Z.; Ji, S.; Liu, Y.; Cao, X.; Tian, S.; Chen, Y.; Niu, Z.; Li, Y. Well-Defined Materials for Heterogeneous Catalysis: From Nanoparticles to Isolated Single-Atom Sites. *Chem. Rev.* **2020**, *120*, 623–682. [CrossRef] [PubMed]
182. Ishida, T.; Murayama, T.; Taketoshi, A.; Haruta, M. Importance of Size and Contact Structure of Gold Nanoparticles for the Genesis of Unique Catalytic Processes. *Chem. Rev.* **2019**, *120*, 464–525. [CrossRef]
183. Shifrina, Z.B.; Matveeva, V.G.; Bronstein, L.M. Role of Polymer Structures in Catalysis by Transition Metal and Metal Oxide Nanoparticle Composites. *Chem. Rev.* **2020**, *120*, 1350–1396. [CrossRef] [PubMed]
184. Cai, J.; Li, C.; Kong, N.; Lu, Y.; Lin, G.; Wang, X.; Yao, Y.; Manners, I.; Qiu, H. Tailored multifunctional micellar brushes via crystallization-driven growth from a surface. *Science* **2019**, *366*, 1095–1098. [CrossRef]
185. Astruc, D. Introduction: Nanoparticles in Catalysis. *Chem. Rev.* **2020**, *120*, 461–463. [CrossRef] [PubMed]
186. Hils, C.; Dulle, M.; Sitaru, G.; Gekle, S.; Schöbel, J.; Frank, A.; Drechsler, M.; Greiner, A.; Schmalz, H. Influence of patch size and chemistry on the catalytic activity of patchy hybrid nonwovens. *Nanoscale Adv.* **2020**, *2*, 438–452. [CrossRef]
187. Schöbel, J.; Burgard, M.; Hils, C.; Dersch, R.; Dulle, M.; Volk, K.; Karg, M.; Greiner, A.; Schmalz, H. Bottom-Up Meets Top-Down: Patchy Hybrid Nonwovens as an Efficient Catalysis Platform. *Angew. Chem. Int. Ed.* **2016**, *56*, 405–408. [CrossRef] [PubMed]
188. Mitschang, F.; Schmalz, H.; Agarwal, S.; Greiner, A. Tea-Bag-Like Polymer Nanoreactors Filled with Gold Nanoparticles. *Angew. Chem. Int. Ed.* **2014**, *53*, 4972–4975. [CrossRef] [PubMed]
189. Taguchi, T.; Isozaki, K.; Miki, K. Enhanced Catalytic Activity of Self-Assembled-Monolayer-Capped Gold Nanoparticles. *Adv. Mater.* **2012**, *24*, 6462–6467. [CrossRef] [PubMed]
190. Raffa, P.; Evangelisti, C.; Vitulli, G.; Salvadori, P. First examples of gold nanoparticles catalyzed silane alcoholysis and silylative pinacol coupling of carbonyl compounds. *Tetrahedron Lett.* **2008**, *49*, 3221–3224. [CrossRef]
191. Mitsudome, T.; Yamamoto, Y.; Noujima, A.; Mizugaki, T.; Jitsukawa, K.; Kaneda, K. Highly Efficient Etherification of Silanes by Using a Gold Nanoparticle Catalyst: Remarkable Effect of O_2. *Chem. Eur. J.* **2013**, *19*, 14398–14402. [CrossRef]

192. Kronawitt, J.; Dulle, M.; Schmalz, H.; Agarwal, S.; Greiner, A. Poly(p-xylylene) Nanotubes Decorated with Nonagglomerated Gold Nanoparticles for the Alcoholysis of Dimethylphenylsilane. *ACS Appl. Nano Mater.* **2020**, *3*, 2766–2773. [CrossRef]
193. Isozaki, K.; Taguchi, T.; Ishibashi, K.; Shimoaka, T.; Kurashige, W.; Negishi, Y.; Hasegawa, T.; Nakamura, M.; Miki, K. Mechanistic Study of Silane Alcoholysis Reactions with Self-Assembled Monolayer-Functionalized Gold Nanoparticle Catalysts. *Catalysts* **2020**, *10*, 908. [CrossRef]
194. Hils, C.; Fuchs, E.; Eger, F.; Schöbel, J.; Schmalz, H. Converting Poly(Methyl Methacrylate) into a Triple-Responsive Polymer. *Chem. Eur. J.* **2020**, *26*, 5611–5614. [CrossRef]

Review

Solution Self-Assembly of Coil-Crystalline Diblock Copolypeptoids Bearing Alkyl Side Chains

Naisheng Jiang [1,*] and Donghui Zhang [2,*]

1. School of Materials Science and Engineering, University of Science and Technology Beijing, Beijing 100083, China
2. Macromolecular Studies Group, Department of Chemistry, Louisiana State University, Baton Rouge, LA 70803, USA
* Correspondence: naishengjiang@ustb.edu.cn (N.J.); dhzhang@lsu.edu (D.Z.)

Abstract: Polypeptoids, a class of synthetic peptidomimetic polymers, have attracted increasing attention due to their potential for biotechnological applications, such as drug/gene delivery, sensing and molecular recognition. Recent investigations on the solution self-assembly of amphiphilic block copolypeptoids highlighted their capability to form a variety of nanostructures with tailorable morphologies and functionalities. Here, we review our recent findings on the solutions self-assembly of coil-crystalline diblock copolypeptoids bearing alkyl side chains. We highlight the solution self-assembly pathways of these polypeptide block copolymers and show how molecular packing and crystallization of these building blocks affect the self-assembly behavior, resulting in one-dimensional (1D), two-dimensional (2D) and multidimensional hierarchical polymeric nanostructures in solution.

Keywords: polypeptoids; diblock copolymers; crystallization; solution self-assembly

1. Introduction

Self-assembly of amphiphilic block copolymers (BCPs) in solution is one of the most fascinating phenomena in polymer physics due to the unique properties and numerous potential applications of the resulting nanostructures. The creation of various well-defined polymeric nanostructures with tailorable size and functionalities via solution self-assembly is not only useful in drug delivery, catalysis, optoelectronics and structured nanomaterials, but also provided unique perspective to understand the structural complexity and assembly rules of biomacromolecules observed in biological systems. To minimize the total free energy of the system, polymeric amphiphiles tend to self-assemble into well-defined morphologies in a selective solvent whenever the polymer concentration is above the critical micelle concentration (CMC). For coil–coil diblock copolymers in selective solvent, the most common self-assembled morphologies include spherical micelles, wormlike micelles and vesicles with a core-shell type of architecture. The thermodynamic equilibrium morphology of these self-assembled structures is described by the so-called dimensionless packing parameter, p, which is defined by $p = v/a_0 l_c$, where v and l_c are the volume and the length of the solvophobic block, respectively, and a_0 is the optimal surface area of the solvophilic block at the core-corona interface [1]. In many cases, the thermodynamic equilibrium with lowest free energy is not readily achieved, as the molecular exchange amongst polymeric aggregates is sluggish relative to their self-assembly process [2]. This in turn opens up opportunities to utilize different self-assembly pathways to attain uncommon solution morphologies that are kinetically trapped.

The early scaling work done by Vilgis and Halperin [3] suggested that by introducing a crystallizable block, which adds an extra driving force into the system, the self-assembly behavior of amphiphilic BCPs in solution can be significantly altered. The crystalline core confined within polymeric micelles may provide novel control options when used as seeds for further crystallization. During the past two decades, solution self-assembly of BCPs

with a crystallizable block has been used as an effective method for the generation of various non-spherical polymeric micelles or nanostructures, including one-dimensional (1D) nanofibers or nanorods [4–6], two-dimensional (2D) nanosheets or platelets [7–9] or even more sophisticated hierarchical nanostructures under specific conditions [10]. This is also known as the so-called crystallization-driven self-assembly (CDSA) process where the aggregate morphology and self-assembly pathways are dominated by the epitaxial crystalline growth of macromolecular building blocks in solution. More importantly, upon crystallization, the molecular exchange is often restricted due to high free energy penalty, resulting in kinetically trapped molecular assemblies in an out-of-equilibrium state with a very long lifetime [2,11–15]. As a result, one can access novel polymeric nanostructures with varying morphology by controlling the self-assembly pathways during sample preparation. It has been shown that CDSA pathways and final solution morphologies of BCPs can be influenced by many factors, such as chemical composition, block ratios, polymer concentration, polymer–solvent interactions, molecular packing of the crystallizable block, annealing condition and other external stimuli [8,16–21]. In some cases, the solution self-assembly of crystallizable BCPs can proceed in a living fashion by gradually adding molecularly dissolved unimers to the pre-existing "seed" crystals, enabling the access to near monodisperse anisotropic nanostructures or hierarchical assemblies with varying levels of structural complexity in solution [5,10,19,22–29].

Non-spherical nanomaterials formed by the solution self-assembly of biocompatible macromolecules exhibit unique properties that are desirable in many biomedical and biotechnological applications, such as drug/gene delivery [30,31] and biomineralization [32]. For example, relative to spherical nanoparticles, elongated filomicelles or nano-disks can either lead to longer blood circulation time [30] or promote cell exterior binding with reduced cell uptake [33]. To date, chemists have used many crystalline polymers as the primary core-forming building blocks to facilitate the CDSA of amphiphilic BCPs in solution, including flexible linear polymers (e.g., polyethylene [15,16], poly(ε-caprolactone) [9,17,34], poly(L-lactide)) [8,21,23] and polymers with either relatively rigid backbones or bulky side groups that exhibit either crystalline or liquid crystalline (LC)-like behaviors (e.g., poly(2-(perfluorooctyl)ethyl methacrylate [35,36], poly(γ-benzyl-L-glutamate) [37], polyferrocenylsilanes [6,10,38,39] and polythiophenes [26,28]). However, considering the potential use of polymeric self-assemblies in biomedical and biotechnological applications, there is also a need of polymers that have desirable bioactivity, cytocompatibility, biodegradability and enzymatically stability.

As a class of bioinspired synthetic polymers, polypeptoids featuring N-substituted polyglycine backbones are structural mimics of polypeptides [40–45]. Due to the absence of hydrogen bonding and stereogenic centers along the backbone (Figure 1), polypeptoids exhibit good thermal processability and solubility in various organic solvents, as well as enhanced protease stability, in sharp contrast to polypeptides. These features make polypeptoid a potential candidate for a wide variety of biomedical and biotechnological applications, such as antifouling coatings [46–51], drug/gene delivery [52–55] and biosensing [56–58]. Recent developments in the controlled polymerization and solid-phase synthesis have enabled access to a variety of polypeptoids with tailorable chain length, N-substituent structures, sequence, and architecture [59–68]. In the past several years, solution self-assembly of block copolypeptoids have especially attracted increasing attention due their capability to form various self-assembled nanostructures with tailorable structure, morphology, and functionality [60,67–85]. Given their molecular tunability, polypeptoid-based BCPs are considered as a promising biomimetic platform for macromolecular and supramolecular engineering for biomedical and biotechnological applications.

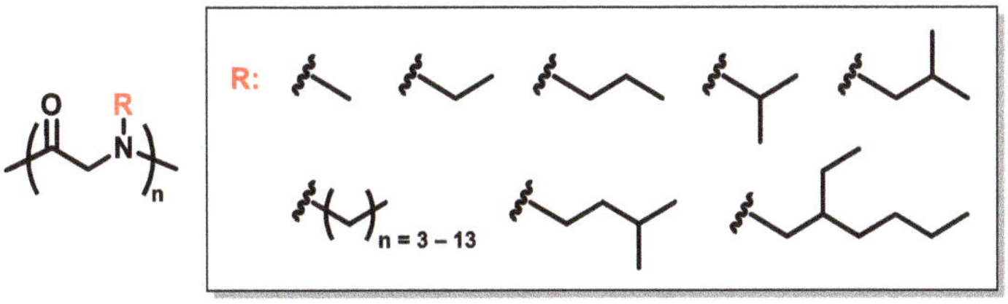

Figure 1. Structures of polypeptoids bearing various alkyl side chains that have been reported.

This review highlights our recent experimental findings on solution self-assembly of coil-crystalline diblock copolypeptoids bearing alkyl side chains. In the next section, we briefly discuss the synthesis and crystallization behavior of polypeptoids bearing alkyl side chains. In Section 3, we focus on the CDSA of coil-crystalline diblock copolypeptoids in solution. The effects of molecular packing, block composition and side chain architecture on CDSA, as well as the self-assembly pathways are presented. Perspectives on the solution self-assembly of coil-crystalline diblock copolypeptoids bearing alkyl side chains are followed by a brief conclusion.

2. Polypeptoids Bearing Alkyl Side Chains: Synthetic Methods and Their Phase Behavior

2.1. Controlled Ring-Opening Polymerization (ROP) for Polypeptoids

Polypeptoid-based polymers are commonly synthesized via two methods: (1) Submonomer solid-phase synthesis, and (2) ring-opening polymerization (ROP) of N-substituted glycine derived N-carboxyanhydride (R-NCA) or N-thiocarboxyanhydride (R-NTA) monomers using nucleophilic initiators (e.g., primary amine). The former method involves the growth of polypeptoid chain from the C-terminus to the N-terminus by alternating attachment of bromoacetic acid and various primary amines on a NH_2-bearing solid support [42,86]. The stepwise synthetic method allows the access to monodispersed polypeptoids with diverse structures and precise control of chain length and sequences, which is advantageous for applications where sequence-defined copolypeptoids with complex side chain functionalities are required [40,68,69,72,73,87–92]. However, long chain polypeptoids with degree of polymerization (DP_n) greater than 50 are difficult to obtain by the sub-monomer method. By contrast, high molecular weight polypeptoids can be synthesized by controlled ROP of R-NCA or R-NTA monomers using nucleophilic initiators such as primary amines [41,62,66,93]. A wide variety of R-NCA or R-NTA monomers have been reported for primary amine-initiated ROP of polypeptoids bearing various N-substituents in a one-pot fashion [44,62,65,66,93–96]. As this review is focused on the solution self-assembly of diblock copolypeptoids bearing alkyl side chains with relatively high molecular weight, we will mainly discuss the polymer synthesis using the controlled ROP method. The readers are referred to the previous reviews [41–45] for a more comprehensive view regarding the synthesis of peptoids and polypeptoids.

As shown in Scheme 1, diblock copolypeptoids can be synthesized by controlled ROP of R-NCA monomers in a sequential manner using nucleophilic initiators such as primary amines. This feature has been attributed to the controlled/living nature of ROP of R-NCAs [59,61,62]. By reaching a complete conversion of the first R-NCA, the second R-NCA can be directly added to the reaction mixture as long as the product maintains a good solubility, where the entire polymerization reaction can be easily monitored by infrared (IR) spectroscopy. The actual molecular weight and block ratio of the final product can be determined by end-group analysis using ^1H NMR spectroscopy in conjunction with matrix-assisted laser desorption/ionization time-of-flight mass spectrometry (MALDI-TOF MS), size-exclusion chromatography (SEC) or viscosity measurements. As R-NCAs are very sensitive to moisture, the use of anhydrous solvents is necessary, and the reactions are normally conducted under anhydrous condition with water content less than 30 ppm to avoid side reactions. Luxenhofer et al. have shown that the benzyl amine-initiated ROP of R-NCA, such as N-methyl N-carboxyanhydride (Me-NCA) and N-butyl N-carboxyanhydride (Bu-NCA) proceeded in a controlled manner without chain transfer or termination events, yielding well-defined homopolypeptoids with controlled molecular weights and narrow molecular weight distributions (PDI < 1.1–1.3) [61,62]. In addition, they also have shown that well-defined block copolypeptoids, e.g., poly(N-methyl glycine)-b-poly(N-butyl glycine) (PNMG-b-PNBG), can be produced by controlled ROP with sequential monomer addition [61,62]. In our previous studies [60,70,79,84], a variety of well-defined amphiphilic diblock copolypeptoids that comprised of at least one crystallizable block with relatively low PDIs have been synthesized via benzyl amine-initiated ROPs of R-NCAs in a sequential manner, allowing us to further investigate their CDSA behaviors in solution. There are several aliphatic N-substituents can be readily used for the molecular design (Figure 1). For coil-crystalline diblock copolypeptoids, Me-NCA is the most commonly used monomer to produce the amorphous PNMG block, whereas R-NCAs bearing long n-alkyl groups (e.g., N-octyl N-carboxyanhydride (Oct-NCA) and N-decyl N-carboxyanhydride (De-NCA)) are often used to produce the crystallizable block. Branching of the alkyl side chains can also be introduced to tune the inter- and intramolecular interactions of polypeptoid, thus allowing their molecular packing and phase behavior to be systematically tailored.

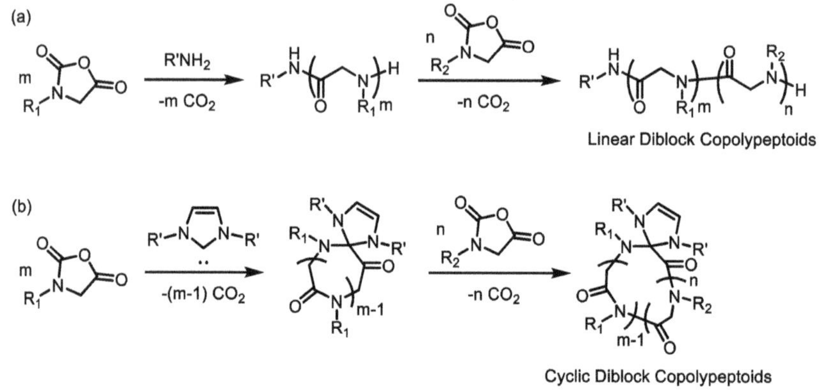

Scheme 1. (a) Synthesis of linear diblock copolypeptoids via primary amine-initiated ROP of R-NCAs in a sequential manner. R'NH$_2$: Primary amine. (b) Synthesis of cyclic diblock copolypeptoids via NHC-mediated ZROP of R-NCAs in a sequential manner.

While primary amine-initiated ROP of R-NCAs yields linear diblock copolypeptoids, recent synthetic developments in the organo-mediated zwitterionic ring-opening polymerization (ZROP) of R-NCAs have also enabled access to polypeptoids with cyclic topology [59]. In previous studies, N-heterocyclic carbenes (NHCs) have been used as initiators/organo-catalysts to initiate/mediate ZROP of R-NCAs (e.g., Me-NCA, Bu-NCA Oct-NCA and De-NCA) for a variety of well-defined cyclic polypepotids with tunable molecular weight and narrow dispersity [59,63,93,97]; 1,8-Diazabicycloundec-7-ene (DBU), a bicyclic amidine that is less sensitive to air and moisture relative to NHC, has also been demonstrated capable of mediating ZROPs of R-NCAs in a similar manner [98,99]. It has been demonstrated that the ZROP proceeded through a zwitterionic propagating intermediate where the two oppositely charged chain ends are held in proximity through electrostatic interaction. Low dielectric solvents, such as tetrahydrofuran (THF) and toluene, were normally used to avoid intramolecular transamidation relative to chain propagation and ensure a controlled ZROP reaction [44]. The quasi-living nature of the organo-mediated ZROP also enable the access to well-defined amphiphilic cyclic diblock copolypeptoids by sequential monomer addition (Scheme 1b), which allow us to exploit the effect of chain topology on solution self-assembly [60].

2.2. Molecular Packing and Phase Behavior of Polypeptoid Homopolymers Bearing Alkyl Side Chains

As aforementioned, the self-assembly pathway and final aggregate morphology of amphiphilic coil-crystalline diblock copolymers in solution depends strongly on the molecular packing and phase behavior of the crystallizable block. Thus, it is important to first gain a thorough understanding on the general packing motifs and crystallization behavior of the corresponding homopolymers. As one would expect, molecular packing, crystallization behaviors, thermal transitions and solubility of polypeptoids highly rely on their N-substituent structures. Here, we briefly summarize the molecular packing and phase behavior of homopolypeptoids bearing alkyl side chains.

Polypeptoids bearing linear aliphatic N-substituents shorter than 2 carbons are amorphous and behave as random coil-like polymers. The most famous example is poly (N-methyl glycine) (PNMG), a.k.a. polysarcosine, the simplest member of the polypeptoid family, which is amorphous and can be readily dissolved in water or alcohol [100–102]. As a promising biodegradable poly (ethylene glycol) (PEG)-alternative, PNMG can be served as hydrophilic component by providing steric stabilization of nanostructures, plasmonic particles or proteins for various biomedical applications, such as drug delivery and theranostics [54,102–107]. On the other hand, it has been found that polypeptoids with relatively long linear n-alkyl side chains ($4 \leq S \leq 14$, where S is the number of carbon atoms in the n-alkyl group) are highly crystalline in the solid state [89,97,108]. Using X-ray diffraction and molecular dynamics simulations, Balsara, Zuckermann and coworkers have revealed their general packing motif: Polypeptoid molecules bearing n-alkyl side chains tend to adopt a board-like structure in the crystalline state, where the backbone is fully extended in an all-cis backbone conformation and is approximately coplanar with the n-alkyl side chains (Figure 2) [108]. The all cis-amide backbone conformation, which is more compact and possesses a higher degree of ordering than the all-trans backbone conformation, allows for more favorable intra- and inter-molecular interactions upon crystallization/supramolecular assembly. With N backbone repeating units and S number of carbons on the n-alkyl side chains, the unit cell dimensions, namely a, b and c, follow a universal relationship, in which $a = 0.455$ nm, $b = (0.298N + 0.035)$ nm and $c = (0.186S + 0.55)$ nm. Note that crystallization of these board-like polypeptoids can occur at relatively low number-average degree of polymerization (e.g., $DP_n = 9$) in both solid and solution states [72,108].

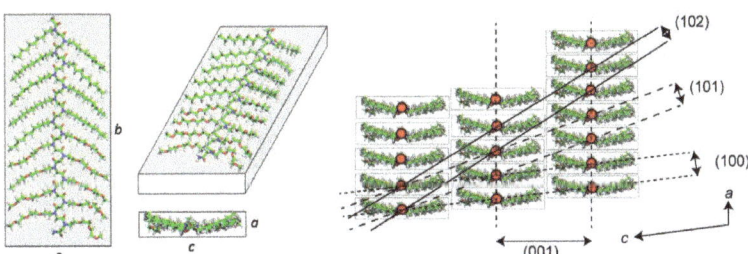

Figure 2. The unit cell dimensions (**left**) and supramolecular assembly of peptoid molecules with all cis-amide backbone conformation in the crystalline state (**right**). Figure reproduced from reference [108] with permission from the American Chemical Society.

By using differential scanning calorimetry (DSC), Luxenhofer showed that the glass transition temperature (T_g) of polypeptoids with $1 \leq S \leq 5$ decreases with increasing n-alkyl side chain length [109], which is likely due to the plasticization effect induced by the flexible n-alkyl side chains [110]. The increase of n-alkyl side chain length also leads to an increasing tendency towards crystallization. Lee et al. found that both linear and cyclic polypeptoids bearing long linear n-alkyl side chains ($4 \leq S \leq 14$) exhibit two phase transitions upon temperature increase (Figure 3a) [97]. These two transition temperatures are strongly coupled and highly depend on the number of carbon atoms in the n-alkyl group. Using temperature-dependent X-ray scattering, Balsara, Zuckermann and coworkers further evidenced the broadening of the (100) peak and the disappearance of higher order peaks of diblock polypeptoid bearing n-decyl side chains during the lower-temperature transition, indicating the diminished ordering for the face-to-face stacking of the board-like polypeptoid molecules (Figure 3b) [111]. Thus, it has been suggested that the lower-temperature transition corresponds to a crystalline phase to a "sanidic" liquid crystalline (LC) mesophase transition, while the higher-temperature transition corresponds to the LC mesophase to isotropic melt transition (Figure 3b) [111]. In our very recent study on poly (N-decyl glycine) (PNDG) thin films prepared on solid substrates, we showed that both linear and cyclic PNDG exhibit two amorphous halos when heating above the isotropic melt transition temperature (Figure 3c) [112]. Interestingly, the d-spacings of these two amorphous halos are in good agreement with the theoretical molecular dimension of PNDG in an extended trans-amide backbone conformation. Therefore, it was proposed that polypeptoid molecules undergo a cis-to-trans amide backbone conformational transition when heating above the isotropic melting temperature, where the long n-alkyl side chains are still nearly coplanar with the polypeptoid backbone [112].

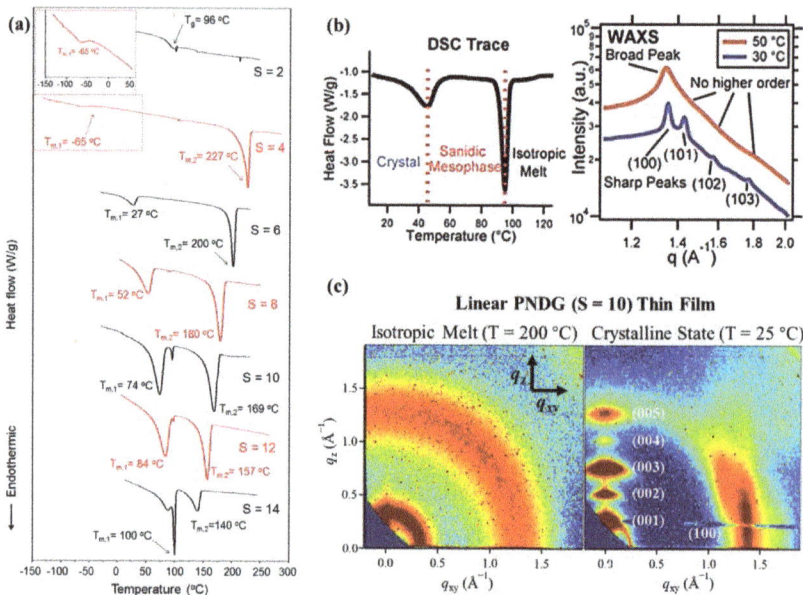

Figure 3. (a) DSC thermograms of linear polypeptoids bearing different n-alkyl side chains with $2 \leq S \leq 14$ during the second heating cycle. (b) DSC and WAXS results of N-acetylated diblock polypeptoid bearing n-decyl side chains, i.e., Ac-pNdc$_9$-b-pNte$_9$. Three different phases can be identified, i.e., crystalline phase, sanidic LC mesophase and isotropic melt. (c) Two-dimensional GIWAXD images of 48 nm thick linear PNDG (DP$_n$ = 52) thin film prepared on Si substrate measured at T = 200 °C and T = 25 °C after cooling from 200 °C, respectively. The out-of-plane (q_z) and in-plane (q_{xy}) directions are indicated by arrows. Figures reproduced from references [97,111,112] with permission from the American Chemical Society.

Side chain engineering has long been served as an effective strategy to modulate inter and intra-molecular interactions and packing of crystallizable polymers, thus allowing their morphology, solubility, and functionality to be systematically tailored [113]. It has been found that the molecular packing and crystallization behavior of polypeptoids are turned into a different scenario when the alkyl side chains are asymmetrically branched. For example, in the case of racemic 2-ethyl-l-hexyl side chains, it was found that relatively short poly

(N-2-ethyl-1-hexyl glycine) (PNEHG) molecules with $DP_n \leq 20$ are amorphous with no first-order transition observed by DSC [90,91]. On the other hand, longer PNEHG chains, e.g., $DP_n \geq 100$, exhibit a single first-order thermal transition with a small enthalpic change, but are still in sharp contrast with poly (N-octyl glycine) (PNOG) homopolymer which also possesses eight carbon atoms on their side chains [84,97]. Our recent WAXD measurements on PNOG and PNEHG homopolymers with similar DP_n ($DP_n \approx 100$) revealed the effect of side chain branching on the molecular geometry and supramolecular assembly [84]. As shown in Figure 4b, $PNOG_{103}$ homopolymer was found to exhibit typical reflection peaks in the WAXD profile due to the side-by-side (along crystallographic c-axis) and face-to-face stackings (along crystallographic a-axis) of the board-like molecules, consistent with those observed for polypeptides bearing linear n-alkyl side chains. By contrast, $PNEHG_{100}$ exhibits a primary diffraction peak at $q^* = 0.50$ Å^{-1} due to the distance (1.26 nm) between adjacent PNEHG backbones that are separated by the interdigitated N-2-ethyl-1-hexyl side chains, along with multiple weak higher order peaks located at $\sqrt{3}q^*$, $\sqrt{4}q^*$, $\sqrt{7}q^*$, respectively. A broad amorphous peak near $q = 1.2$ Å^{-1} is also discernible, which is likely to arise from the interchain distance among the 2-ethyl-1-hexyl side chains. This result indicates that the PNEHG molecules are rod-like and packed into a hexagonal mesophase upon cooling from isotropic melt. Unlike PNOG chains that preferentially adopt a board-like geometry with all n-octyl side chains aligned in the same plane, the greater steric hindrance of the bulky racemic 2-ethyl-1-hexyl side chains makes it energetically unfavorable for the PNEHG to adopt a planar geometry. Instead, the PNEHG molecules adopt a rod-like geometry with an extended backbone conformation, which allow the side chains to orient outwardly along the backbone, thereby minimizing steric repulsion amongst the bulky branched alkyl substituents. Therefore, asymmetric branching of the aliphatic N-substituents (e.g., 2-ethyl-l-hexyl) not only suppresses the degree of crystallization, but also changes the packing motif of polypeptides. As we will show, such feature allows us to modulate the solution self-assembly of amphiphilic diblock copolypeptoids by side chain branching pattern.

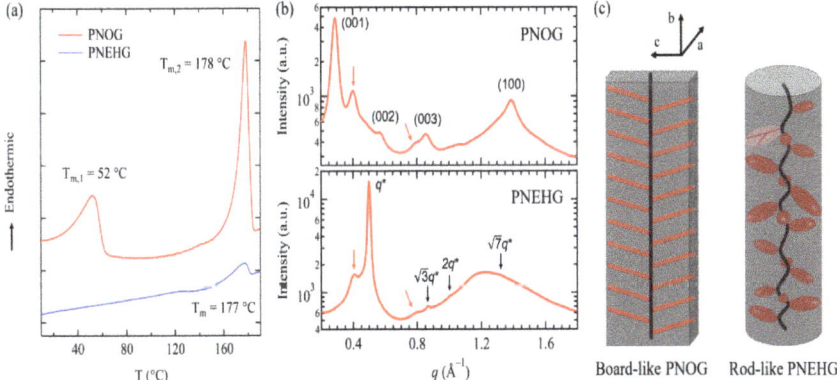

Figure 4. (a) DSC thermograms of PNOG ($DP_n = 103$) and PNEHG ($DP_n = 100$) homopolymers during the second heating cycle. (b) One-dimensional WAXD profiles of the PNOG and PNEHG homopolymers measured at T = 45 °C. The primary and secondary peaks associated to the Kapton windows on the sample cell are indicated by the red arrows. (c) Proposed molecular geometries of PNOG and PNEHG at crystalline/liquid-crystalline state. Figure reproduced from reference [84] with permission from the American Chemical Society.

3. Solution Self-Assembly of Coil-Crystalline Diblock Copolypeptoids Bearing Alkyl Side Chains

3.1. Sample Preparation and Characterization of Coil-Crystalline Diblock Copolypeptoid Solutions

For long chain macromolecules, crystallization is a kinetically controlled process, implying that sample preparation and processing pathways have profound impacts on the final structure and crystalline morphology. For solution self-assembly of BCPs that comprised of a crystallizable block, size, shape and morphology of the final nanostructures strongly rely on the self-assembly pathways, which can be affected by many preparation/processing factors, including initial polymer concentration, solvent quality, impurity and thermal history. The interplay between aggregation and crystallization also plays a key role in determining the final solution morphology. As crystallization is strongly temperature and concentration dependent, it can be manipulated to control the self-assembly pathway. The simplest way to trigger a CDSA process is by first dissolve the BCPs molecularly in good or nonselective solvent, then changing the temperature or solvent quality to induce crystallization.

It should be noted that the solubility of individual blocks of BCPs in a given solvent is important when planning a CDSA experiment. For polypeptoids, the polymer–solvent interaction may vary significantly depending on the number of carbon atoms in their alkyl side groups (S). For example, while both PNMG (S = 1) and poly (N-ethyl glycine) (S = 2) have good solubility in water [107], poly (N-propyl glycine) with S = 3 was found to exhibit lower critical solution temperature (LCST) in water, which can phase separate when heated above the cloud point temperature (i.e., 15–25 °C) [114]. As the S value further increases, polypeptoid molecules become increasingly more hydrophobic and exhibited diminished solubility in water. Similar trend of solubility versus side chain length was also observed for polypeptoids in methanol, which showed that PNDG with S = 10 is more solvophobic than PNMG at room temperature, as evidenced by liquid contact angle measurements [79]. It was found that a wide range of polypeptoids bearing n-alkyl side chains with $1 \leq S \leq 12$ can be readily solubilized in dichoromethane and chloroform [97]. Toluene and DMF can also dissolve polypeptoids with relatively short n-alkyl side chains (e.g., S = 4) [59,63], whereas for crystallizable polypeptoids bearing longer n-alkyl side chains, THF can provide good solubility at high temperatures (e.g., 50 °C or above) [97,112]. Note that the polypeptoid solubility may also depends on molecular weight and preparation history of the polymer samples [60,97,107].

Our previous studies mainly focused on the solution self-assembly of coil-crystalline diblock copolypeptoids bearing alkyl side chains in methanol. Using scattering techniques, we found that most coil-crystalline diblock copolypeptoids, e.g., PNMG-b-PNOG with $f_{PNOG} \leq 0.73$ and PNMG-b-PNDG with $f_{PNDG} \leq 0.44$ (where f is the volume fraction of the crystalline block), can be molecularly dissolved (i.e., forming unimers) at dilute concentrations in methanol by heating at high temperatures [79,84]. Cooling the methanol solution down to room temperature induces the recrystallization of PNOG or PNDG blocks. Therefore, CDSA of diblock copolypeptoids can be triggered by first heating the solution at high temperature, subsequently cooling to desired temperature and keep the solution under isothermal condition for prolonged time. Note that multiple structural changes and phase transitions can take place during these steps, thus, the solution samples must be carefully characterized with good spatial and temporal resolutions.

Various in situ and ex situ techniques can be used to monitor the structural evolution of coil-crystalline diblock copolypeptoids during the solution self-assembly, such as static light scattering (SLS), small-/wide-angle X-ray scattering (SAXS/WAXS), small-angle neutron scattering (SANS), (cryo-)transmission electron microscopy (TEM or cryo-TEM) and atomic force microscopy (AFM). While TEM and AFM can "visually" characterize the polymeric nanostructures self-assembled in real space, reciprocal scattering techniques using X-ray or neutron sources are more powerful tools in terms of providing global averaged structural information at length scales ranging from micrometers to angstroms [115]. In addition, scattering techniques are generally non-destructive and can directly characterize the self-assembled structures in solution environments. Modern synchrotron X-ray sources with high flux and small beam divergence, in conjunction with the use of hybrid pixel array detectors that allow direct photon detection (e.g., PILATUS detectors by Dectris Ltd.), also enable fast data collection (down to milliseconds per data) for dilute samples, which makes SAXS/WAXS ideally suited for probing multiscale structural evolution of BCPs as a function of time during solution self-assembly [116–119]. It should be mentioned that polypeptoid self-assemblies induced by CDSA often possess multiple levels of structural hierarchy and heterogeneity in solution (see examples below). Therefore, extreme care should be taken when interpreting the scattering data, especially for those collected from the small-angle regime. In this regard, SAXS and SANS are often complimentarily used in conjunction with other microscopic techniques to provide a compelling characterization of the solution CDSA process over a wide range of length- and time-scales.

3.2. Crystallization-Driven Self-Assembly of 1D nanofibrils

Long, wormlike 1D nanofibrils can be generated by the CDSA of coil-crystalline diblock copolypeptoids with relatively low volume fraction of the crystalline block in methanol. Lee et al. previously showed that $PNMG_{112}$-b-$PNDG_{16}$ comprised of a soluble PNMG block and a much shorter crystallizable PNDG slowly self-assembled into long, wormlike nanofibrils in dilute methanol with a polymer concentration of $c = 1$ mg/mL [60]. Cryo-TEM imaging revealed a morphological transition from spherical micelles to micrometer length nanofibrils during the course of seven days after the solution was cooled down to room temperature (Figure 5a–c). Similar self-assembly behavior was also found for the cyclic counterpart, i.e., cyclic-$PNMG_{105}$-b-$PNDG_{10}$, except for its relatively slower self-assembly kinetics. At higher concentrations (e.g., 10 wt%), these linear and cyclic PNMG-b-PNDGs form free-standing gels after the solution was cooled, which is attributed to the formation of gel network that comprised of entangled crystalline nanofibrils [70].

To better understand the crystalline packing of PNDG segment in the wormlike micelles, X-ray scattering experiments of the worm-micelle solution was conducted under unidirectional flow. Figure 5d–f shows the 2D SAXS, MAXS and WAXS data of the 5 mg/mL methanol solution containing the $PNMG_{105}$-b-$PNDG_{20}$ long, wormlike 1D nanofibrils under unidirectional flow in a capillary flow cell with a constant shear rate of ~25.6 s^{-1} near the wall. Under the influence of the flow field, the wormlike micelles were preferentially aligned parallelly with the flow direction, evidenced by the increasing anisotropy of the 2D SAXS patterns with increasing flow rate [79,120–122]. Meanwhile, M/WAXS analysis (Figure 5g,h) has revealed a significantly more pronounced scattering peak due to the (001) reflection in the $q_\perp$ direction relative to that in the $q_{//}$ direction, indicating that the (001) molecular packing separated by the n-decyl side chains (i.e., side-by-side stacking) with a d-spacing of $d_{001} = 2.4$ nm was aligned in the direction perpendicular to the long axis of the $PNMG_{105}$-b-$PNDG_{20}$ nanofibrils. Consistently, the scattering peaks due to (100) reflection and the associated higher order reflections from (101) and (102) planes are more notable along the $q_{//}$ direction as compared to those in the $q_\perp$ direction. According to Figure 2, the result indicates that the adjacent cis-amide backbones with a d-spacing of $d_{100} = 0.46$ nm due to the face-to-face stacking was aligned in a direction parallel to the long axis of the nanofibrils. These combined results support an anisotropic crystalline core structure for the long, wormlike 1D nanofibrils, as depicted in the inset of Figure 5c.

Time/temperature-dependent synchrotron X-ray/neutron scattering experiments were performed to further investigate the self-assembly pathway and formation mechanism of these 1D nanofibrils. High-temperature solution SAXS result shows that the $PNMG_{105}$-b-$PNDG_{20}$ polymers are well-dissolved and exist as unimers in methanol at 65 °C (Figure 6a,b), which gives a radius of gyration (R_g) of 2.3 nm by Guinier plot analysis. Upon cooling the solution to room temperature, a drastic change of the scattering profile was observed (Figure 5i). At time $t = 0$ min, i.e., immediately after the solution was cooled down from 65°C to room temperature, the SAXS profile shows a noticeable upturn at the low q region, while the overall intensity is still relatively weak. Such intensity upturn with a $q^{-2.5}$

dependence is attributed to the formation of polymer aggregates, i.e., the "seeds", at the very early stage after the solution was cooled down to room temperature. With increasing of time, (001), (100) and their higher order reflections started to appear after ~100 min and intensify over time. Meanwhile, at the low q regime in SAXS spectra, the dependence of intensity over q gradually changed over from I ~ $q^{-2.5}$ to $q^{-1.5}$, and eventually to q^{-1} after ~400 min, while the overall absolute SAXS intensity continue to increase until a final state was reached. We attribute this increase in the SAXS intensity mainly due to the one-dimensional elongation of the nanofibrils, consistent with the time-dependent cryo-TEM results shown in Figure 5a–c. Note that the SAXS profile of the final PNMG$_{105}$-b-PNDG$_{20}$ nanofibrils at t = 15 days can be well-fitted by using the scattering model for core-corona cylindrical micelles developed by Pedersen and co-workers [123,124], which gives a core radius (R_c) of 3.3 ± 0.2 nm and a radius of gyration of the corona chains of 3.8 ± 0.2 nm. The average diameter of the nanofibrils is then estimated to be approximately 23.4 nm, in good agreement with the cryo-TEM result.

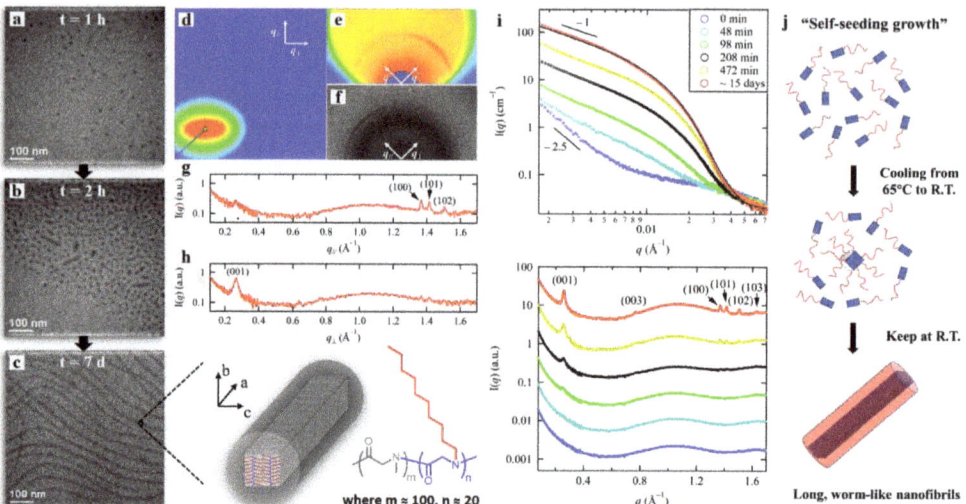

Figure 5. (a–c) Representative cryo-TEM images of 1 mg/mL PNMG$_{112}$-b-PNDG$_{16}$ methanol solution at different waiting time (t) after been cooled down to room temperature. The chemical structure of PNMG-b-PNDG block copolypeptoid and the schematic illustration of molecular packings of the core-forming PNDG blocks within a nanofibril are shown in the inset of (c), where the backbones and n-decyl side chains of PNDG are indicated in blue and red, respectively. (d–f) Two-dimensional SAXS (d), MAXS (e) and WAXS (f) images for 5 mg/mL PNMG$_{105}$-b-PNDG$_{20}$ methanol solution measured during unidirectional flow at room temperature. The directions parallel ($q_{//}$) and perpendicular ($q_\perp$) to the flow direction are indicated by arrows. The corresponding one-dimensional profiles of the MAXS/WAXS results along the $q_{//}$ and $q_\perp$ directions are plotted in (g) and (h), respectively. (i) SAXS (top) and MAXS/WAXS (bottom) profiles of 5 mg/mL PNMG$_{105}$-b-PNDG$_{20}$ methanol solution measured at static state at different t. The solid line in the SAXS profile at t = ~15 days corresponds to the best-fit to the data based on the cylindrical-shaped micelle model. (j) Schematic illustration of the proposed self-assembly mechanisms for the 1D nanofibrils (e.g., PNMG$_{105}$-b-PNDG$_{20}$) via CDSA. Figures reproduced from references [60,79] with permission from the American Chemical Society.

The above results clearly show that the elongation of PNMG$_m$-b-PNDG$_n$ nanofibrils with relatively low volume fraction of PNDG segments (i.e., m ≈ 100 and n ≈ 20) is induced by the face-to-face stacking of PNDG backbones along the crystallographic a-axis (i.e., the (100) packing), while the cross-sectional dimension (or lateral diameter) of the nanofibrils is determined by the backbone length of PNDG and the (001) packing along the crystallographic c-axis. Based on the time-dependent SAXS/WAXS and cryo-TEM results, it is reasonable to conclude that the 1D nanofibrils formation is mainly governed by the so-called "self-seeding growth" mechanism [6,9,13,19,28,125,126], which involves the initial formation of a few small "seed" crystals followed by the preferential addition of the unimers to the crystalline front (Figure 5j). Apparently, for board-like PNDG molecules, the creation of anisotropic crystalline core requires preferential addition of the unimers to a certain crystallographic facet, instead of adding unimers equally in all directions. As we found, the face-to-face stacking of the PNDG segment is more favored over the side-by-side packing during the seeded growth process, resulting in the unidirectional elongation of long wormlike micelles.

3.3. Effect of Block Composition on the Solution Self-Assemblies

For amphiphilic coil-crystalline BCPs, increasing the volume fraction or DP_n of the solvophobic crystalline block often leads to drastic changes in the self-assembly pathway and final aggregate morphology in solution (Figure 6d,e) show the cryo-TEM images for the self-assembled $PNMG_{121}$-b-$PNDG_{46}$ (f_{PNDG} = 0.61) and $PNMG_{124}$-b-$PNDG_{63}$ (f_{PNDG} = 0.68) nanostructures in 5 mg/mL methanol solution. Unlike the long, wormlike nanofibrils formed by CDSA of PNMG-b-PNDGs with relatively low PNDG volume fractions (i.e., f_{PNDG} = 0.44), the $PNMG_{121}$-b-$PNDG_{46}$ having an intermediate PNDG volume fraction self-assembled into rigid rod-like structures with much shorter length (~100–400 nm) under identical sample preparation condition. The best-fit to the SANS data of $PNMG_{121}$-b-$PNDG_{46}$ nanorods (Figure 6f) using cylindrical-shaped polymer micelle model gives a PNDG core radius of R_c = 6.8 ± 0.2 nm, which is two times larger than that for $PNMG_{105}$-b-$PNDG_{20}$ nanofibrils. The discrete reflection peaks due to the face-to-face and side-by-side packing of PNDG were also observed by WAXS, suggesting the occurrence of highly ordered crystalline structure of core-forming PNDG blocks. Assuming the crystalline packing of the PNDG segments in the 1D nanorods are identical to that in the 1D nanofibrils, the cross section of the former core would have at least 4 PNDG molecules stacked side-by-side in a fully extended cis-amide backbone conformation.

As the volume fraction of PNDG segments is further increased, cryo-TEM revealed the predominant presence of 2D nanosheets in addition to some short nanorods for the $PNMG_{124}$-b-$PNDG_{63}$ methanol solution. The majority of the nanosheets exhibit a rectangular shape that is ~100 nm in width and several hundreds of nm in length. The average thickness of the $PNMG_{124}$-b-$PNDG_{63}$ nanosheets was estimated to be ~14 nm based on AFM analysis. WAXS analysis of the $PNMG_{124}$-b-$PNDG_{63}$ solution (Figure 6g) revealed notable diffraction peaks due to side-by-side packing of PNDG. However, the (100) reflection and the associated higher order (101) and (102) reflections are barely discernible in the WAXS region, indicating the relatively poor molecular ordering of adjacent PNDG backbones along the crystallographic a-axis. This result suggests that the formation of $PNMG_{124}$-b-$PNDG_{63}$ nanosheets is mainly driven by the side-by-side molecular packing along the crystallographic c-axis, while the face-to-face molecular packing along the a-axis is significantly diminished. We postulate that the length of the nanosheets is determined the side-by-side packing of PNDG segments along the crystallographic c-axis, whereas the width of the nanosheets is resulted from the face-to-face stacking of the PNDG segments. This picture is consistent with recent cryo-electron microscopy and molecular dynamic simulation studies on crystallizable diblock copolypeptoid nanosheets [127]. From WAXS analysis, the distance of adjacent PNDG backbones separated by the long n-decyl side chains was found to be slightly increased to 2.5 nm for the $PNMG_{124}$-b-$PNDG_{63}$ nanosheets as compared to that of the PNMG-b-PNDG nanofibrils and nanorods. Meanwhile, the (001) peak is slightly broader than those observed from $PNMG_{105}$-b-$PNDG_{20}$ nanofibrils and $PNMG_{121}$-b-$PNDG_{46}$ nanorods (2.4 nm), implying the more disordered molecular packing of PNDG backbones along the crystallographic c-axis in the $PNMG_{124}$-b-$PNDG_{63}$ nanosheets presumably due to the backbone folding.

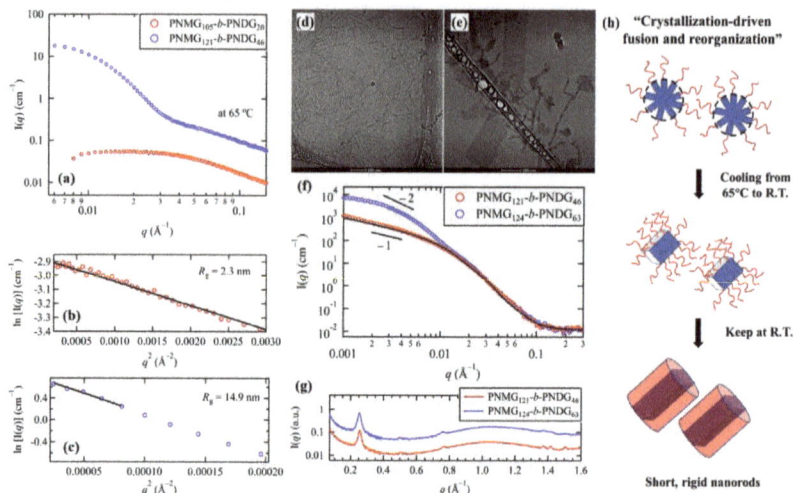

Figure 6. (a) SAXS profiles of the 5 mg/mL PNMG$_{105}$-b-PNDG$_{20}$ and PNMG$_{121}$-b-PNDG$_{46}$ methanol solutions measured at 65 °C. The corresponding Guinier plot analysis of PNMG$_{105}$-b-PNDG$_{20}$ (R_g = 2.3 nm) and PNMG$_{121}$-b-PNDG$_{46}$ (R_g = 14.9 nm) based on the criteria of $qR_g < 1.3$ were shown in (b) and (c), respectively. Representative cryo-TEM images for the self-assembled (d) PNMG$_{121}$-b-PNDG$_{46}$ and (e) PNMG$_{124}$-b-PNDG$_{63}$ nanostructures in methanol. (f) SANS intensity profile (open circles) for the self-assembled PNMG$_{121}$-b-PNDG$_{46}$ and PNMG$_{124}$-b-PNDG$_{63}$ nanostructures in deuterated methanol. The solid line in (f) corresponds to the best-fit to the data based on the cylindrical-shaped micelle model. (g) WAXS intensity profiles for the 5 mg/mL PNMG$_{121}$-b-PNDG$_{46}$ and PNMG$_{124}$-b-PNDG$_{63}$ in methanol. (h) Schematic illustration of the proposed self-assembly mechanisms for the 1D nanorods (e.g., PNMG$_{121}$-b-PNDG$_{46}$) via CDSA. Figure reproduced from reference [79] with permission from the American Chemical Society.

The above results show that the solution self-assembly of coil-crystalline diblock copolypeptides highly relies on the volume fraction of the crystallizable PNDG segment relative to that of the solvophilic PNMG segment. With increasing volume fraction of the PNDG block, the final aggregate morphology gradually transits from long wormlike nanofibrils, to short rigid nanorods and then to 2D nanosheets. Here, the self-assembly pathway of PNMG-b-PNDG plays a key role in determining the final aggregate morphology. SAXS (Figure 6a–c) and cryo-TEM analysis [79] revealed the formation of spherical micelles of PNMG$_{121}$-b-PNDG$_{46}$ in methanol at 65 °C with a R_g of 14.9 nm and an aggregation number of ~175. This is in sharp contrast to the PNMG$_{105}$-b-PNDG$_{20}$ methanol solution in which all polymers exist as unimers (R_g = 2.3 nm) at 65 °C. Note that no crystallization of PNDG block was observed at this temperature. We attributed the large difference in the initial association state to the solubility difference of these polymers in methanol: With longer PNDG segments, the stronger solvophobic interaction drives the micellation of diblock copolypeptides at high temperature (i.e., T > T$_m$), forming amorphous spherical micelles. Using SAXS/WAXS, we also found that subsequent cooling the solution to room temperature immediately induces the crystallization of PNDG segments within the micellar core. As dissociation of micelles upon cooling to room temperature is highly unlikely, the pre-formed spherical micelles of PNMG$_{121}$-b-PNDG$_{46}$ must undergo confined crystallization of PNDG within the micellar core upon cooling [15,16]. If we consider the length of the PNMG$_{121}$-b-PNDG$_{46}$ nanorods to be ~200 nm in average at the final state, as evidenced by cryo-TEM (Figure 6d), the aggregation number of the PNMG$_{121}$-b-PNDG$_{46}$ nanorods is then estimated to be ~1834. This number is about 10 times larger than that of the amorphous spherical micelle precursor (~175) at 65 °C prior to the onset of crystallization, which clearly indicates that the crystallization-induced fusion and structural rearrangement of preformed spherical micelles must have occurred to yield the final nanorods. Therefore, the self-assembly pathway for the $PNMG_{121}$-b-$PNDG_{46}$ nanorods, as depicted in Figure 6h, is distinctly different from the aforementioned self-seeding growth of $PNMG_{105}$-b-$PNDG_{20}$ nanofibrils (Figure 5j).

We shall also mention that the scenario becomes more complicated for the $PNMG_{124}$-b-$PNDG_{63}$ nanosheets and the detail formation mechanism of these nanosheets remains somewhat ambiguous. Based on SAXS/WAXS and cryo-TEM results, $PNMG_{124}$-b-$PNDG_{63}$ molecules aggregate into large non-spherical, amorphous clusters even at high temperature presumably due to the strong solvophobic effect. Upon cooling, PNDG segments are recrystallized within these large aggregates, which introduces an additional crystallization driving force for subsequent fusion/reorganization of preexisting aggregates. However, fusion/reorganization can be a rare event if the number of preexisting micelles per unit volume is too low. The existence of large aggregates due to strong solvophobic effect may thus poses an obstacle to the formation of well-defined polypeptoid nanostructures via CDSA. As will be described in the following section, the self-assembly pathway and final aggregate morphology of diblock copolypeptoids become more defined when the PNDG is replaced by PNOG, i.e., a crystallizable but less solvophobic block.

3.4. Effect of Side Chain Branching on the Solution Self-Assemblies

As we mentioned in Section 2.2, asymmetric branching of the aliphatic N-substituents has profound impact on the molecular packing and phase behavior of polypeptoid homopolymers. To understand how side chain branching further influences the solution self-assembly of amphiphilic diblock copolypeptoids, two types of diblock copolypeptoids, i.e., $PNMG_{116}$-b-$PNOG_{94}$ and $PNMG_{121}$-b-$PNEHG_{101}$ (Figure 7a), with nearly identical molecular weight (i.e., $M_n \approx 25$ kDa) and volume fraction of solvophobic block (f_{PNOG} = 0.73, f_{PNEHG} = 0.74) were recently investigated for their solution self-assembly behavior in methanol. Both samples were first heated at high temperatures, allowing polymers to be fully dissolved and existed as unimers in methanol. The solution self-assembly were then triggered by cooling of the respective methanol solution from high temperature to room temperature.

As seen in Figure 7b,c, $PNMG_{116}$-b-$PNOG_{94}$ molecules bearing linear *n*-octyl side chains self-assembled into large hierarchical microflowers that comprised of radially arranged nanoribbon subunits (i.e., flower petals) after ~24 h of assembly time. Such morphological feature at the final stage of self-assembly has also been captured by solution SAXS, which gives $I(q) \sim q^{-2}$ power law behavior at the intermediate q range and followed by an intensity minimum at $q \sim 0.045$ Å^{-1} (Figure 7d). The highly ordered crystalline structure of PNOG blocks with a typical board-like molecular geometry was also revealed by WAXS, where the molecules are fully extended in an all cis-amide conformation and are stacked side by side and face to face simultaneously. Time-dependent SAXS/WAXS (Figure 7d–g) and AFM results [84] show that the overall self-assembly process is relatively sluggish and involves the assembly of multilevel building blocks in a stepwise fashion: Upon cooling, the $PNMG_{116}$-b-$PNOG_{94}$ unimers first associate to form amorphous spherical micelles owing to the high solvophobic content; These amorphous micelles are further aggregated via a nucleation-and-growth mechanism, resulting in the formation of flower petal junction; Finally, the growth of the flower petals (i.e., nanoribbon sub-units) occurs by the continuous addition of the amorphous $PNMG_{116}$-b-$PNOG_{94}$ materials to the crystallization front following a 2D crystallization kinetic, evidenced by the Avrami analysis (Figure 7g). For a 5 mg/mL solution, the entire self-assembly process takes few hundreds of minutes to complete. The relative sluggish self-assembly process (compared to $PNMG_{121}$-b-$PNEHG_{101}$ counterpart) is mainly attributed to the slow epitaxial 2D crystalline growth of board-like PNOG segments during micellar fusion/reorganization process, resulting in the formation of long-range ordered crystal lattice at molecular level.

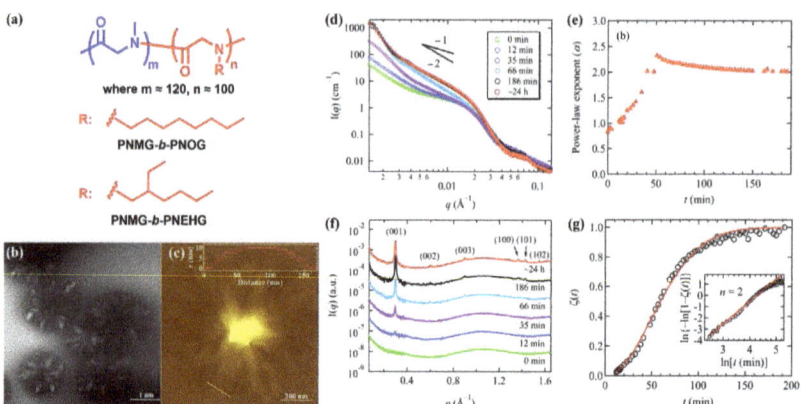

Figure 7. (**a**) The chemical structures of PNMG-*b*-PNOG and PNMG-*b*-PNEHG used for comparison purpose. (**b**) Cryo-TEM and (**c**) AFM images for PNMG$_{116}$-*b*-PNOG$_{94}$ in a diluted methanol solution. (**d**) Representative SAXS profiles of the 5 mg/mL PNMG$_{116}$-*b*-PNOG$_{94}$ methanol solution at different waiting times (*t*) after being cooled down to room temperature. (**e**) Plot of the exponent (*a*) values of $I(q) \sim q^{-a}$ near $q = 0.006$ Å^{-1} as a function of *t*. (**f**) The corresponding WAXS profiles at different *t*, where the data have been shifted vertically for clarity. (**g**) The $\zeta(t)$ values (black circles) obtained from the normalized integrated intensity of the (001) peak at different *t*. Inset of (**g**) shows the corresponding Sharp–Hancock plot. The red solid lines in (**g**) correspond to the best-fits to the data using the Avrami–Erofeev expression, $\zeta(t) = 1-\exp[-(kt)^n]$, with $n = 2$. Figure reproduced from reference [84] with permission from the American Chemical Society.

By contrast, the PNMG$_{121}$-*b*-PNEHG$_{101}$ molecules bearing bulky branched racemic 2-ethyl-1-hexyl side chains were found to self-assemble into symmetric 2D hexagonal nanosheets (Figure 8a,b). The best-fit to the SAXS profile of the final PNMG$_{121}$-*b*-PNEHG$_{101}$ nanosheets using the scattering form factor for disk-shaped polymer micelles [123,124] gives a core thickness of 11.3 ± 0.2 nm and a radius of gyration of the corona chains of 2.7 ± 0.2 nm. The total thickness of the hexagonal nanosheets is then estimated to be 23.1 ± 1.0 nm, which is larger than that (16 ± 1 nm) estimated at the dry state by AFM analysis. Meanwhile, the WAXS profile of PNMG$_{121}$-*b*-PNEHG$_{101}$ nanosheets (Figure 8d) shows a single diffraction peak at $q^* = 0.50$ Å^{-1}, corresponding to the distance (1.26 nm) between adjacent PNEHG segments that are separated by the bulky branched N-2-ethyl-1-hexyl side chains [97]. Grazing-incidence wide-angle X-ray diffraction (GI-WAXD) measurements were performed to unveil the molecular packing and orientation inside of the PNMG$_{121}$-*b*-PNEHG$_{101}$ nanosheets. Due to geometric confinement, the large 2D hexagonal nanosheets were laid flat on the Si substrate (inset of Figure 8f), allowing the molecular orientation within the dried hexagonal nanosheets to be resolved. Aside from the primary diffraction peak at $q^* = 0.50$ Å^{-1}, multiple higher order peaks located at $\sqrt{3}q^*$, $\sqrt{4}q^*$, $\sqrt{7}q^*$ along the in-plane (q_{xy}) direction were also observed by GIWAXD (Figure 8e–f), indicating the rod-like PNEHG molecules are packed into a hexagonal lattice with the long axis of the rods aligned normal to the substrate (and the surface of hexagonal nanosheets). We speculate that the absence of these higher order peaks in Figure 8d is due to the relatively strong incoherent scattering from solution WAXS measurements using a capillary cell. There are also two discrete off-axis streaks with low intensity that are aligned parallel to q_{xy}, as indicated by the red arrows in the GIWAXD (Figure 8e), possibly due to the presence of short and non-continuous helix-like segments along the backbone in low abundance.

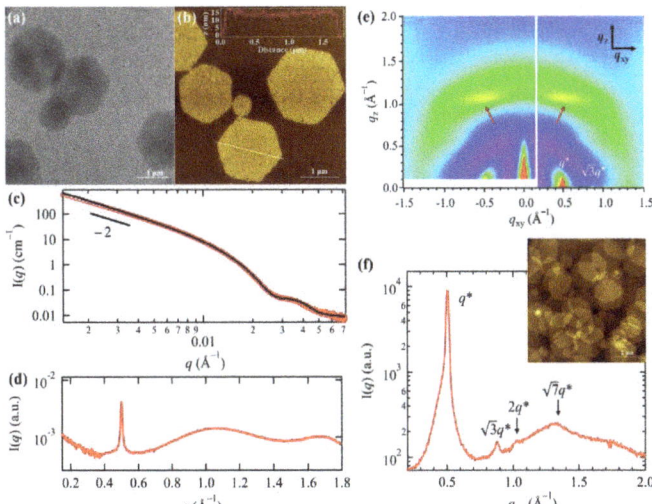

Figure 8. (a) Cryo-TEM and (b) AFM images for PNMG$_{121}$-b-PNEHG$_{101}$ in diluted methanol solution. (c) SAXS and (d) WAXS intensity profiles for the 5 mg/mL PNMG$_{121}$-b-PNEHG$_{101}$ in methanol. The solid line in (c) corresponds to the best-fit to the data based on the scattering form factor for disk-shaped polymer micelles. (e) Two-dimensional GIWAXD image for the PNMG$_{121}$-b-PNEHG$_{101}$ hexagonal nanosheets deposited onto a Si substrate. The two off-axis streaks are indicated by red arrows in (e). (f) One-dimensional GIWAXD profile along the q_{xy} (in-plane) direction. The corresponding AFM image for the GIWAXD sample is shown in the inset of (f). Figure reproduced from reference [84] with permission from the American Chemical Society.

Hence, it is evident that the lateral dimension of the PNMG$_{121}$-b-PNEHG$_{101}$ nanosheet is governed by the preferential packing of rod-like PNEHG molecules in a columnar hexagonal lattice in 2D, whereas the thickness of the nanosheet core is determined by the height of hexagonal columns (Figure 9a). Within the columnar hexagonal mesophase, PNEHG blocks adopt a rod-like molecular geometry with an extended backbone conformation with the bulky racemic N-2-ethyl-1-hexyl side chains radially and outwardly displayed along the backbone. Interestingly, it was also found that the final PNMG$_{121}$-b-PNEHG$_{101}$ nanosheets were formed immediately (i.e., within seconds) after the solution was cooled below the clearing temperature, as evidenced by the little change of scattering profiles with time. Differs from the long-range ordered PNOG crystals, the intermolecular packing of the rod-like PNEHG blocks favors the formation of a columnar hexagonal LC mesophase which may not be very long ranged, evidenced by the weak higher order diffraction peaks even at the dried state (Figure 8f). This is consistent with DSC and WAXD results for the bulk PNEHG homopolymer with similar DP$_n$ (Figure 4a,b). Thus, the formation of the mesophase within the PNEHG micellar core occurs much more rapidly relative to that of the crystalline PNOG micellar core presumably due to the less defined molecular packing structure in the former relative to the latter [37,128].

The correlations among aggregate morphology, molecular packing and N-substituent architecture (i.e., linear versus branched) of diblock copolypeptoid can be now rationalized. According to the general packing motif, a single PNOG board-like molecule in the crystalline lattice contains three different facets: The main face of PNOG that comprised of both backbone and N-aliphatic side chain, the surface comprised of PNOG backbone chain ends and the surface comprised of only N-aliphatic side chain ends, which are perpendicular to the crystallographic a-, b- and c-axes of the PNOG molecule, respectively (Figure 4c). Since PNOG blocks are covalently linked with the corona-forming PNMG blocks in the diblock copolypeptoids, it is reasonable that the PNOG crystals cannot grow along crystallographic b-axis. The Avrami analysis on the time-dependent WAXS results also showed that the crystalline growth of the core-forming PNOGs is two-dimensional (Figure 7g). We therefore postulate that the core thickness of the PNMG-b-PNOG nanoribbons is determined by the crystalline dimension along b-axis, while the other two axes determine the lateral dimension of the nanoribbon core (Figure 9b). Based on AFM analysis, the average length of the nanoribbons is approximately 4–5 times larger than their width, which suggests that the tendency for the core-forming PNOG block to grow along one axis is 4–5 times higher than the other axis. This is consistent with other previous studies [77–79,127] which show that the 2D nanosheets assembled from the diblock copolypeptoids bearing board-like crystallizable blocks from solution usually appears non-symmetrical, such as ribbon-like or rectangular shapes, rather than forming symmetrical 2D geometry. Yet, how these elongated 2D nanoribbons are stacked radially along flower petal junction and the detailed hierarchical self-assembly mechanism of the microflowers remain somewhat ambiguous. However, it is clear that the nanoribbon formation is due to the favorable molecular packing along one of the crystallographic axes during the 2D crystalline growth, which is dictated by the disparate inter- or intramolecular interactions and polymer–solvent interactions along different crystallographic axes.

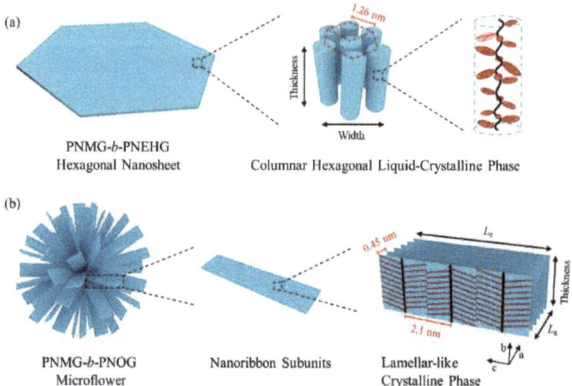

Figure 9. Schematic illustration of the molecular arrangement inside the (**a**) PNMG$_{121}$-b-PNEHG$_{101}$ hexagonal nanosheet and (**b**) PNMG$_{116}$-b-PNOG$_{94}$ microflower. The corona-forming PNMG blocks and possible chain folding of PNOG were omitted for clarity. Figure reproduced from reference [84] with permission from the American Chemical Society.

By contrast, as the bulky racemic 2-ethyl-1-hexyl side chains are randomly distributed around the PNEHG backbone (Figure 4c), the rod-like PNEHG blocks would afford isotropic inter- or intramolecular interactions and polymer–solvent interfacial interaction in the radial direction of the rods. As the solvophilic PNMG block is chemically linked with PNEHG block, columnar hexagonal packing of PNEHG becomes energetically favored, resulting in the formation of large hexagonal nanosheets that possess a symmetrical 2D geometry (Figure 9a). We also found that the lateral dimension of PNMG$_{121}$-b-PNEHG$_{101}$ hexagonal nanosheets can be manipulated from nano-size (e.g., ~200 nm) to micro-size (e.g., ~2 μm) by tuning the initial polymer concentration within the dilute regime, while the thickness of hexagonal nanosheets remains unaffected by the concentration [84]. It should be noted that the symmetrical hexagonal nanosheets formed by PNMG$_{121}$-b-PNEHG$_{101}$ mesogens are highly unusual and rarely observed by crystallization or CDSA of typical crystalline polymers (e.g., polyethylene and polycaprolactone) [129–131]. This finding, which uses rod-like mesogens as the primary building block to induce the formation of 2D hexagonal nanosheets via solution self-assembly, sheds new light on the creation of highly symmetric 2D nano-/micro-scale materials for a wide range of applications.

4. Conclusions and Outlook

In this article, we review our recent experimental studies on the crystallization-driven solution self-assembly of amphiphilic diblock copolypeptoids bearing alkyl side chains. It has been found that supramolecular self-assembly and aggregate morphology of diblock copolypeptoids in solution are extremely sensitive to their molecular characteristics, such as block composition, molecular weight and N-substituent architecture. This is because micellation (solvophilic/solvophobic interaction), crystallization and the interplay between these two driving forces are directly linked to the detailed molecular characteristics of the diblock copolypeptoids. Because crystallization is more of a kinetically controlled process that can lead to non-equilibrium assembly, the spatiotemporal evolution of crystallizable BCPs during solution self-assembly must be carefully characterized in situ in order to gain a comprehensive understanding on the assembly pathway. Here, we highlight the use of in situ small-/wide-angle X-ray/neutron scattering in conjunction with other microscopic techniques in probing the molecular packing, hierarchical structure, and self-assembly pathways of crystallizable diblock copolypeptoids in solution, which allow us to better understand their multiscale structural evolution and self-assembly pathways. However, in priori design of solution self-assembly process to arrive at a targeted nanostructure/morphology still remain challenging.

Here, we listed several fundamental questions regarding diblock copolypeptoids bearing alkyl side chains that remain unsolved: (i) How do the interplay among inter-/intra-molecular interactions of board-like polypeptoids along different crystallographic axes dictate the anisotropic growth or the aspect ratio of 1D or 2D nanostructures. Can these interactions be mediated through chemical design or sample preparation protocols to achieve a tunable morphology? (ii) If chain folding is inevitable upon crystallization, in what direction (a or c-axis) do long polypeptoid segments prefer to fold inter-/intra-molecularly within the nanostructures? How this affects the chain conformation (or packing density) of solvophilic block, micellar fusion/reorganization, dispersibility and final aggregate morphology in solution? (iii) What lead to the radial stacking of nanoribbons to form hierarchical microflowers? How can we tune the structure and level of hierarchy based on the kinetic and thermodynamic behaviors of primary building blocks? How can we control each growth step that constitute the hierarchical assembly process? (iv) What is the self-assembly pathway and formation mechanism of hexagonal nanosheets composed of diblock copolypeptoids bearing asymmetrically branched alkyl side chains, e.g., PNMG-*b*-PNEHG? Can the use of mesogenic building blocks with less ordered LC-like packing serve as a new paradigm for the design of solution self-assemblies with unique structures and properties? Future efforts that incorporate predictive theoretical tools and advanced structural characterization methods would be helpful to address these issues.

Nevertheless, it is exciting to see that even for the simplest AB-type diblock copolypeptoids with alkyl side chains, a variety of well-defined non-spherical nanostructures with diverse morphology and hierarchy can be readily fabricated by CDSA, ranging from 1D nanofibrils, to nanorods, to 2D nanosheets and to hierarchical nanostructures. Owing to the recent advances in controlled synthetic methods, such as the submonomer solid-phase synthesis and sequential ROP of R-NCAs or R-NTAs, well-defined crystallizable block copolypeptoids with diverse N-substituent structure and tunable molecular sequences can now be produced with high efficiency, providing seemingly unlimited choices of polypeptoid building blocks for solution self-assembly. Besides, by thinking solution CDSA as a reaction process, recent findings on the living CDSA of several other BCPs (in particular polyferrocenylsilane-based polymers) that utilizes the "seeded-growth" protocol have envisioned a more precise control of size dispersity, complexity and hierarchy of polymeric self-assemblies [5,10,19,22–29]. Considering the unique biological properties of polypeptoid, these advancements would open new opportunities for the future design of novel polypeptoid nanomaterials with tailorable structure, property and functionality, which are potentially useful in molecular biomimicry and biomedical/biotechnological applications.

Author Contributions: N.J. and D.Z. contributed to writing (review and editing). All authors have read and agreed to the published version of the manuscript.

Funding: This research was funded by the National Natural Science Foundation of China (52073025), the Fundamental Research Funds for the Central Universities (FRF-IDRY-20-003, Interdisciplinary Research Project for Young Teachers of USTB) and the National Science Foundation (CHE 2003458).

Institutional Review Board Statement: Not applicable.

Informed Consent Statement: Not applicable.

Data Availability Statement: The data are available from the corresponding author upon request.

Acknowledgments: N.J. acknowledges the financial support by the National Natural Science Foundation of China (52073025) and the Fundamental Research Funds for the Central Universities (FRF-IDRY-20-003, Interdisciplinary Research Project for Young Teachers of USTB). D.Z. acknowledges the financial support by the National Science Foundation (CHE 2003458).

Conflicts of Interest: The authors declare no conflict of interest.

References

1. Blanazs, A.; Armes, S.P.; Ryan, A.J. Self-Assembled Block Copolymer Aggregates: From Micelles to Vesicles and their Biological Applications. *Macromol. Rapid Commun.* **2009**, *30*, 267–277. [CrossRef]
2. Hayward, R.C.; Pochan, D.J. Tailored Assemblies of Block Copolymers in Solution: It Is All about the Process. *Macromolecules* **2010**, *43*, 3577–3584. [CrossRef]
3. Vilgis, T.; Halperin, A. Aggregation of coil-crystalline block copolymers: Equilibrium crystallization. *Macromolecules* **1991**, *24*, 2090–2095. [CrossRef]
4. Massey, J.A.; Temple, K.; Cao, L.; Rharbi, Y.; Raez, J.; Winnik, M.A.; Manners, I. Self-Assembly of Organometallic Block Copolymers: The Role of Crystallinity of the Core-Forming Polyferrocene Block in the Micellar Morphologies Formed by Poly(ferrocenylsilane-b-dimethylsiloxane) in n-Alkane Solvents. *J. Am. Chem. Soc.* **2000**, *122*, 11577–11584. [CrossRef]
5. Gilroy, J.B.; Gädt, T.; Whittell, G.R.; Chabanne, L.; Mitchels, J.M.; Richardson, R.M.; Winnik, M.A.; Manners, I. Monodisperse cylindrical micelles by crystallization-driven living self-assembly. *Nat. Chem.* **2010**, *2*, 566–570. [CrossRef]
6. Wang, X.; Guerin, G.; Wang, H.; Wang, Y.; Manners, I.; Winnik, M.A. Cylindrical Block Copolymer Micelles and Co-Micelles of Controlled Length and Architecture. *Science* **2007**, *317*, 644–647. [CrossRef]
7. He, X.; He, Y.; Hsiao, M.-S.; Harniman, R.L.; Pearce, S.; Winnik, M.A.; Manners, I. Complex and Hierarchical 2D Assemblies via Crystallization-Driven Self-Assembly of Poly(l-lactide) Homopolymers with Charged Termini. *J. Am. Chem. Soc.* **2017**, *139*, 9221–9228. [CrossRef]
8. Inam, M.; Cambridge, G.; Pitto-Barry, A.; Laker, Z.P.L.; Wilson, N.R.; Mathers, R.T.; Dove, A.P.; O'Reilly, R.K. 1D vs. 2D shape selectivity in the crystallization-driven self-assembly of polylactide block copolymers. *Chem. Sci.* **2017**, *8*, 4223–4230. [CrossRef] [PubMed]
9. Arno, M.C.; Inam, M.; Coe, Z.; Cambridge, G.; Macdougall, L.J.; Keogh, R.; Dove, A.P.; O'Reilly, R.K. Precision Epitaxy for Aqueous 1D and 2D Poly(ε-caprolactone) Assemblies. *J. Am. Chem. Soc.* **2017**, *139*, 16980–16985. [CrossRef]
10. Qiu, H.; Hudson, Z.M.; Winnik, M.A.; Manners, I. Multidimensional hierarchical self-assembly of amphiphilic cylindrical block comicelles. *Science* **2015**, *347*, 1329–1332. [CrossRef]
11. Nicolai, T.; Colombani, O.; Chassenieux, C. Dynamic polymeric micelles versus frozen nanoparticles formed by block copolymers. *Soft Matter* **2010**, *6*, 3111–3118. [CrossRef]
12. Jain, S.; Bates, F.S. Consequences of Nonergodicity in Aqueous Binary PEO–PB Micellar Dispersions. *Macromolecules* **2004**, *37*, 1511–1523. [CrossRef]
13. Tritschler, U.; Pearce, S.; Gwyther, J.; Whittell, G.R.; Manners, I. 50th Anniversary Perspective: Functional Nanoparticles from the Solution Self-Assembly of Block Copolymers. *Macromolecules* **2017**, *50*, 3439–3463. [CrossRef]
14. Lunn, D.J.; Finnegan, J.R.; Manners, I. Self-assembly of "patchy" nanoparticles: A versatile approach to functional hierarchical materials. *Chem. Sci.* **2015**, *6*, 3663–3673. [CrossRef]
15. Yin, L.; Lodge, T.P.; Hillmyer, M.A. A Stepwise "Micellization–Crystallization" Route to Oblate Ellipsoidal, Cylindrical, and Bilayer Micelles with Polyethylene Cores in Water. *Macromolecules* **2012**, *45*, 9460–9467. [CrossRef]
16. Schmelz, J.; Karg, M.; Hellweg, T.; Schmalz, H. General Pathway toward Crystalline-Core Micelles with Tunable Morphology and Corona Segregation. *ACS Nano* **2011**, *5*, 9523–9534. [CrossRef]
17. He, W.-N.; Zhou, B.; Xu, J.-T.; Du, B.-Y.; Fan, Z.-Q. Two Growth Modes of Semicrystalline Cylindrical Poly(ε-caprolactone)-b-poly(ethylene oxide) Micelles. *Macromolecules* **2012**, *45*, 9768–9778. [CrossRef]
18. Pitto-Barry, A.; Kirby, N.; Dove, A.P.; O'Reilly, R.K. Expanding the scope of the crystallization-driven self-assembly of polylactide-containing polymers. *Polym. Chem.* **2014**, *5*, 1427–1436. [CrossRef]

19. Qian, J.; Lu, Y.; Chia, A.; Zhang, M.; Rupar, P.A.; Gunari, N.; Walker, G.C.; Cambridge, G.; He, F.; Guerin, G.; et al. Self-Seeding in One Dimension: A Route to Uniform Fiber-like Nanostructures from Block Copolymers with a Crystallizable Core-Forming Block. *ACS Nano* **2013**, *7*, 3754–3766. [CrossRef]
20. Hsiao, M.-S.; Yusoff, S.F.M.; Winnik, M.A.; Manners, I. Crystallization-Driven Self-Assembly of Block Copolymers with a Short Crystallizable Core-Forming Segment: Controlling Micelle Morphology through the Influence of Molar Mass and Solvent Selectivity. *Macromolecules* **2014**, *47*, 2361–2372. [CrossRef]
21. Sun, L.; Petzetakis, N.; Pitto-Barry, A.; Schiller, T.L.; Kirby, N.; Keddie, D.J.; Boyd, B.J.; O'Reilly, R.K.; Dove, A.P. Tuning the Size of Cylindrical Micelles from Poly(l-lactide)-b-poly(acrylic acid) Diblock Copolymers Based on Crystallization-Driven Self-Assembly. *Macromolecules* **2013**, *46*, 9074–9082. [CrossRef]
22. Presa Soto, A.; Gilroy, J.B.; Winnik, M.A.; Manners, I. Pointed-Oval-Shaped Micelles from Crystalline-Coil Block Copolymers by Crystallization-Driven Living Self-Assembly. *Angew. Chem. Int. Ed.* **2010**, *49*, 8220–8223. [CrossRef]
23. Petzetakis, N.; Dove, A.P.; O'Reilly, R.K. Cylindrical micelles from the living crystallization-driven self-assembly of poly(lactide)-containing block copolymers. *Chem. Sci.* **2011**, *2*, 955–960. [CrossRef]
24. Hudson, Z.M.; Boott, C.E.; Robinson, M.E.; Rupar, P.A.; Winnik, M.A.; Manners, I. Tailored hierarchical micelle architectures using living crystallization-driven self-assembly in two dimensions. *Nat. Chem.* **2014**, *6*, 893–898. [CrossRef]
25. Finnegan, J.R.; Lunn, D.J.; Gould, O.E.C.; Hudson, Z.M.; Whittell, G.R.; Winnik, M.A.; Manners, I. Gradient Crystallization-Driven Self-Assembly: Cylindrical Micelles with "Patchy" Segmented Coronas via the Coassembly of Linear and Brush Block Copolymers. *J. Am. Chem. Soc.* **2014**, *136*, 13835–13844. [CrossRef] [PubMed]
26. Patra, S.K.; Ahmed, R.; Whittell, G.R.; Lunn, D.J.; Dunphy, E.L.; Winnik, M.A.; Manners, I. Cylindrical Micelles of Controlled Length with a π-Conjugated Polythiophene Core via Crystallization-Driven Self-Assembly. *J. Am. Chem. Soc.* **2011**, *133*, 8842–8845. [CrossRef]
27. Qiu, H.; Gao, Y.; Du, V.A.; Harniman, R.; Winnik, M.A.; Manners, I. Branched Micelles by Living Crystallization-Driven Block Copolymer Self-Assembly under Kinetic Control. *J. Am. Chem. Soc.* **2015**, *137*, 2375–2385. [CrossRef]
28. Qian, J.; Li, X.; Lunn, D.J.; Gwyther, J.; Hudson, Z.M.; Kynaston, E.; Rupar, P.A.; Winnik, M.A.; Manners, I. Uniform, High Aspect Ratio Fiber-like Micelles and Block Co-micelles with a Crystalline π-Conjugated Polythiophene Core by Self-Seeding. *J. Am. Chem. Soc.* **2014**, *136*, 4121–4124. [CrossRef]
29. Tritschler, U.; Gwyther, J.; Harniman, R.L.; Whittell, G.R.; Winnik, M.A.; Manners, I. Toward Uniform Nanofibers with a π-Conjugated Core: Optimizing the "Living" Crystallization-Driven Self-Assembly of Diblock Copolymers with a Poly(3-octylthiophene) Core-Forming Block. *Macromolecules* **2018**, *51*, 5101–5113. [CrossRef]
30. Geng, Y.; Dalhaimer, P.; Cai, S.; Tsai, R.; Tewari, M.; Minko, T.; Discher, D.E. Shape effects of filaments versus spherical particles in flow and drug delivery. *Nat. Nanotechnol.* **2007**, *2*, 249–255. [CrossRef]
31. Kim, Y.; Dalhaimer, P.; Christian, D.A.; Discher, D.E. Polymeric worm micelles as nano-carriers for drug delivery. *Nanotechnology* **2005**, *16*, S484–S491. [CrossRef]
32. Hartgerink, J.D.; Beniash, E.; Stupp, S.I. Self-Assembly and Mineralization of Peptide-Amphiphile Nanofibers. *Science* **2001**, *294*, 1684–1688. [CrossRef]
33. Zhang, Y.; Tekobo, S.; Tu, Y.; Zhou, Q.; Jin, X.; Dergunov, S.A.; Pinkhassik, E.; Yan, B. Permission to enter cell by shape: Nanodisk vs nanosphere. *ACS Appl. Mater. Interfaces* **2012**, *4*, 4099–4105. [CrossRef] [PubMed]
34. Rizis, G.; van de Ven, T.G.M.; Eisenberg, A. "Raft" Formation by Two-Dimensional Self-Assembly of Block Copolymer Rod Micelles in Aqueous Solution. *Angew. Chem. Int. Ed.* **2014**, *53*, 9000–9003. [CrossRef]
35. Li, X.; Jin, B.; Gao, Y.; Hayward, D.W.; Winnik, M.A.; Luo, Y.; Manners, I. Monodisperse Cylindrical Micelles of Controlled Length with a Liquid-Crystalline Perfluorinated Core by 1D "Self-Seeding". *Angew. Chem. Int. Ed.* **2016**, *55*, 11392–11396. [CrossRef]
36. Jin, B.; Sano, K.; Aya, S.; Ishida, Y.; Gianneschi, N.; Luo, Y.; Li, X. One-pot universal initiation-growth methods from a liquid crystalline block copolymer. *Nat. Commun.* **2019**, *10*, 2397. [CrossRef] [PubMed]
37. Gao, L.; Gao, H.; Lin, J.; Wang, L.; Wang, X.-S.; Yang, C.; Lin, S. Growth and Termination of Cylindrical Micelles via Liquid-Crystallization-Driven Self-Assembly. *Macromolecules* **2020**, *53*, 8992–8999. [CrossRef]
38. Rider, D.A.; Manners, I. Synthesis, Self-Assembly, and Applications of Polyferrocenylsilane Block Copolymers. *Polym. Rev.* **2007**, *47*, 165–195. [CrossRef]
39. Hailes, R.L.N.; Oliver, A.M.; Gwyther, J.; Whittell, G.R.; Manners, I. Polyferrocenylsilanes: Synthesis, properties, and applications. *Chem. Soc. Rev.* **2016**, *45*, 5358–5407. [CrossRef]
40. Kirshenbaum, K.; Barron, A.E.; Goldsmith, R.A.; Armand, P.; Bradley, E.K.; Truong, K.T.V.; Dill, K.A.; Cohen, F.E.; Zuckermann, R.N. Sequence-specific polypeptoids: A diverse family of heteropolymers with stable secondary structure. *Proc. Natl. Acad. Sci. USA* **1998**, *95*, 4303–4308. [CrossRef]
41. Zhang, D.; Lahasky, S.H.; Guo, L.; Lee, C.-U.; Lavan, M. Polypeptoid Materials: Current Status and Future Perspectives. *Macromolecules* **2012**, *45*, 5833–5841. [CrossRef]
42. Sun, J.; Zuckermann, R.N. Peptoid Polymers: A Highly Designable Bioinspired Material. *ACS Nano* **2013**, *7*, 4715–4732. [CrossRef] [PubMed]
43. Gangloff, N.; Ulbricht, J.; Lorson, T.; Schlaad, H.; Luxenhofer, R. Peptoids and Polypeptoids at the Frontier of Supra- and Macromolecular Engineering. *Chem. Rev.* **2016**, *116*, 1753–1802. [CrossRef] [PubMed]

44. Chan, B.A.; Xuan, S.; Li, A.; Simpson, J.M.; Sternhagen, G.L.; Yu, T.; Darvish, O.A.; Jiang, N.; Zhang, D. Polypeptoid polymers: Synthesis, characterization, and properties. *Biopolymers* **2017**, *109*, e23070.
45. Xuan, S.; Zuckermann, R.N. Diblock copolypeptoids: A review of phase separation, crystallization, self-assembly and biological applications. *J. Mater. Chem. B* **2020**, *8*, 5380–5394. [CrossRef] [PubMed]
46. Statz, A.R.; Meagher, R.J.; Barron, A.E.; Messersmith, P.B. New Peptidomimetic Polymers for Antifouling Surfaces. *J. Am. Chem. Soc.* **2005**, *127*, 7972–7973. [CrossRef] [PubMed]
47. Statz, A.R.; Barron, A.E.; Messersmith, P.B. Protein, cell and bacterial fouling resistance of polypeptoid-modified surfaces: Effect of side-chain chemistry. *Soft Matter* **2008**, *4*, 131–139. [CrossRef]
48. Lau, K.H.A.; Ren, C.; Sileika, T.S.; Park, S.H.; Szleifer, I.; Messersmith, P.B. Surface-Grafted Polysarcosine as a Peptoid Antifouling Polymer Brush. *Langmuir* **2012**, *28*, 16099–16107. [CrossRef]
49. Leng, C.; Buss, H.G.; Segalman, R.A.; Chen, Z. Surface Structure and Hydration of Sequence-Specific Amphiphilic Polypeptoids for Antifouling/Fouling Release Applications. *Langmuir* **2015**, *31*, 9306–9311. [CrossRef] [PubMed]
50. Patterson, A.L.; Wenning, B.; Rizis, G.; Calabrese, D.R.; Finlay, J.A.; Franco, S.C.; Zuckermann, R.N.; Clare, A.S.; Kramer, E.J.; Ober, C.K.; et al. Role of Backbone Chemistry and Monomer Sequence in Amphiphilic Oligopeptide- and Oligopeptoid-Functionalized PDMS- and PEO-Based Block Copolymers for Marine Antifouling and Fouling Release Coatings. *Macromolecules* **2017**, *50*, 2656–2667. [CrossRef]
51. Gao, Q.; Li, P.; Zhao, H.; Chen, Y.; Jiang, L.; Ma, P.X. Methacrylate-ended polypeptides and polypeptoids for antimicrobial and antifouling coatings. *Polym. Chem.* **2017**, *8*, 6386–6397. [CrossRef]
52. Li, A.; Zhang, D. Synthesis and Characterization of Cleavable Core-Cross-Linked Micelles Based on Amphiphilic Block Copolypeptoids as Smart Drug Carriers. *Biomacromolecules* **2016**, *17*, 852–861. [CrossRef]
53. Zhu, L.P.; Simpson, J.M.; Xu, X.; He, H.; Zhang, D.H.; Yin, L.C. Cationic Polypeptoids with Optimized Molecular Characteristics toward Efficient Nonviral Gene Delivery. *ACS Appl. Mater. Interfaces* **2017**, *9*, 23476–23486. [CrossRef] [PubMed]
54. Deng, Y.; Chen, H.; Tao, X.; Cao, F.; Trépout, S.; Ling, J.; Li, M.-H. Oxidation-Sensitive Polymersomes Based on Amphiphilic Diblock Copolypeptoids. *Biomacromolecules* **2019**, *20*, 3435–3444. [CrossRef] [PubMed]
55. Song, Y.; Wang, M.; Li, S.; Jin, H.; Cai, X.; Du, D.; Li, H.; Chen, C.-L.; Lin, Y. Efficient Cytosolic Delivery Using Crystalline Nanoflowers Assembled from Fluorinated Peptoids. *Small* **2018**, *14*, 1803544. [CrossRef] [PubMed]
56. Luo, Y.; Song, Y.; Wang, M.; Jian, T.; Ding, S.; Mu, P.; Liao, Z.; Shi, Q.; Cai, X.; Jin, H.; et al. Bioinspired Peptoid Nanotubes for Targeted Tumor Cell Imaging and Chemo-Photodynamic Therapy. *Small* **2019**, *15*, 1902485. [CrossRef] [PubMed]
57. Murray, D.J.; Kim, J.H.; Grzincic, E.M.; Kim, S.C.; Abate, A.R.; Zuckermann, R.N. Uniform, Large-Area, Highly Ordered Peptoid Monolayer and Bilayer Films for Sensing Applications. *Langmuir* **2019**, *35*, 13671–13680. [CrossRef]
58. Tao, X.; Chen, H.; Trépout, S.; Cen, J.; Ling, J.; Li, M.-H. Polymersomes with aggregation-induced emission based on amphiphilic block copolypeptoids. *Chem. Commun.* **2019**, *55*, 13530–13533. [CrossRef]
59. Guo, L.; Zhang, D. Cyclic Poly(α-peptoid)s and Their Block Copolymers from N-Heterocyclic Carbene-Mediated Ring-Opening Polymerizations of N-Substituted N-Carboxylanhydrides. *J. Am. Chem. Soc.* **2009**, *131*, 18072–18074. [CrossRef]
60. Lee, C.-U.; Smart, T.P.; Guo, L.; Epps, T.H.; Zhang, D. Synthesis and Characterization of Amphiphilic Cyclic Diblock Copolypeptoids from N-Heterocyclic Carbene-Mediated Zwitterionic Polymerization of N-Substituted N-Carboxyanhydride. *Macromolecules* **2011**, *44*, 9574–9585. [CrossRef] [PubMed]
61. Fetsch, C.; Grossmann, A.; Holz, L.; Nawroth, J.F.; Luxenhofer, R. Polypeptoids from N-Substituted Glycine N-Carboxyanhydrides: Hydrophilic, Hydrophobic, and Amphiphilic Polymers with Poisson Distribution. *Macromolecules* **2011**, *44*, 6746–6758. [CrossRef]
62. Fetsch, C.; Luxenhofer, R. Highly Defined Multiblock Copolypeptoids: Pushing the Limits of Living Nucleophilic Ring-Opening Polymerization. *Macromol. Rapid Commun.* **2012**, *33*, 1708–1713. [CrossRef]
63. Guo, L.; Lahasky, S.H.; Ghale, K.; Zhang, D. N-Heterocyclic Carbene-Mediated Zwitterionic Polymerization of N-Substituted N-Carboxyanhydrides toward Poly(α-peptoid)s: Kinetic, Mechanism, and Architectural Control. *J. Am. Chem. Soc.* **2012**, *134*, 9163–9171. [CrossRef]
64. Robinson, J.W.; Schlaad, H. A versatile polypeptoid platform based on N-allyl glycine. *Chem. Commun.* **2012**, *48*, 7835–7837. [CrossRef]
65. Tao, X.; Du, J.; Wang, Y.; Ling, J. Polypeptoids with tunable cloud point temperatures synthesized from N-substituted glycine N-thiocarboxyanhydrides. *Polym. Chem.* **2015**, *6*, 3164–3174. [CrossRef]
66. Tao, X.; Deng, Y.; Shen, Z.; Ling, J. Controlled Polymerization of N-Substituted Glycine N-Thiocarboxyanhydrides Initiated by Rare Earth Borohydrides toward Hydrophilic and Hydrophobic Polypeptoids. *Macromolecules* **2014**, *47*, 6173–6180. [CrossRef]
67. Xuan, S.; Lee, C.-U.; Chen, C.; Doyle, A.B.; Zhang, Y.; Guo, L.; John, V.T.; Hayes, D.; Zhang, D. Thermoreversible and Injectable ABC Polypeptoid Hydrogels: Controlling the Hydrogel Properties through Molecular Design. *Chem. Mater.* **2016**, *28*, 727–737. [CrossRef] [PubMed]
68. Sternhagen, G.L.; Gupta, S.; Zhang, Y.; John, V.; Schneider, G.J.; Zhang, D. Solution Self-Assemblies of Sequence-Defined Ionic Peptoid Block Copolymers. *J. Am. Chem. Soc.* **2018**, *140*, 4100–4109. [CrossRef] [PubMed]
69. Nam, K.T.; Shelby, S.A.; Choi, P.H.; Marciel, A.B.; Chen, R.; Tan, L.; Chu, T.K.; Mesch, R.A.; Lee, B.C.; Connolly, M.D.; et al. Free-floating ultrathin two-dimensional crystals from sequence-specific peptoid polymers. *Nat. Mater.* **2010**, *9*, 454–460. [CrossRef] [PubMed]

70. Lee, C.-U.; Lu, L.; Chen, J.; Garno, J.C.; Zhang, D. Crystallization-Driven Thermoreversible Gelation of Coil-Crystalline Cyclic and Linear Diblock Copolypeptides. *ACS Macro Lett.* **2013**, *2*, 436–440. [CrossRef]
71. Sanii, B.; Haxton, T.K.; Olivier, G.K.; Cho, A.; Barton, B.; Proulx, C.; Whitelam, S.; Zuckermann, R.N. Structure-Determining Step in the Hierarchical Assembly of Peptoid Nanosheets. *ACS Nano* **2014**, *8*, 11674–11684. [CrossRef] [PubMed]
72. Sun, J.; Jiang, X.; Lund, R.; Downing, K.H.; Balsara, N.P.; Zuckermann, R.N. Self-assembly of crystalline nanotubes from monodisperse amphiphilic diblock copolypeptoid tiles. *Proc. Natl. Acad. Sci. USA* **2016**, *113*, 3954–3959. [CrossRef] [PubMed]
73. Jin, H.; Jiao, F.; Daily, M.D.; Chen, Y.; Yan, F.; Ding, Y.-H.; Zhang, X.; Robertson, E.J.; Baer, M.D.; Chen, C.-L. Highly stable and self-repairing membrane-mimetic 2D nanomaterials assembled from lipid-like peptoids. *Nat. Commun.* **2016**, *7*, 12252. [CrossRef] [PubMed]
74. Fetsch, C.; Gaitzsch, J.; Messager, L.; Battaglia, G.; Luxenhofer, R. Self-Assembly of Amphiphilic Block Copolypeptoids—Micelles, Worms and Polymersomes. *Sci. Rep.* **2016**, *6*, 33491. [CrossRef]
75. Robertson, E.J.; Battigelli, A.; Proulx, C.; Mannige, R.V.; Haxton, T.K.; Yun, L.S.; Whitelam, S.; Zuckermann, R.N. Design, Synthesis, Assembly, and Engineering of Peptoid Nanosheets. *Acc. Chem. Res.* **2016**, *49*, 379–389. [CrossRef]
76. Ni, Y.; Sun, J.; Wei, Y.; Fu, X.; Zhu, C.; Li, Z. Two-Dimensional Supramolecular Assemblies from pH-Responsive Poly(ethyl glycol)-b-poly(l-glutamic acid)-b-poly(N-octylglycine) Triblock Copolymer. *Biomacromolecules* **2017**, *18*, 3367–3374. [CrossRef]
77. Shi, Z.; Wei, Y.; Zhu, C.; Sun, J.; Li, Z. Crystallization-Driven Two-Dimensional Nanosheet from Hierarchical Self-Assembly of Polypeptoid-Based Diblock Copolymers. *Macromolecules* **2018**, *51*, 6344–6351. [CrossRef]
78. Wei, Y.; Tian, J.; Zhang, Z.; Zhu, C.; Sun, J.; Li, Z. Supramolecular Nanosheets Assembled from Poly(ethylene glycol)-b-poly(N-(2-phenylethyl)glycine) Diblock Copolymer Containing Crystallizable Hydrophobic Polypeptoid: Crystallization Driven Assembly Transition from Filaments to Nanosheets. *Macromolecules* **2019**, *52*, 1546–1556. [CrossRef]
79. Jiang, N.; Yu, T.; Darvish, O.A.; Qian, S.; Mkam Tsengam, I.K.; John, V.; Zhang, D. Crystallization-Driven Self-Assembly of Coil–Comb-Shaped Polypeptoid Block Copolymers: Solution Morphology and Self-Assembly Pathways. *Macromolecules* **2019**, *52*, 8867–8877. [CrossRef]
80. Sun, J.; Wang, Z.; Zhu, C.; Wang, M.; Shi, Z.; Wei, Y.; Fu, X.; Chen, X.; Zuckermann, R.N. Hierarchical supramolecular assembly of a single peptoid polymer into a planar nanobrush with two distinct molecular packing motifs. *Proc. Natl. Acad. Sci. USA* **2020**, *117*, 31639–31647. [CrossRef]
81. Gangloff, N.; Höferth, M.; Stepanenko, V.; Sochor, B.; Schummer, B.; Nickel, J.; Walles, H.; Hanke, R.; Würthner, F.; Zuckermann, R.N.; et al. Linking two worlds in polymer chemistry: The influence of block uniformity and dispersity in amphiphilic block copolypeptoids on their self-assembly. *Biopolymers* **2019**, *110*, e23259. [CrossRef] [PubMed]
82. Wang, Z.; Lin, M.; Bonduelle, C.; Li, R.; Shi, Z.; Zhu, C.; Lecommandoux, S.; Li, Z.; Sun, J. Thermoinduced Crystallization-Driven Self-Assembly of Bioinspired Block Copolymers in Aqueous Solution. *Biomacromolecules* **2020**, *21*, 3411–3419. [CrossRef]
83. Xuan, S.; Jiang, X.; Balsara, N.P.; Zuckermann, R.N. Crystallization and self-assembly of shape-complementary sequence-defined peptoids. *Polym. Chem.* **2021**, *12*, 4770–4777. [CrossRef]
84. Kang, L.; Chao, A.; Zhang, M.; Yu, T.; Wang, J.; Wang, Q.; Yu, H.; Jiang, N.; Zhang, D. Modulating the Molecular Geometry and Solution Self-Assembly of Amphiphilic Polypeptoid Block Copolymers by Side Chain Branching Pattern. *J. Am. Chem. Soc.* **2021**, *143*, 5890–5902. [CrossRef] [PubMed]
85. Wei, Y.; Liu, F.; Li, M.; Li, Z.; Sun, J. Dimension control on self-assembly of a crystalline core-forming polypeptoid block copolymer: 1D nanofibers versus 2D nanosheets. *Polym. Chem.* **2021**, *12*, 1147–1154. [CrossRef]
86. Simon, R.J.; Kania, R.S.; Zuckermann, R.N.; Huebner, V.D.; Jewell, D.A.; Banville, S.; Ng, S.; Wang, L.; Rosenberg, S.; Marlowe, C.K.; et al. Peptoids: A modular approach to drug discovery. *Proc. Natl. Acad. Sci. USA* **1992**, *89*, 9367–9371. [CrossRef]
87. Chen, C.L.; Qi, J.H.; Tao, J.H.; Zuckermann, R.N.; DeYoreo, J.J. Tuning calcite morphology and growth acceleration by a rational design of highly stable protein-mimetics. *Sci. Rep.* **2014**, *4*, 6266. [CrossRef]
88. Rosales, A.M.; Murnen, H.K.; Zuckermann, R.N.; Segalman, R.A. Control of Crystallization and Melting Behavior in Sequence Specific Polypeptoids. *Macromolecules* **2010**, *43*, 5627–5636. [CrossRef]
89. Sun, J.; Teran, A.A.; Liao, X.X.; Balsara, N.P.; Zuckermann, R.N. Crystallization in Sequence-Defined Peptoid Diblock Copolymers Induced by Microphase Separation. *J. Am. Chem. Soc.* **2014**, *136*, 2070–2077. [CrossRef]
90. Sun, J.; Teran, A.A.; Liao, X.X.; Balsara, N.P.; Zuckermann, R.N. Nanoscale Phase Separation in Sequence-Defined Peptoid Diblock Copolymers. *J. Am. Chem. Soc.* **2013**, *135*, 14119–14124. [CrossRef]
91. Sun, J.; Jiang, X.; Siegmund, A.; Connolly, M.D.; Downing, K.H.; Balsara, N.P.; Zuckermann, R.N. Morphology and Proton Transport in Humidified Phosphonated Peptoid Block Copolymers. *Macromolecules* **2016**, *49*, 3083–3090. [CrossRef]
92. Chen, C.L.; Zuckermann, R.N.; DeYoreo, J.J. Surface-Directed Assembly of Sequence Defined Synthetic Polymers into Networks of Hexagonally Patterned Nanoribbons with Controlled Functionalities. *ACS Nano* **2016**, *10*, 5314–5320. [CrossRef]
93. Guo, L.; Li, J.; Brown, Z.; Ghale, K.; Zhang, D. Synthesis and characterization of cyclic and linear helical poly(α-peptoid)s by N-heterocyclic carbene-mediated ring-opening polymerizations of N-substituted N-carboxyanhydrides. *Peptide Sci.* **2011**, *96*, 596–603. [CrossRef]
94. Guo, L.; Zhang, D. Synthesis and Characterization of Helix-Coil Block Copoly(α-peptoid)s. In *Non-Conventional Functional Block Copolymers*; ACS Symposium Series; American Chemical Society: Washington, DC, USA, 2011; Volume 1066, pp. 71–79.

Article

Self-Seeding Procedure for Obtaining Stacked Block Copolymer Lamellar Crystals in Solution

Brahim Bessif [1], Thomas Pfohl [1] and Günter Reiter [1,2,*]

[1] Physikalisches Institut, Albert-Ludwigs-Universität, 79104 Freiburg, Germany; brahimbessif@gmail.com (B.B.); thomas.pfohl@physik.uni-freiburg.de (T.P.)
[2] Freiburg Materials Research Center (FMF), Albert-Ludwigs-Universität, 79104 Freiburg, Germany
* Correspondence: guenter.reiter@physik.uni-freiburg.de

Abstract: We examined the formation of self-seeded platelet-like crystals from polystyrene-*block*-polyethylene oxide (PS-*b*-PEO) diblock copolymers in toluene as a function of polymer concentration (c), crystallization temperature (T_C), and self-seeding temperature (T_{SS}). We showed that the number (N) of platelet-like crystals and their mean lateral size (L) can be controlled through a self-seeding procedure. As (homogeneous) nucleation was circumvented by the self-seeding procedure, N did not depend on T_C. N increased linearly with c and decayed exponentially with T_{SS} but was not affected significantly by the time the sample was kept at T_{SS}. The solubility limit of PS-*b*-PEO in toluene (c^*), which was derived from the linear extrapolation of $N(c) \to 0$ and from the total deposited mass of the platelets per area ($M_C(c) \to 0$), depended on T_C. We have also demonstrated that at low N, stacks consisting of a (large) number (η) of uniquely oriented lamellae can be achieved. At a given T_C, L was controlled by N and η as well as by $\Delta c = c - c^*$. Thus, besides being able to predict size and number of platelet-like crystals, the self-seeding procedure also allowed control of the number of stacked lamellae in these crystals.

Keywords: crystal morphologies; polymer crystallization; nucleation mechanism; scaling relations

1. Introduction

Polymer crystallization can be initiated by homogenous nucleation. However, at temperatures close to the melting point this process is extremely slow and is often competing with heterogonous nucleation through foreign substances (nucleating agents, surfaces, or "dirt") [1,2]. Furthermore, homogeneous nucleation is a statistical process, which continuously initiates (with a decreasing probability in time) the growth of additional crystals [3]. Correspondingly, after a given crystallization time, the resulting crystals will have a distribution in size. The control of the starting time of nucleation and the number density of nuclei allows the tuning of crystalline structures of organic and inorganic materials [1,4–6]. Without such control, crystalline structures are often the result of multiple steps of nucleation, yielding complex morphologies such as spherulites with no direct relation to the symmetry of the crystal unit cell. By contrast, if many crystals were nucleated simultaneously, each from a single nucleus, and if they do not contain grain boundaries or major defects, we can deduce parameters of the crystal unit cell and its symmetry directly from the well-ordered and often simple crystalline morphology [7–9].

The kinetics of the growth of single crystals was explored in-situ in thin polymer films through various microscopy techniques [5,10]. The observed morphologies and the size of these crystalline structures depended on thermal conditions, film thickness, molecular weight and volume fraction of the crystallizable polymer [5,7,9–12]. Furthermore, similar single crystal structures were formed in polymer solutions, where solvent-polymer interactions and the solubility limit represent additional key parameters [8,11,13,14]. At concentrations below the solubility limit, the polymer solution is homogenous and no crystalline structures form [15]. Above the solubility limit, polymer–polymer interactions become more frequent allowing the formation of ordered structures, which, however, typically is accompanied by a nucleation barrier [3]. Crystals only form when this energy barrier for nucleation is overcome. Nucleating agents, surfaces or "dirt" may help to lower this barrier [8,13,16].

Self-seeding approaches represent a way to circumvent nucleation. Such approaches have been widely investigated in thin polymer films and polymer solutions. Self-seeding allows growing large and almost defect-free polymer single crystals at low super-cooling or low super-saturation. Such would not be possible if one must rely on homogeneous nucleation as under such conditions, nucleation events would not occur within acceptable time intervals. As shown by Blundell, Keller, and Kovacs [17,18], crystalline structures, in particular their shape and morphology, obtained via a self-seeding procedure allowed to infer parameters of the crystal unit cell. In one of their experiments, crystallization of poly-ethylene (PE) in dilute xylene solutions was studied [17]. First, thermal history of the PE sample was removed by completely dissolving the polymer above the clearing temperature of PE in the xylene solution (this temperature was in the range of $T_D = 97\ °C$ to 110 °C). Subsequently, the sample was quenched to a low crystallization temperature ($T_C = 80\ °C$) where the polymer rapidly crystallized, resulting in a large number of rather small PE crystals of various degrees of order [17]. In the next step, these crystals were heated to the self-seeding temperature T_{SS}. For dilute solutions, besides varying T_C and the time spent at T_{SS} of different values, the influence of polymer concentration (c) on the number (N) of resulting crystals was examined [17].

Similar studies of self-seeding approaches were also performed in thin polymer films [19,20]. Kovacs and Gonthier investigated the decoration of crystals via self-seeding in thin (ca. 1 μm) films of polyethylene oxide (PEO) [21]. There, the polymer film was first completely melted at 80 °C, i.e., well above the equilibrium melting temperature ($T_M \approx 70\ °C$) and then quenched to a low crystallization temperature to ensure fast growth and a high nucleation density [21,22]. These crystalline PEO films were heated to T_{SS}, slightly lower than T_M, yielding a controlled number of remaining crystalline seeds. Furthermore, using a sequential crystallization procedure, these crystals were decorated with many smaller ones [21]. Interestingly, the properties of the initial crystals from which the seeds originated allowed controlling characteristic features of the resulting seeded crystals such as their orientation [19]. In a set of self-seeding experiments, Xu et al. employed a specific thermal protocol combined with a systematic variation of T_{SS} [19]. They demonstrated that seeds, and the subsequently formed crystals, preserved the orientation of the initial single crystal from which they were originated [19,23].

As summarized by Sekerka [24], three main factors control the morphological evolution of crystals. Crystal growth is controlled by (i) the transport of molecules (diffusion of chains to the front of the growing crystal), (ii) the probability of attachment and detachment (interfacial kinetics) and (iii) the minimization of interfacial energy (capillarity) [12,24–27]. Growth of polymer crystals is controlled by the same processes, even if due to chain folding, they mainly grow in two-dimensions only, i.e., form lamellar crystals. The probably most widely studied polymer single crystals are based on PE and are often diamond-shaped, as independently shown by Till, Keller, and Fischer already in 1957 [14,28–30]. Based on such studies, Keller introduced the concept of a lamellar crystal consisting of folded polymer chains [13,29–31]. To reduce the degree of chain folding, i.e., to increase the lamellar thickness, to improve crystallinity and thermal stability of the polymer crystals [31–34], crystals had to be grown slowly at high T_C [5,12].

In solutions of block copolymers (PS-*b*-PEO) consisting of polystyrene (PS) and polyethylene oxide (PEO), Lotz and Kovacs observed similar crystalline lamellae in the early 1960s [35]. They found that the resultant crystalline structures had features of PEO single crystals [36]. The morphology of these PS-*b*-PEO crystals was that of square-shaped platelets, where a crystalline PEO lamella was sandwiched between two glassy PS layers [35]. Changes in the ratio of the molecular weight of PS and PEO led to changes in morphology and thickness of the platelets [37,38]. PS-*b*-PEO block copolymers have been employed in various applications, ranging from semiconductors and in microelectronics, micro purification, surface treatments to medical systems [7,39,40].

To allow for the observation of individual lamellar crystals and to avoid aggregation of crystalline structures, a rather low nucleation density is required. Thus, in the present work, we used self-seeding to control the number of crystal nuclei of PS-b-PEO in a toluene solution. Furthermore, as all seeds (nuclei) were already present before lowering the temperature to T_C, all resulting crystalline PS-b-PEO platelets started to grow at the same time and thus always had the same size [17]. Here, we present basic studies of self-seeding in solution [41], focusing on the influence of parameters such as self-seeding temperature, crystallization temperature, and polymer concentration. Besides the formation of mono-lamellar platelets, we also demonstrate that under certain conditions stacks of uniquely oriented PS-b-PEO lamellae can form.

2. Materials and Methods

In our experiments, we have used a symmetric diblock copolymer of poly(styrene)-block-poly(ethylene oxide), purchased from the Advanced Polymer Materials Inc., Montreal, Canada. The copolymers consist of a polystyrene block (with a number-average molecular weight M_n = 60 kg/mol) and a PEO block (M_n = 61 kg/mol) with a dispersity Đ = 1.10.

We have prepared solutions of PS-b-PEO in toluene at various concentrations ranging from 5 to 40 mg/mL. A Lauda thermostat (water tank) with a temperature precision of ±0.5 °C was used to control the desired temperatures, i.e., T_{SS} for self-seeding and T_C for crystallization from solution. First, the polymer powder was dispersed in toluene at T_{D_p} = 24 °C for 30 min using a rotating vortex mixer with 2500 rpm. The obtained polymer solution was transparent without any observable aggregates, indicating that most of PS-b-PEO was dissolved. However, some (small) aggregates must have existed in this supersaturated polymer solution, which was aged (stabilized) by keeping the toluene dispersion at room temperature (T_N = RT = 22 °C) for at least 24 h. After this protocol, the resulting polymer solution was slightly turbid, indicating the presence of suspended aggregates and possibly crystalline structures. To generate PS-b-PEO lamellar crystals under controlled conditions, the dispersion was subsequently heated to a seeding temperature (T_{SS} was varied from 35 °C to 60 °C) and then quenched to the crystallization temperature (T_C was varied from 10 °C to 22 °C). We kept the solution at T_C for 24 h to reach equilibrium, i.e., all polymers above the solubility limit were included in crystalline structures [16,42]. The thermal protocol shown in Figure 1 was employed for the crystallization of PS-b-PEO in solution.

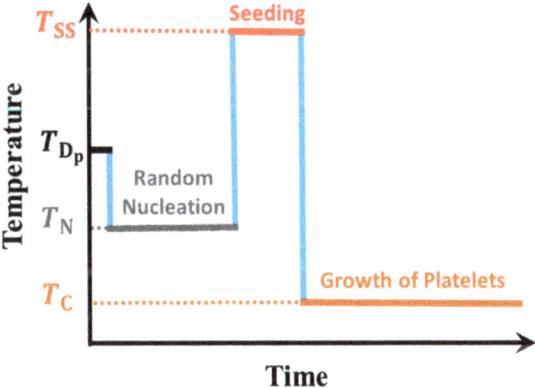

Figure 1. Thermal protocol used for crystallization of polymers in solution: Temperature defined the various stages of the polymer dispersion: partial dissolution (at T_{D_p}), high nucleation probability and rapid growth (at T_N), self-seeding (at T_{SS}) and slow growth of uniform PS-b-PEO platelets (at T_C).

To visualize the crystalline structures, 0.1 mL of the polymer solution containing suspended crystals was spin-casted at 2000 rpm onto UV-ozone treated silicon substrates (1.5 cm × 1.5 cm). Besides crystalline structures, also dissolved (non-crystallized) polymer chains were deposited. During the fast evaporation of toluene, these polymers could crystallize (rapidly). Thus, the resulting films contained large platelet-like polymer crystals surrounded by rapidly crystallized or amorphous polymers. We removed the latter by washing the samples, i.e., by putting the film for 5 to 30 s in a bath of toluene at room temperature.

The randomly distributed and washed crystals deposited on a silicon substrate were analyzed at ambient temperature using optical microscopy (OM, Olympus, Hamburg, Germany) and atomic force microscopy (AFM, JPK, Berlin, Germany) in the tapping mode [43]. Through optical microscopy, we analyzed the size and the density of these crystalline square-shaped and platelet-like crystals on multiple, randomly selected areas of the sample (ranging from 143 μm × 106 μm up to 10-times larger areas). AFM was used to examine smaller crystals on smaller areas.

On highly reflecting Si-substrates, white light from the optical microscope is reflected from both interfaces of thin polymer films, leading to interference colors. These colors represent the thickness of the polymer film and the embedded structures [44]. After washing the films, we could observe only platelet-like objects in various colors on a white background.

For samples with a high nucleation density and thus small platelet-like crystals, the lateral resolution of the optical microscope was not sufficient. Therefore, we used AFM to determine the size distribution and the nucleation density of small platelets. For higher polymer concentrations and higher number density of seeds, crystals might overlap during deposition. In such cases, before the deposition, we diluted the crystallized solution. The observed number of crystals was then multiplied by the dilution factor.

3. Results and Discussion

To allow for the growth of large platelet-like crystals, we aimed to keep the nucleation density at a low level. To this end, we employed different seeding temperatures (T_{SS}) and crystallization temperatures (T_C) for crystallization of PS-b-PEO in toluene solutions of various polymer concentrations (c). In an additional series of experiments, we varied c systematically for given values of T_{SS} = 40° (where the sample was kept for 5 min) and T_C = 20 °C (there, the sample was kept for 24 h). We present typically observed crystal morphologies obtained in PS-b-PEO solutions with c being 5, 10, and 20 mg/mL, respectively (Figure 2). Increasing c led to an increase of the number density (N) of square-shaped crystalline platelets, which all had approximately the same lateral size (L).

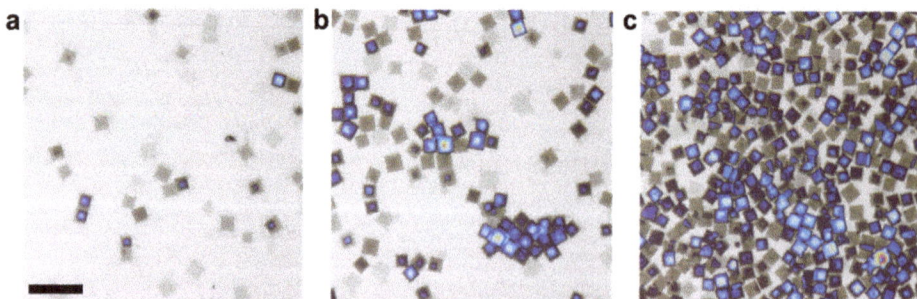

Figure 2. Optical micrographs demonstrating the influence of the concentration (c) of PS-b-PEO in toluene on the number (N) of resulting platelet-like crystals on silicon wafers. c varied from (**a**) 5 mg/mL, (**b**) 10 mg/mL, and (**c**) 20 mg/mL. The polymer was crystallized at T_C = 20 °C for 24 h after seeding at T_{SS} = 40 °C for 30 min. The scale bar represents 15 μm.

The square shape of the platelets reflects the symmetry of the crystal unit cell of PEO and was observed previously for single crystals of PEO and PS-b-PEO [14,17,18,21,36,45]. From the optical micrographs, we deduced N, the number of the platelet-like crystals per unit area, and assumed that each crystal resulted from a single nucleation site provided by a seed surviving at T_{SS}. Thus, we interpret N also as the number density reflecting the density of seeds.

From the optical micrographs represented in Figure 2, one can notice differences in the colors of the crystals, which originated from the interference of the white light (Figure S1). The colors indicate the thickness of the obtained crystalline objects. Interestingly, besides N also the number (per area) of crystals exhibiting a blue color increased with c. As indicated by the change of the interference color from light brown to light blue, we can conclude that the thickness of the platelets increased from ca. 20 nm to ca. 100 nm.

We quantified the changes of the thickness (h), as well as L and N of the PS-b-PEO crystals shown in the optical micrographs of Figure 2, which were formed during 24 h at $T_C = 20\,°C$ after being self-seeded at $T_{SS} = 40\,°C$ for 30 min. The results are shown in Figure 3. We found that N increased approximately linearly with c while L basically did not change with c. Furthermore, we observed that some crystals exhibited a thickness (h) larger than $h_0 \approx 20$ nm of a mono-lamellar crystal. $h > h_0$ resulted from stacking of several uniquely oriented lamellae. The fraction of crystals with $h > h_0$ increased with increasing c, as can be seen from the increase in the fraction of "blue" platelets in Figure 2.

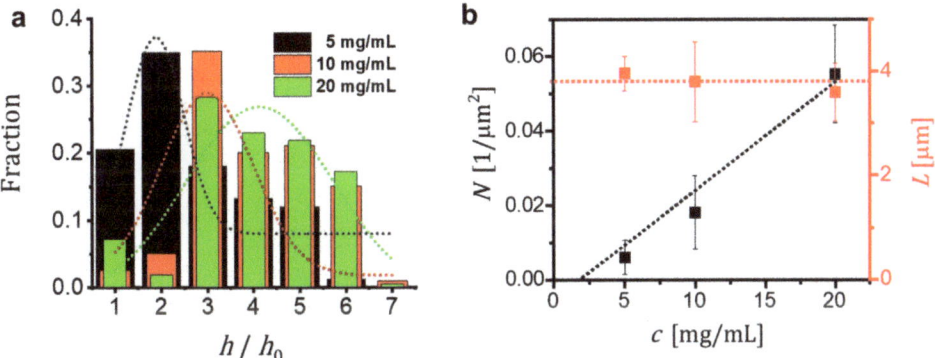

Figure 3. Influence of polymer concentration (c) on the number density and size of the crystals. Samples were crystallized at $T_C = 20\,°C$ for 24 h after self-seeding $T_{SS} = 40\,°C$ for 30 min. (**a**) Fraction of crystals as a function of their thickness (h) normalized by the thickness $h_0 \approx 20$ nm of a mono-lamellar crystal formed at $T_C = 20\,°C$. The black, red and green bars represent the results obtained for c being 5, 10 and 20 mg/mL, respectively. The dotted lines represent the corresponding fits assuming a Gaussian distribution; (**b**) Number density (N) and side length (L) of the obtained crystals.

Consistent with previous observations [4,17], the time the sample was kept at T_{SS} did not have a major impact on N and L. For controlling N and L, T_{SS} turned out to be the most relevant parameter. To reach this conclusion, we performed a series of experiments with solutions of a given c and varied T_{SS} systematically from 35 °C to 55 °C. After 5 min at T_{SS}, all solutions were quenched to a chosen T_C where they were kept for 24 h. The optical micrographs shown in Figure 4 represent N for a 10 mg/mL solution crystallized at $T_C = 13\,°C$ after seeding at various values of T_{SS}.

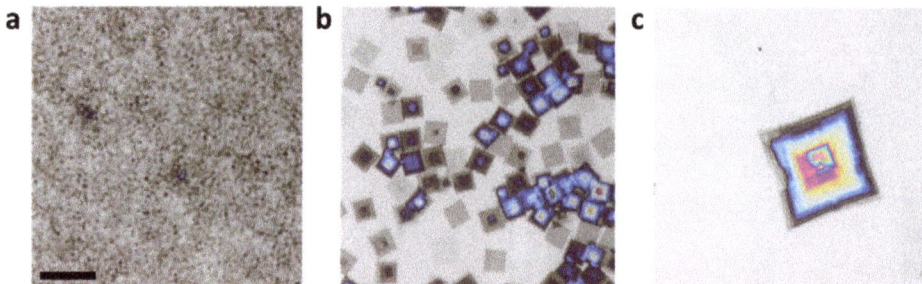

Figure 4. Influence of T_{SS} on N and L. The optical micrographs show crystals obtained in a 10 mg/mL PS-b-PEO solution in toluene crystallized at $T_C = 13\,°C$ for 24 h after being 5 min at a self-seeding temperature (**a**) $T_{SS} = 35\,°C$; (**b**) $T_{SS} = 40\,°C$, and (**c**) $T_{SS} = 45\,°C$. The scale bar represents 15 µm.

Independent of T_{SS}, the distribution in size of the platelets was very narrow, indicating that all crystals started to grow simultaneously and grew at a constant rate, as expected for a self-seeding procedure [17]. Figure 4 also shows that with increasing T_{SS}, N and L of the obtained crystals changed. We found that for a given c, N and L are related (increase of N lead to a decrease of L). Interestingly, in addition to N and L, also h of the crystalline objects, i.e., the number of stacked and uniquely oriented lamellae, increased with T_{SS} (Figures S1 and S3).

We measured N and L of the obtained crystals for many systematically repeated experiments such as the one shown in Figure 4 for different c. For increasing T_{SS} and a given c, Figure 5 displays that a decrease in N was accompanied by an increase of L. Accordingly, as shown also in Figure 4, very large crystals were obtained for high T_{SS}. Consistent with the graph shown in Figure 3b, N was proportional to c but L showed rather small changes with c (see Figure 5).

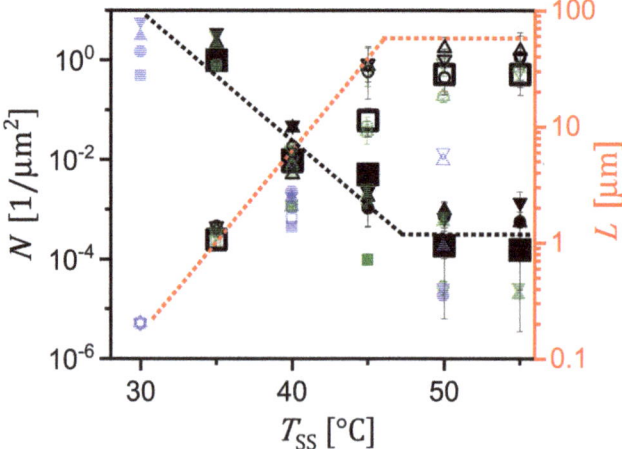

Figure 5. The effect of self-seeding temperature (T_{SS}) on the size (mean value of the side length L, open symbols) of the crystals and their number density (N, closed symbols), shown for various concentrations (c), crystallized at $T_C = 13\,°C$ (black), 20 °C (green) and 22 °C (blue) for 24 h, respectively. Squares, circles, up triangles and down triangles represent data points obtained for $c = 5$ mg/mL, $c = 10$ mg/mL, $c = 20$ mg/mL and $c = 40$ mg/mL, respectively.

For high $T_{SS} \geq 50\,°C$, L and N became rather constant and did not further decrease exponentially with T_{SS}. This behaviour may indicate that at these temperatures self-seeding was outpaced by heterogeneous nucleation through some rare but highly active substances within the solution.

Interestingly, comparing results for different c and various T_C (Figure 5 and Figure S2), we can confirm that the total volume (or total mass) of all crystals is proportional to the number of polymers above the solubility limit of the solution (represented by the deposited mass of the platelets per area, M_C). From Figure 5, we can deduce that for $T_{SS} < 50\,°C$, N is (roughly) proportional to $1/L^2$. For a constant T_C, we can assume that the mean thickness of the platelets $\bar{h} = \eta \cdot h_0$ consisting on average of a number η of lamellar crystals with a thickness h_0 each, was constant, i.e., independent of T_{SS}. Thus, the mass (m_C) of an individual platelet is proportional to its volume and the density of the polymer ($\rho \approx 1\,g/cm^3$). Therefore, the amount of polymer in solution above the solubility limit, deduced from the total volume per area of all the crystals obtained for various conditions, can be approximated by $V_C \approx N \cdot L^2 \cdot \eta \cdot h_0$, where η represents the average number of lamellae in a stack. Hence, we obtain the total mass per area of the crystallized polymer $M_C = N \cdot m_C \approx \rho \cdot V_C = N \cdot L^2 \cdot \bar{h}$. Furthermore, for a given c and at constant T_C, the value of $\bar{h}$ is well defined, as observed experimentally. Thus, the product $N \cdot L^2$ should be constant. Blundell and Keller found that the $N_{BK} \cdot m_{C,BK} = 1$ where N_{BK} is the number of nuclei per gram of polymer and $m_{C,BK}$ the mass of each crystal [17].

In Figure 6, we show that especially for $T_C = 13\,°C$, the product $N \cdot L^2 \cdot h_0$ did not vary significantly with T_{SS}. Some deviations from a constant value may be attributed to the formation of stacks of lamellar crystals, i.e., not all platelets had the same thickness (Figure S3), and to possible homogeneous nucleation events occurring at $T_C = 13\,°C$. However, at low T_{SS}, the values of $N \cdot L^2$ varied linearly with c (L is independent of c and N varied linearly as shown in Figure 3). We concluded that for high densities of seeds the formation of stacks of lamellae was unlikely and almost all platelets consisted of mono-lamellar crystals ($\bar{h} \approx h_0$, $\eta = 1$) (see Figure 6b).

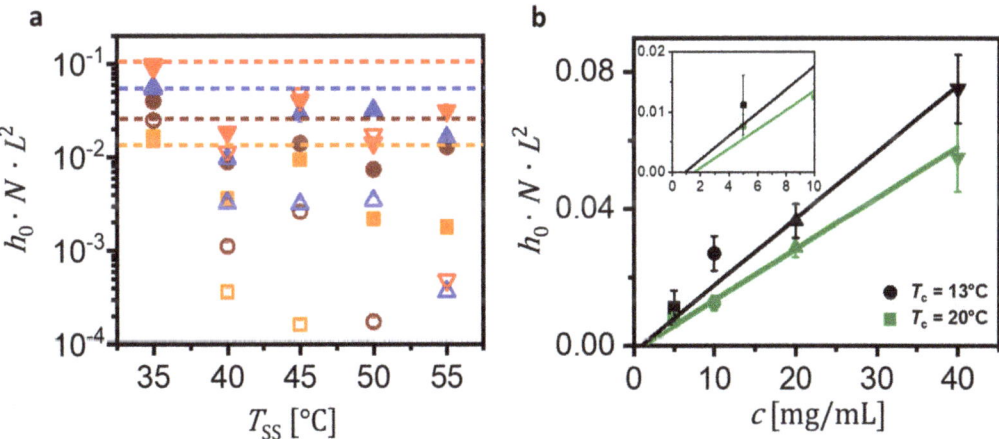

Figure 6. The effect of T_{SS} on the volume of the obtained crystals, assumed to be mono-lamellar, i.e., $\bar{h} \approx h_0$, $\eta = 1$; Amount (V_C) of polymer in solution above the solubility limit, deduced from the total volume per area of all the crystals (assuming $V_C \approx N \cdot L^2 \cdot h_0$) obtained for various conditions. (**a**) Closed and open symbols represent data points of $h_0 \cdot N \cdot L^2$ as a function of T_{SS} of crystals formed at $T_C = 13\,°C$ and $T_C = 20\,°C$, respectively. Orange, brown, blue and red symbols represent results for $c = 5$ mg/mL, $c = 10$ mg/mL, $c = 20$ mg/mL and $c = 40$ mg/mL, respectively. The corresponding dotted lines indicate the values of $h_0 \cdot N \cdot L^2$ which are expected to be independent of T_{SS}. Observed data points below these lines may indicate $\eta > 1$, especially for higher values of T_{SS}; (**b**) Plot of $N \cdot L^2 \cdot h_0$ (i.e., we assumed $\eta = 1$) for $T_{SS}= 35\,°C$ as function of c for $T_C = 13\,°C$ (black circles) and $T_C = 20\,°C$ (green squares), respectively. Linear fits to these data points extrapolated to $N \cdot L^2 \cdot h_0 = 0$ (magnified in the inset) yielded values for the solubility limit c^*. We obtained $c^* \approx (1 \pm 0.3)$ mg/mL and $c^* \approx (1.5 \pm 0.3)$ mg/mL for $T_C = 13\,°C$ and $T_C = 20\,°C$, respectively.

We observed that V_C increased with polymer concentration c (Figure 6,). We note that for concentrations less than $c \cong 1$ mg/mL, no crystals were observed. As can be observed from the graphs in Figures 3b and 6b, the linear fits extrapolated to $N \cdot L^2 \cdot h_0 = 0$ yielded, depending on T_C, values for the solubility limit c^*. That means that for $c < c^*$, all polymers were dissolved. We obtained $c^* \approx (1 \pm 0.3)$ mg/mL and $c^* \approx (1.5 \pm 0.3)$ mg/mL for the data points obtained for $T_C = 13\,°C$ and $T_C = 20\,°C$, respectively. Our results are on good agreement with results published by Keller and co-workers [17]. Due to self-seeding, N does not vary with T_C but c^* does and thus $N \cdot L^2$ depends on T_C. For the explored values of T_{SS} ranging from 35 °C to 55 °C, our results for N, the distribution of L and c^* were reproducible with an uncertainty of less than ca. 10%.

Besides the dependence of N and L on T_{SS}, we observed that L, but also c^*, depended on T_C. M_C can also be related to the super-saturation $\Delta c = c - c^*$: $M_C \approx N \cdot L^2 \cdot \bar{h} = V_S \cdot \Delta c$, where V_S is volume of the deposited polymer solution per unit area and $(\bar{h} = \eta \cdot h_0)$. For example, in $V_S = 0.1$ µL of polymer solution deposited on an area of 1 µm^2. Assuming that all platelets were mono-lamellar crystals, we can predict L of the formed platelets, controlling N through T_{SS} and M_C through Δc, by following relation: $L^2 \approx \Delta c / N \cdot \bar{h}$. We concluded that for a given c, N depended only on T_{SS}. Thus, for a constant T_{SS} but varying T_C, L can be expressed as $L^2 \approx 1/h \cdot \Delta c$. Increasing T_C leads to a reduction of M_C and thus to platelets with smaller L. If η is increasing with T_C, as observed, L decreases even more [17,36,46,47].

4. Conclusions

The number (N) of platelet-like PS-b-PEO block copolymer crystals and their mean lateral size (L) in a toluene solution can be well controlled via a self-seeding procedure. We examined the influence of polymer concentration (c), the crystallization temperature (T_C), the self-seeding temperature (T_{SS}) on N and L. We concluded that the volume of the crystallized polymer in solution (V_C) is determined by the solubility limit (c^*). Accordingly, the product $N \cdot L^2 \cdot \bar{h}$ depended on T_C. As (homogeneous) nucleation was circumvented through the self-seeding procedure, N did not depend on T_C but was mainly controlled through T_{SS} and c. Consistent with previous reports [17,23,35,36], for a given T_{SS}, N was not affected significantly by the time the sample was kept at T_{SS}. On the other hand, for a given T_{SS}, L depended on T_C. Therefore, L decreased when T_C increased, often accompanied by the formation of stacked lamellar crystals, especially for low values of N. To summarize, by determining the crucial parameters of the self-seeding procedure, we can predict the size of platelet-like crystals, their number density, and the number of uniquely oriented lamellae stacked in such crystals. Therefore, we conclude that self-seeding represents a highly suitable means for a controlled and predictable growth of well-defined polymer crystals in a polymer solution.

Supplementary Materials: The following are available online at https://www.mdpi.com/article/10.3390/polym13111676/s1, Figure S1: Height of stacks of lamellar crystals obtained by calibration of interference colors with AFM measurements, Figure S2: The influence of seeding temperature T_{SS} on side length (L) and number density (N), Figure S3: Number density and its consequence on the formation stacks of lamellar crystals.

Author Contributions: Conceptualization, methodology, validation, B.B., T.P. and G.R.; formal analysis, data curation, figure preparation, writing—original draft preparation, B.B.; writing—review and editing, T.P. and G.R.; supervision, G.R. All authors have read and agreed to the published version of the manuscript.

Funding: This research was support by "Photo-Emulsion", a European Innovative Training Network (ITN) project of the H2020 program (MSCA-ITN2017 n°: 765341) and by the Deutsche Forschungsgemeinschaft (RE 2273/18-1).

Institutional Review Board Statement: Not applicable.

Informed Consent Statement: Not applicable.

Acknowledgments: Financial support by "Photo-Emulsion", a European Innovative Training Network (ITN) project of the H2020 program (MSCA-ITN2017 n°: 765341) and by the Deutsche Forschungsgemeinschaft (RE 2273/18-1) are acknowledged. We thank Jun Xu (Tsinghua University, Beijing, China) for stimulating discussions.

Conflicts of Interest: The authors declare no conflict of interest.

References

1. Thanh, N.T.K.; Maclean, N.; Mahiddine, S. Mechanisms of Nucleation and Growth of Nanoparticles in Solution. *Chem. Rev.* **2014**, *114*, 7610–7630. [CrossRef] [PubMed]
2. Zhong, G.; Wang, K.; Zhang, L.; Li, Z.-M.; Fong, H.; Zhu, L. Nanodroplet Formation and Exclusive Homogenously Nucleated Crystallization in Confined Electrospun Immiscible Polymer Blend Fibers of Polystyrene and Poly(Ethylene Oxide). *Polymer* **2011**, *52*, 5397–5402. [CrossRef]
3. Xu, J.; Reiter, G.; Alamo, R.G. Concepts of Nucleation in Polymer Crystallization. *Crystals* **2021**, *11*, 304. [CrossRef]
4. Alfonso, G.C.; Valenti, B.; Saccone, A.; Pedemonte, E. Nature of Nuclei in Seeded Crystallization. *Polym. J.* **1974**, *6*, 322–325. [CrossRef]
5. Reiter, G.; Strobl, G.R. *Progress in Understanding of Polymer Crystallization*; Springer: Berlin/Heidelberg, Germany, 2007; ISBN 978-3-540-47307-7.
6. Zhang, G.; Roy, B.K.; Allard, L.F.; Cho, J. Titanium Oxide Nanoparticles Precipitated from Low-Temperature Aqueous Solutions: I. Nucleation, Growth, and Aggregation. *J. Am. Ceram. Soc.* **2008**, *91*, 3875–3882. [CrossRef]
7. Agbolaghi, S.; Abbaspoor, S.; Abbasi, F. A Comprehensive Review on Polymer Single Crystals—From Fundamental Concepts to Applications. *Prog. Polym. Sci.* **2018**, *81*, 22–79. [CrossRef]
8. Zhang, B.; Chen, J.; Baier, M.C.; Mecking, S.; Reiter, R.; Mülhaupt, R.; Reiter, G. Molecular-Weight-Dependent Changes in Morphology of Solution-Grown Polyethylene Single Crystals. *Macromol. Rapid Commun.* **2015**, *36*, 181–189. [CrossRef] [PubMed]
9. Hamie, H. Morphology and Thermal Behavior of Single Crystals of Polystyrene-Poly(Ethylene Oxide) Block Copolymers. Ph.D. Thesis, Université de Haute Alsace, Mulhouse, France, 2010; p. 252.
10. Majumder, S.; Poudel, P.; Zhang, H.; Xu, J.; Reiter, G. A Nucleation Mechanism Leading to Stacking of Lamellar Crystals in Polymer Thin Films. *Polym. Int.* **2020**, *69*, 1058–1065. [CrossRef]
11. Naga, N.; Yoshida, Y.; Noguchi, K. Crystallization of Poly(L-Lactic Acid)/Poly(D-Lactic Acid) Blend Induced by Organic Solvents. *Polym. Bull.* **2019**, *76*, 3677–3691. [CrossRef]
12. Reiter, G. Some Unique Features of Polymer Crystallisation. *Chem. Soc. Rev.* **2014**, *43*, 2055–2065. [CrossRef]
13. Welch, P.; Muthukumar, M. Molecular Mechanisms of Polymer Crystallization from Solution. *Phys. Rev. Lett.* **2001**, *87*, 218302. [CrossRef]
14. Keller, A. A note on single crystals in polymers: Evidence for a folded chain configuration. *Philos. Mag.* **1957**, *2*, 1171–1175. [CrossRef]
15. Blanks, R.F.; Prausnitz, J.M. Thermodynamics of Polymer Solubility in Polar and Nonpolar Systems. *Ind. Eng. Chem. Fund.* **1964**, *3*, 1–8. [CrossRef]
16. Wu, T.; Pfohl, T.; Chandran, S.; Sommer, M.; Reiter, G. Formation of Needle-like Poly(3-Hexylthiophene) Crystals from Metastable Solutions. *Macromolecules* **2020**, *53*, 8303–8312. [CrossRef]
17. Blundell, D.J.; Keller, A. Nature of Self-Seeding Polyethylene Crystal Nuclei. *J. Macromol. Sci. Part B* **1968**, *2*, 301–336. [CrossRef]
18. Vidotto, G.; Levy, D.; Kovacs, A.J. Cristallisation et fusion des polymères autoensemencés. *Colloid Polym. Sci.* **1969**, *230*, 289–305. [CrossRef]
19. Xu, J.; Ma, Y.; Hu, W.; Rehahn, M.; Reiter, G. Cloning Polymer Single Crystals through Self-Seeding. *Nat. Mater.* **2009**, *8*, 348–353. [CrossRef]
20. Poudel, P.; Chandran, S.; Majumder, S.; Reiter, G. Controlling Polymer Crystallization Kinetics by Sample History. *Macromol. Chem. Phys.* **2018**, *219*, 1700315. [CrossRef]
21. Kovacs, A.J.; Gonthier, A. Crystallization and fusion of self-seeded polymers. *Colloid Polym. Sci.* **1972**, *250*, 530–552. [CrossRef]
22. Murphy, C.J.; Henderson, G.V.S.; Murphy, E.A.; Sperling, L.H. The Relationship between the Equilibrium Melting Temperature and the Supermolecular Structure of Several Polyoxetanes and Polyethylene Oxide. *Polym. Eng. Sci.* **1987**, *27*, 781–787. [CrossRef]
23. Sangroniz, L.; Cavallo, D.; Müller, A.J. Self-Nucleation Effects on Polymer Crystallization. *Macromolecules* **2020**, *53*, 4581–4604. [CrossRef]
24. Sekerka, R.F. Role of Instabilities in Determination of the Shapes of Growing Crystals. *J. Cryst. Growth* **1993**, *128*, 1–12. [CrossRef]
25. Xiao, R.-F.; Alexander, J.I.D.; Rosenberger, F. Growth Morphology with Anisotropic Surface Kinetics. *J. Cryst. Growth* **1990**, *100*, 313–329. [CrossRef]
26. Gránásy, L.; Pusztai, T.; Tegze, G.; Warren, J.A.; Douglas, J.F. Growth and Form of Spherulites. *Phys. Rev. E* **2005**, *72*, 011605. [CrossRef] [PubMed]
27. Brener, E.; Müller-Krumbhaar, H.; Temkin, D. Structure formation and the morphology diagram of possible structures in two-dimensional diffusional growth. *Phys. Rev. E* **1996**, *54*, 2714–2722. [CrossRef]
28. Till, P.H. The Growth of Single Crystals of Linear Polyethylene. *J. Polym. Sci.* **1957**, *24*, 301–306. [CrossRef]

29. Fischer, E.W. Notizen: Stufen- Und Spiralförmiges Kristallwachstum Bei Hochpolymeren. *Z. Naturforsch. A* **1957**, *12*, 753–754. [CrossRef]
30. Fischer, E.W. Effect of annealing and temperature on the morphological structure of polymers. *Pure Appl. Chem.* **1972**, *31*, 113–132. [CrossRef]
31. Luo, C.; Han, X.; Gao, Y.; Liu, H.; Hu, Y. Aggregate Morphologies of PS-b-PEO-b-PS Copolymer in Mixed Solvents. *J. Dispers. Sci. Technol.* **2011**, *32*, 159–166. [CrossRef]
32. DiMarzio, E.A.; Guttman, C.M.; Hoffman, J.D. Calculation of Lamellar Thickness in a Diblock Copolymer, One of Whose Components Is Crystalline. *Macromolecules* **1980**, *13*, 1194–1198. [CrossRef]
33. Weeks, J.J. Melting Temperature and Change of Lamellar Thickness with Time for Bulk Polyethylene. *J. Res. Natl. Bur. Stan. Sect. A* **1963**, *67*, 441. [CrossRef]
34. Hsiao, M.-S.; Zheng, J.X.; Leng, S.; Van Horn, R.M.; Quirk, R.P.; Thomas, E.L.; Chen, H.-L.; Hsiao, B.S.; Rong, L.; Lotz, B.; et al. Crystal Orientation Change and Its Origin in One-Dimensional Nanoconfinement Constructed by Polystyrene-Block-Poly(Ethylene Oxide) Single Crystal Mats. *Macromolecules* **2008**, *41*, 8114–8123. [CrossRef]
35. Lotz, B.; Kovacs, A.J. Propriétés des copolymères biséquencés polyoxyéthylène-polystyrène. *Colloid Polym. Sci.* **1966**, *209*, 97–114. [CrossRef]
36. Lotz, B.; Kovacs, A.J.; Bassett, G.A.; Keller, A. Properties of copolymers composed of one poly-ethylene-oxide and one polystyrene block. *Colloid Polym. Sci.* **1966**, *209*, 115–128. [CrossRef]
37. Savariar, E.N.; Krishnamoorthy, K.; Thayumanavan, S. Molecular Discrimination inside Polymer Nanotubules. *Nat. Nanotechnol.* **2008**, *3*, 112–117. [CrossRef]
38. Yang, S.Y.; Yang, J.-A.; Kim, E.-S.; Jeon, G.; Oh, E.J.; Choi, K.Y.; Hahn, S.K.; Kim, J.K. Single-File Diffusion of Protein Drugs through Cylindrical Nanochannels. *ACS Nano* **2010**, *4*, 3817–3822. [CrossRef]
39. Karunakaran, M.; Nunes, S.P.; Qiu, X.; Yu, H.; Peinemann, K.-V. Isoporous PS-b-PEO Ultrafiltration Membranes via Self-Assembly and Water-Induced Phase Separation. *J. Membr. Sci.* **2014**, *453*, 471–477. [CrossRef]
40. Phan, T.N.; Issa, S.; Gigmes, D. Poly(Ethylene Oxide)-Based Block Copolymer Electrolytes for Lithium Metal Batteries. *Polymer Int.* **2019**, *68*, 7–13. [CrossRef]
41. Marra, J.; Hair, M.L. Interactions between Two Adsorbed Layers of Poly(Ethylene Oxide)/Polystyrene Diblock Copolymers in Heptane—Toluene Mixtures. *Colloids Surf.* **1988**, *34*, 215–226. [CrossRef]
42. Wu, T.; Valencia, L.; Pfohl, T.; Heck, B.; Reiter, G.; Lutz, P.J.; Mülhaupt, R. Fully Isotactic Poly(p-Methylstyrene): Precise Synthesis via Catalytic Polymerization and Crystallization Studies. *Macromolecules* **2019**, *52*, 4839–4846. [CrossRef]
43. Eaton, P.; West, P. *Atomic Force Microscopy*; Oxford University Press: Oxford, UK, 2010.
44. Kitagawa, K. Thin-Film Thickness Profile Measurement by Three-Wavelength Interference Color Analysis. *Appl. Opt.* **2013**, *52*, 1998. [CrossRef] [PubMed]
45. Keller, A. Crystal configurations and their relevance to the crystalline texture and crystallization mechanism in polymers. *Colloid Polym. Sci.* **1964**, *197*, 98–115. [CrossRef]
46. Majumder, S.; Busch, H.; Poudel, P.; Mecking, S.; Reiter, G. Growth Kinetics of Stacks of Lamellar Polymer Crystals. *Macromolecules* **2018**, *51*, 8738–8745. [CrossRef]
47. Guo, Z.; Yan, S.; Reiter, G. Formation of Stacked Three-Dimensional Polymer "Single Crystals". *Macromolecules* **2021**. [CrossRef]

Article

In-Depth Analysis of the Effect of Fragmentation on the Crystallization-Driven Self-Assembly Growth Kinetics of 1D Micelles Studied by Seed Trapping

Gerald Guerin [1,2,*], Paul A. Rupar [3] and Mitchell A. Winnik [2,4,*]

1. Shanghai Key Laboratory of Advanced Polymeric Materials, Key Laboratory for Ultrafine Materials of Ministry of Education, School of Materials Science and Engineering, East China University of Science and Technology, Shanghai 200237, China
2. Department of Chemistry, University of Toronto, 80 St. George Street, Toronto, ON M5S 3H6, Canada
3. Department of Chemistry, University of Alabama, Tuscaloosa, AL 35487, USA; parupar@ua.edu
4. Department of Chemical Engineering and Applied Chemistry, University of Toronto, Toronto, ON M5S 3E2, Canada
* Correspondence: gguerin@ecust.edu.cn (G.G.); mwinnik@utoronto.ca (M.A.W.)

Abstract: Studying the growth of 1D structures formed by the self-assembly of crystalline-coil block copolymers in solution at elevated temperatures is a challenging task. Like most 1D fibril structures, they fragment and dissolve when the solution is heated, creating a mixture of surviving crystallites and free polymer chains. However, unlike protein fibrils, no new nuclei are formed upon cooling and only the surviving crystallites regrow. Here, we report how trapping these crystallites at elevated temperatures allowed us to study their growth kinetics at different annealing times and for different amounts of unimer added. We developed a model describing the growth kinetics of these crystallites that accounts for fragmentation accompanying the 1D growth process. We show that the growth kinetics follow a stretched exponential law that may be due to polymer fractionation. In addition, by evaluating the micelle growth rate as a function of the concentration of unimer present in solution, we could conclude that the micelle growth occurred in the mononucleation regime.

Keywords: block copolymers; crystallization-driven self-assembly; kinetics; fragmentation; growth

1. Introduction

"Seeing is believing" is a well-known idiom that has found an echo in science, where many breakthroughs involved microscopy techniques. In polymer science, one of the major discoveries was made in 1957 by Andrew Keller [1], while studying homopolymer single crystals by transmission electron microscopy (TEM). Keller confirmed that the thin lamellae of these crystals consisted of folded chains, as first suggested by Storks [2], and then showed that the lamellar thickness of a homopolymer single crystal depends directly on the temperature at which it was grown [3]. Since that time, microscopy techniques have become an important tool to further investigate crucial aspects of polymer crystallization, e.g., crystal morphology [4–7] and crystal growth kinetics [8–11].

Several groups have examined the self-assembly in solution of crystalline-coil block copolymers (BCPs) that crystallize to form 1-dimensional (1D) micelles [12–18]. In several cases, the length and the composition of these micelles can be controlled precisely by adding free block copolymer ("unimer") with the same crystalline block, leading to the formation of elongated micelles or more complex structures referred to as block co-micelles [19–21]. Among these BCPs, those with a polyferrocenyldimethylsilane (PFS) core-forming block have been the most intensely studied. PFS BCPs have allowed the formation of the most advanced structures via stepwise hierarchical assembly [22–28]. Such sophisticated structures are, however, difficult to create with other crystalline-coil BCPs, a situation compounded by our limited fundamental understanding of the crystallization and growth of these 1D micelles.

To follow the growth kinetics of homopolymer single crystals in solution by microscopy, the crystals need to be isolated from their supersaturated unimer solution at different annealing times. For 2D single crystals, one can sediment the crystals [29] or transfer a crystal suspension quickly from one thermostated bath to another to surround the original growth front with a crystalline layer of different thickness [8]. For 1D core crystalline micelles, Boott et al. [11] followed the growth kinetic of PFS-b-PDMS (where PDMS stands for polydimethylsiloxane) core-crystalline micelles by depositing a drop of the micelle solution onto a TEM grid at a given annealing time and measuring the micelle lengths. They evaluated the effect of solvent, unimer concentration and block ratio on the micelle growth rate of PFS-b-PDMS micelles at temperatures where micelle fragmentation and dissolution could be ignored.

Furthermore, valuable information about the formation and growth of core-crystalline micelles can be obtained by studying their growth kinetics as a function of polymer concentration. Indeed, crystal growth in solution is considered to be interface-controlled, and the growth-determining step is the attachment of straight-chain segments ("stems") to the growth front and their rearrangement to form a surface nucleus [30–33]. Modern theory identifies different growth regimes depending on the number and rate at which stems spread on the crystal face [29,34]. These regimes are predicted to exhibit different dependences on polymer concentration, c, and can be identified experimentally by studying the polymer concentration dependence of crystal growth rates, G. This relationship is given by $G \propto c^\gamma$, where γ is the concentration exponent. For polymer single crystals, γ ranges from 0.2 to 2 [8,35]. A value of γ lower than 1 implies the presence of a barrier to chain deposition at the crystal growth front, while γ larger than 1 suggests cooperation between several unimers in solution to form a stable nucleus that would grow on the crystal face.

More recently, we have been particularly interested in understanding the main factors that affect micelles dissolution [36,37]. We noticed that when a seed solution was heated at a temperature where most of the seed dissolved, the surviving seeds could broaden and extend [38]. To further investigate this phenomenon, here we report the growth kinetics study of core-crystalline micelles in the presence of different amounts of unimer, at a temperature where the micelles could both fragment and dissolve. For this purpose, we pre-heated solutions of PFS$_{53}$-b-PI$_{637}$ (where PI stands for polyisoprene, and the subscripts represent the degree of polymerization of each block) crystallites at 75 °C, and added different amounts of unimer of the same BCP. After different annealing times, we injected a large excess of PFS$_{60}$-b-PDMS$_{660}$ to trap the growing crystallites and measured their length for each annealing time, following a seed-trapping approach developed previously [36]. In-depth analysis of the lengths of the trapped seeds and that of the control samples (without trapping the seeds) allowed us to develop a kinetic model that accounts for the fragmentation of the seeds during their growth. We also showed that the growth kinetics could be well described by a stretched exponential, in agreement with the kinetics study of Boott et al. [11]. To explain these results, we hypothesize that the stretched exponential is caused by polymer fractionation. Finally, from the concentration dependency of the seed growth rates, we showed that the 1D growth occurs via the successive addition of polymer chains to the exposed crystal faces.

2. Materials and Method

Decane (99+%) and Karstedt's catalyst were purchased from Sigma-Aldrich (Oakville, ON, Canada) and used without further purification.

The PFS$_{53}$-b-PI$_{637}$ (M_n, GPC = 56,300, Đ = 1.01) and PFS$_{60}$-b-PDMS$_{660}$ were synthesized by one of us and have been reported in ref [39].

2.1. Transmission Electron Microscopy

Bright-field transmission electron microscopy (TEM) images were taken at the nanoimaging facility of the chemistry department of the University of Toronto using a Hitachi H-7000 instrument (Hitachi High-Tech Corporation, Tokyo, Japan). Samples were prepared by placing one drop of solution on a Formvar carbon-coated grid, touching the edge of the droplet with a filter paper to remove excess liquid and allowing the grid to dry.

For each sample, micelle length distributions were determined by tracing more than 200 micelles using the software ImageJ (NIH, Laboratory for Optical and Computational Instrumentation, LOCI, University of Wisconsin, Madison, WI, US). Error bars were calculated using the standard error of the mean, s.e.m., obtained with a 99% confidence interval.

2.2. Sample Preparation

Six vials, each containing 4 mL of the same seed solution (c = 0.02 mg/mL) were heated at 75 °C in a heating bath. After 40 min of heating, different aliquots (0, 16, 33, 52, 76 and 82 µL) of PFS_{53}-b-PI_{637} unimer heated in decane (c = 4.8 mg/mL) at 100 °C were added to these solutions. Those solutions were then further annealed at 75 °C for 100, 420, 1200 and 2640 min. After each annealing time, 1 mL of each solution was transferred into empty vials that were also pre-heated to 75 °C. Half (0.5 mL) of each of these solutions were then injected in an empty vial and let to cool to room temperature (23 °C), following the usual self-seeding procedure, while to the second half we added a 5 times excess of PFS_{60}-b-$PDMS_{660}$ unimers that was pre-heated in decane at 100 °C. This second set of samples was briefly swirled and let at 75 °C for 5 more minutes to fully mix PFS_{60}-b-$PDMS_{660}$ with the PFS_{53}-b-PI_{637} unimers remaining in solution. The samples were then removed from the heating bath to cool to room temperature. This procedure allowed us to trap the PFS_{53}-b-PI_{637} surviving seeds with the large excess of PFS_{60}-b-$PDMS_{660}$ unimer. After two days of aging at room temperature, we added 0.5 mL of decane to each trapped seed solution, followed by 0.1 mL of Karstedt's catalyst. The samples were let to age one more day and then studied by TEM. In parallel, the control samples were aged at room temperature for two days prior to be imaged by TEM. This led to a total of 48 samples that were imaged by TEM.

Note that to follow the micelle growth kinetics as a function of unimer concentration, we used a PFS_{53}-b-PI_{637} seed solution in decane prepared for a previous study [36] and that was carefully stored in a sealed container and aged for one year at 23 °C. Over this time, the size (number average length, $L_{seed,RT}$, equal to 43.5 nm) and concentration of seed crystallites remained constant (Figure S1), but aging decreased the number of crystallites that dissolved at a given annealing temperature (Figure S2). We attribute the enhanced robustness of the one-year-old seeds to an increase in the crystallinity of the 1D PFS core. To confirm that this long aging time did not affect the mechanism of the seed dissolution, we plotted the length distribution of one-year aged seeds annealed and trapped at 75 °C, and compared it with the length distribution of the freshly prepared seeds annealed and trapped in similar conditions (Figure S3).

2.3. Definition of Key Parameters

The parameter p represents the ratio of the amount of PFS_{53}-b-PI_{637} unimer added to the solution at 75 °C ($m_{uni,added}$) to the mass of seed crystallites present in the seed solution at 23 °C ($m_{seeds,RT}$) prior to heating the solution to 75 °C. $L_{ts}(p,t)$ is the number average length of the trapped seeds annealed at 75 °C for different annealing times, while the number average length of the untrapped micelles cooled to 23 °C is defined as $L_{mic}(p,t)$.

Seeded growth can be well described by a simple equation [19]:

$$L_{mic} = \left(\frac{m_{uni}}{m_{seed}} + 1\right) L_{seed} \quad (1)$$

where L_{mic} and L_{seed} are the number average lengths of the micelles and the seeds, respectively, while m_{uni} is the mass of unimer added to the solution and m_{seed} is the mass of seeds. Despite its apparent simplicity, one can extract a large amount of information from Equation (1), helping us understand some key phenomena related to seeded growth and self-seeding.

3. Results and Discussion

The surviving seeds could easily be delineated by our seed-trapping protocol (Figure 1) [36]. Figure 1f,g shows representative TEM images of the stained trapped seeds after the solution was annealed at 75 °C, for 2640 min without unimer added ($p = 0$, Figure 1f) and with the largest amount of PFS$_{53}$-b-PI$_{637}$ unimer added ($p = 4.9$, Figure 1g).

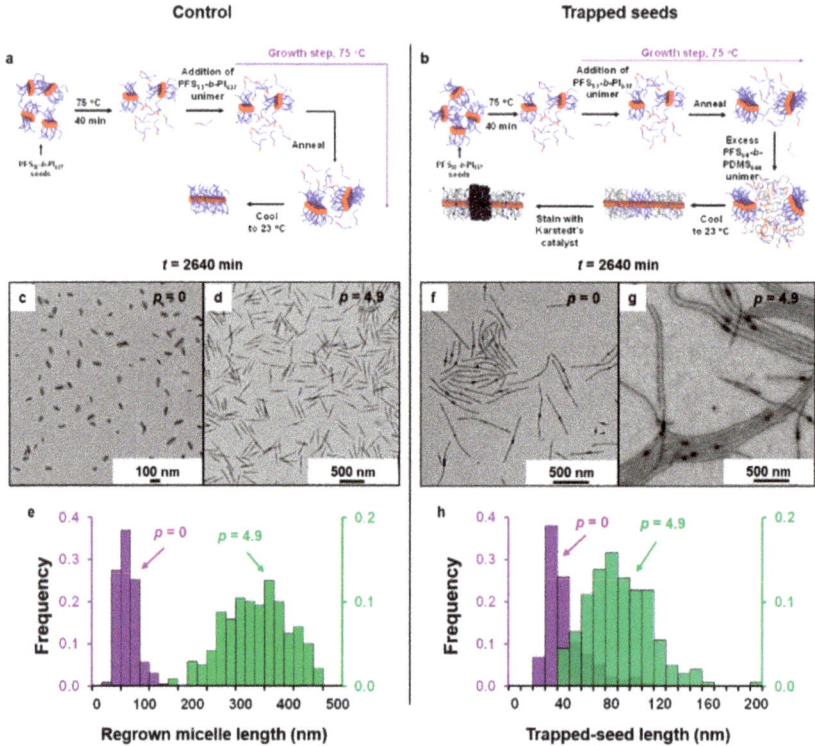

Figure 1. Seed growth kinetics studied by seed trapping. Schematic diagram describing (**a**) the control and (**b**) the seed-trapping experiments performed to study the growth kinetics of PFS$_{53}$-b-PI$_{637}$ crystallites annealed at 75 °C for different times in the presence of added PFS$_{53}$-b-PI$_{637}$ unimer. In this scheme we use a color code to represent different chemical species: Red represents polyferrocenyldimethylsilane (PFS) (either as the PFS block of a unimer or the crystalline core of a micelle), blue represents polyisoprene (PI), grey represents polydimethylsiloxane (PDMS), while the black spheres represent the platinum nanoparticles from the Karstedt's catalyst used to stain PI. (**c**,**d**) TEM images of PFS$_{53}$-b-PI$_{637}$ micelles obtained by heating seeds at 75 °C for 2640 min, and letting the solution cool to 23 °C, (**c**) without PFS$_{53}$-b-PI$_{637}$ unimer added ($p = 0$), and (**d**) in the presence of an initial mass ratio ($p = 4.9$) of PFS$_{53}$-b-PI$_{637}$ unimer added to PFS$_{53}$-b-PI$_{637}$ seeds. (**e**) Respective histograms of the length distributions of the regrown micelles without ($p = 0$, purple columns), and with unimer added ($p = 4.9$, green columns). (**f**,**g**) TEM images of PFS$_{53}$-b-PI$_{637}$ seeds trapped after 2640 min of annealing at 75 °C, (**f**) without PFS$_{53}$-b-PI$_{637}$ unimer added ($p = 0$), and (**g**) in the presence of an initial mass ratio of PFS$_{53}$-b-PI$_{637}$ unimer added to PFS$_{53}$-b-PI$_{637}$ seeds, $p = 4.9$. (**h**) Respective histograms of the length distribution of the surviving PFS$_{53}$-b-PI$_{637}$ seeds without ($p = 0$, purple columns), and with unimer added ($p = 4.9$, green columns). Samples (**f**,**g**) were stained with Karstedt's catalyst to highlight the PI rich regions.

It is important to note that, as shown in a previous study [36], the trapped seeds are ca. 3.5 nm longer than the stained seeds as observed by TEM due to the shrinkage of the corona block induced by the cross-linking of the corona by the karsted's catalyst.

The effect of the addition of PFS_{53}-b-PI_{637} unimer on PFS_{53}-b-PI_{637} seeds can be seen in Figure 1f–h, in Figures S4–S15 and Tables S1 and S2. $L_{ts}(0,2640)$ is similar to that of the un-aged seeds trapped at room temperature (Figure S16), suggesting that the dissolution of some of the seeds did not induce an obvious growth of the surviving seeds at 75 °C. In the presence of extra unimer (p = 4.9, Figure 1g), and after the same annealing time (2640 min), the length distribution of the seeds shifted to larger values, broadened and became Gaussian-like (Figure 1h). However, one sees that even after 2 days of annealing, the length of the trapped seeds remained much smaller than those of the corresponding control samples that were fully regrown at 23 °C (Figure 1c–e), a clear indication that most of the unimer remained in solution at 75 °C.

Figure 2a shows the evolution of micelles that regrow after annealing (control experiments), $L_{mic}(p,t)$, as a function of annealing time for each amount of unimer added to the solution. For all the amounts of unimer added, one sees that the lengths of the micelles regrown at 23 °C decreased with annealing time. This result was rather surprising since one would intuitively expect the lengths of the micelles regrown at 23 °C, $L_{mic}(p,t)$ to be independent of annealing time. For each annealing time, however, the lengths of the micelles subsequently regrown at room temperature, $L_{mic}(p,t)$ versus p obeys Equation (1), still increased linearly as a function of the amount of unimer present in solution (Figure 2b).

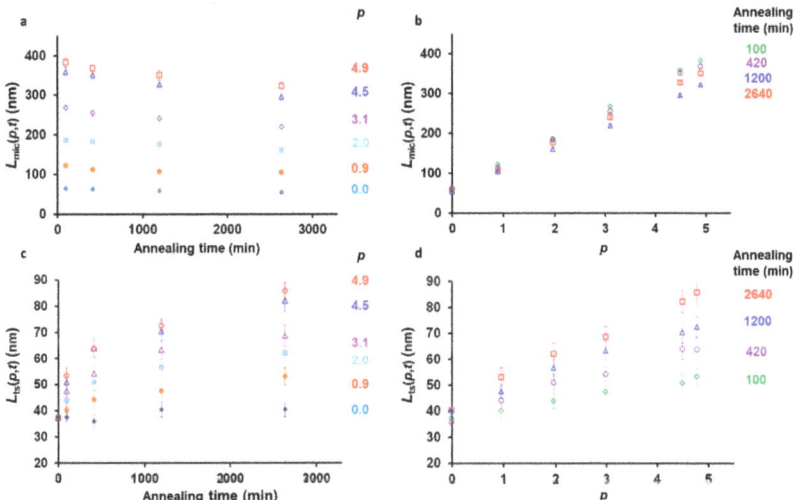

Figure 2. Evolution of the number average lengths of micelle annealed at 75 °C and regrown at room temperature solutions as a function of (a) time and (b) ratios p. Evolution of the number of average lengths of trapped seeds as a function of (c) time and (d) ratios p. p is the ratio of the mass of unimer added to the hot seed solution, $m_{uni,added}$, to the mass of seeds originally present in the solution, $m_{seeds,RT}$. Error bars correspond to the s.e.m. of the length distributions.

In contrast, the lengths of the seeds trapped at 75 °C, $L_{ts}(p,t)$ increased as a function of time (Figure 2c), following much more conventional behavior. Seeded growth was, however, extremely slow since even after two days of annealing, the trapped seeds remained much smaller than the micelles at room temperature, reaching only 90 nm for the largest amount of unimer injected in the solution. The growth kinetics were also non-linear, slowing down with time (Figure S17). Micelle growth, although extremely slow, was still noticeable since, for p = 4.9, the trapped seeds were ca. twice longer after 2640 min of annealing than after 100 min. Finally, in Figure 3d we show the plot of the length of the trapped seeds as a function of the different amount of unimer added to the solution for each annealing time. In this plot, one observes that $L_{ts}(p,t)$ increased linearly as a function of p, as already seen for $L_{mic}(p,t)$ versus p.

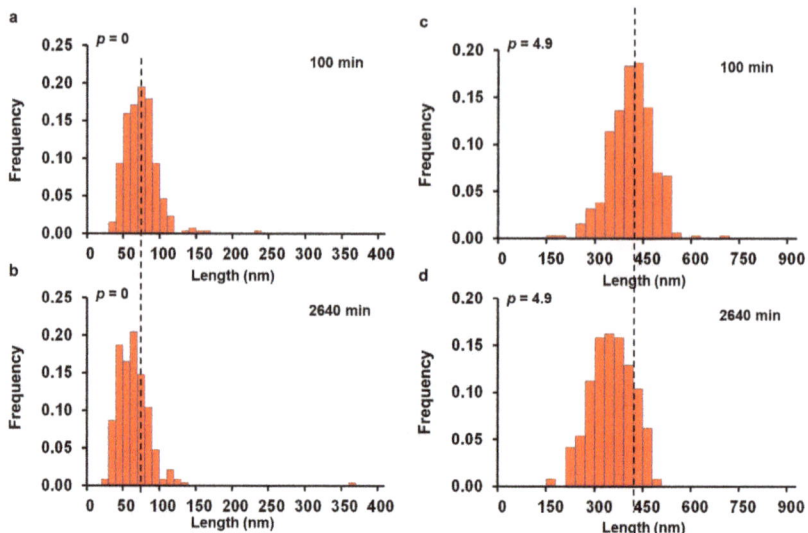

Figure 3. Histograms of the length distributions of PFS$_{53}$-b-PI$_{637}$ micelles regrown at room temperature for $p = 0$ after (**a**) 100 min, (**b**) 2640 min, and for $p = 4.9$ after (**c**) 100 min and (**d**) 2640 min of annealing at 75 °C. The vertical dashed lines highlight the shift of the micelle length distributions towards shorter values.

The seed-trapping and control experiments were rather simple to perform, leaving little doubts about the validity of the results obtained. Interestingly, their combination coupled with the use of Equation (1) (which is also highly straightforward) can unravel complex phenomena, as shown in the following sections.

3.1. Control Experiments: Lateral Growth versus Fragmentation during Annealing

The number average lengths of the seeds annealed at 75 °C for different times (100, 420, 1200 and 2640 min) (Figure S18) and cooled to room temperature decreased monotonically from ca. 64.4 nm for the sample annealed for 100 min down to 55.4 nm when annealed for 2640 min. Although the length distributions of the sample shifted toward lower values (Figure 3a,b), the variation was quite small and could be considered, a priori, as part of the experimental error. However, similar behavior was observed for all the control samples, becoming increasingly noticeable as the amount of unimer added to the solution increased (Figure 3c,d), pointing toward a systematic effect of the annealing time on the final length of the micelles regrown at room temperature.

Previous works have shown that two main phenomena could explain this behavior:

(a) Unimer chains could add laterally onto the micelles during annealing. These chains would thus not participate to the elongation of the micelles. Since lateral growth is expected to be time dependent, the amount of unimer in solution that could participate to the elongation of the micelle at 23 °C would decrease with time.

(b) The micelles could fragment with time, increasing the number of seeds in solution which would lead to a decrease in the lengths of the micelles once they regrow at 23 °C.

The one-year aged seed crystallites were much more stable towards annealing at 75 °C than their freshly prepared counterpart (Figure S2). Therefore, one can conclude that this long aging time favored the packing of the crystalline core and decreased the distance between two grafting points. As a result, the densification of the micelle crystalline core would lead to an increase in their grafting density, hindering the lateral growth of the micelles [38]. We thus assume that the decrease in $L_{mic}(p,t)$ as a function of annealing time is mainly due to seed fragmentation, while lateral growth can be neglected. We verified these assumptions by evaluating the ratio $L_{mic}(p,t)/L_{mic}(0,t)$ as a function of annealing time for both scenarios, and comparing it with the experimental plot of $L_{mic}(p,t)/L_{mic}(100,t)$ versus time shown in Figure 2b.

The lateral growth of the micelles during annealing would lead to an increase in the linear aggregation number of the micelles in the section that was regrown. Therefore, we can rewrite Equation (1) to express $L_{mic}(p,t)$ as a function of both p and t:

$$L_{mic}(p,t) = \left(\frac{N_{agg/L}(p,t)}{N_{agg/L,RT}} \frac{m_{uni}(p,t)}{m_{ts}(p,t)} + 1 \right) L_{ts}(p,t) \qquad (2)$$

where $N_{agg/L}(p,t)$, is the linear aggregation number of the trapped seeds, $N_{agg/L,RT}$ is the linear aggregation number of the seeds at room temperature. $m_{uni}(p,t)$ is the amount of unimer that is still present in solution after it was annealed at 75 °C for a time t.

We recall that $m_{seeds,RT} = m_{uni}(0,0) + m_{ts}(0,0)$, while by definition, $m_{added}(p,0) = p \, m_{seeds,RT}$, where $m_{uni}(0,0)$ is the mass of unimer coming from the dissolution of some of the starting seeds, $m_{added}(p,0)$ is the mass of unimer added to the solution and $m_{ts}(0,0)$ is the mass of the surviving seeds just after heating. Equation (2) thus gives (see Supporting discussion, Section I):

$$L_{mic}(p,t) = L_{ts}(0,0) \left[(p+1) \frac{m_{seeds,RT}}{m_{ts}(0,0)} \right] + L_{ts}(p,t) \left[1 - \frac{N_{agg/L}(p,t)}{N_{agg/L,RT}} \right] \qquad (3)$$

For $t = 0$, $N_{agg/L}(p,0) = N_{agg/L,RT}$ (the linear aggregation number does not change), leading to:

$$L_{mic}(p,0) = \frac{m_{seed,RT}}{m_{ts}(0,0)} (p+1) L_{ts}(0,0) \qquad (4)$$

Here, we note that for the specific case where p and t are equal to 0, Equation (4) gives:

$$\frac{m_{seeds,RT}}{m_{ts}(0,0)} = \frac{L_{mic}(0,0)}{L_{ts}(0,0)} \qquad (5)$$

Therefore, Equation (3) becomes:

$$\frac{L_{mic}(p,t)}{L_{mic}(p,0)} = 1 - \frac{L_{ts}(p,t)}{L_{mic}(0,0)} \left(\frac{N_{agg}(p,t) - N_{agg,RT}}{N_{agg,RT}} \right) \frac{1}{(p+1)} \qquad (6)$$

On the other hand, if the decrease of $L_{mic}(p,t)$ as a function of time is solely due to the fragmentation of the seeds as a function of time (and, thus to an increase in the number of seeds), we can show (Supporting discussion, Section II) that:

$$L_{mic}(p,t) = \frac{m_{seed,RT}}{m_{ts}(0,0)} (p+1) L_{ts,f}(p,t) \qquad (7)$$

where $L_{ts,f}(p,t)$ is the length of the fragmented seeds as a function of annealing time.

We note that Equation (7) is equivalent to Equation (4), using $L_{ts,f}(p,t)$ instead of $L_{ts}(p,t)$. At time, $t = 0$, just after the seed solution reached 75 °C, although the seeds may have fragmented, they did not grow, thus, the length of the fragmented seeds at $t = 0$ is the same as the length of the trapped seeds, i.e., $L_{ts,f}(p,0) = L_{ts}(p,0) = L_{ts}(0,0)$. We can thus write:

$$\frac{L_{mic}(p,t)}{L_{mic}(p,0)} = \frac{L_{ts,f}(p,t)}{L_{ts}(0,0)} \qquad (8)$$

Equations (6) and (8) indicate that the evolution of $L_{mic}(p,t)/L_{mic}(100,t)$ versus time for different values of p strongly depends on the annealing history. If lateral growth occurs during annealing (Equation (6)) then, one should expect $L_{mic}(p,t)/L_{mic}(100,t)$ to vary with the amount of unimer added to the solution, while in the case of seed fragmentation (Equation (8)), $L_{mic}(p,t)/L_{mic}(100,t)$ would be independent of p if the seed fragmentation does not depend on the amount of unimer added to the solution.

To evaluate which equation better describe the experimental results, we plotted $L_{mic}(p,t)/L_{mic}(100,t)$ versus time for the different amount of unimer added to the solution (Figure 4). In this plot, $L_{mic}(p,t)/L_{mic}(100,t)$ appears to be independent on the amount of unimer injected into the seed crystallite solution for the annealing times investigated, a strong indication that the change in the final lengths of the micelles for the control experiment is mainly due to seed fragmentation (additional discussion can be found in Supporting Information).

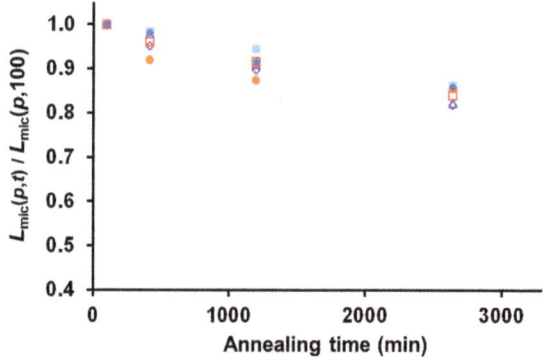

Figure 4. Plot of $[L_{mic}(p,t)/L_{mic}(p,100)]$ as a function of annealing time for the different ratios of the mass of unimer added to the hot seed solution, $m_{uni,added}$, to the mass of seeds originally present in the solution, $m_{seeds,RT}$ ($p = 0, 0.9, 2, 3.1, 4.5, 4.9$).

3.2. Fragmentation versus Annealing Time

The presence of fragmentation could strongly affect the micelle growth kinetics. Indeed, the impact of fragmentation on the micelle lengths during seeded growth is multifold. It decreases the length of the micelles that fragment, and it increases the number of seeds the unimer can add on. Fragmentation could also influence the growth kinetics. For example, if seeded growth was a diffusion limited process, the increase in the number of seeds in solution would decrease the diffusion time between the unimer and the seed ends, which would lead to an increase in growth kinetics. It might thus appear difficult to quantify the effect of fragmentation on micelle growth.

Fortunately, we can use some key observations that were made in previous reports as well as in the present work to simplify the equations. Boott et al. [11] have shown that the seeded growth of PFS_{63}-b-$PDMS_{513}$ (a system similar to that presented in this study) was not diffusion-limited. We can thus assume that the growth kinetics is not influenced by the seed fragmentation. In addition, the overlapping plots of $L_{mic}(p,t)/L_{mic}(p,100)$ indicate that micelle fragmentation does not depend on the amount of unimer present in solution. Finally, we distinguish dissolution from fragmentation, in the sense that fragmentation leads to the formation of more micelles/seeds, but does not add any unimer in solution.

The number average length of a population of N seeds is given by:

$$L_n = \frac{\sum_{i=1}^{N} L_i}{N} \tag{9}$$

where L_i is the length of seed i, and N is the total number of seeds. If z new seeds are formed via fragmentation, the number average length of the seed solution, $L_{n,f}$ becomes:

$$L_{ts,f}(t) = \frac{\sum_{i=1}^{N+f} L'_i(t)}{N+z(t)} = \frac{\sum_{i=1}^{N} L_i}{N+z(t)} = \frac{\sum_{i=1}^{N} L_i}{N} \frac{1}{1+z(t)/N} = L_{ts}(0,0) f(t) \qquad (10)$$

where L'_i is the length of seed i, after fragmentation. It is important to note that Equation (10) is only correct in absence of dissolution during the fragmentation, since in this case, the total length of the seeds is unchanged and $z(t)/N$ increases with time, from 0 to a finite positive value.

In the present work, seed annealing can be schematized as a two-step process. First, as soon as the annealing temperature was reached, the seeds with the lowest crystallinity dissolved [36], while the rest of the seeds survived. The dissolution step can be considered instantaneous. In the second step, the seeds fragmented. As shown in Figures 2a, 3 and 4, this step is time dependent.

The difficulty here resides in choosing a physically meaningful equation that would describe the fragmentation of the trapped seeds during annealing. In a previous work [36], we showed that short seeds, such as those used in this study, would mainly fragment in the center until they reach a critical length, $L_{c,f}$ [40–43], that was estimated to be close to 32 nm. Since we are using the same seed solution, we would expect the critical length to also be close to 32 nm.

It is also important to note that the increase in length of the seeds during annealing at 75 °C (Figure 2c,d), did not affect their fragmentation, since the plots of $L_{mic}(p,t)/L_{mic}(100,t)$ versus time (Figure 4) overlap for all the values of p. This result, in apparent contradiction with previous observations [44], suggests that the original seeds were more fragile than their extended counterparts. This phenomenon may find its origin in the fact that the original seeds were grown at room temperature, while the extended parts were grown at 75 °C, which could facilitate a better packing of the crystalline block, strengthening the micelle core.

From this description, we infer that the evolution of the number of fragmentation events as a function of time could be approximated by a normal distribution centered at $t = 0$. Indeed, at short time, the number of seeds that would fragment might be relatively large, but as the annealing time increases, the seeds would strengthen, and the number of fragmentation events would slowly decrease. Since the number of seeds increases at each fragmentation event, one needs to consider the sum of all the events as a function of time, leading to a cumulative distribution function:

$$z(t) = 2 \int_0^t \frac{\exp\left(-\frac{x^2}{2\sigma^2}\right)}{\sqrt{2\pi\sigma^2}} dx [\alpha - 1] N \qquad (11)$$

and

$$f(t) = \frac{1}{1+z(t)/N} = \frac{1}{1+2\int_0^t \frac{\exp\left(-\frac{x^2}{2\sigma^2}\right)}{\sqrt{2\pi\sigma^2}} dx [\alpha - 1]} \qquad (12)$$

where σ can be related to the rate of fragmentation, since fast fragmentation would lead to a low value of σ, while α is a normalization factor defined as:

$$\alpha = \frac{L_{ts}(0,0)}{L_{c,f}} \qquad (13)$$

If the length of the surviving seeds at $t = 0$ is equal to the critical length, $L_{c,f}$, then $\alpha = 1$, and the seeds will not fragment. However, if α is large, then the decrease in seed lengths due to fragmentation will also be large.

From Equation (10), the change of the trapped seed lengths as a function of fragmentation is thus:

$$L_{ts,f}(0,t) = \frac{L_{ts}(0,0)}{2\int_0^t \frac{\exp\left(-\frac{x^2}{2\sigma^2}\right)}{\sqrt{2\pi\sigma^2}}dx[\alpha-1]+1} \quad (14)$$

Since the amount of unimer added to the seed solution did not affect their fragmentation, we can write the more general equation:

$$L_{ts,f}(p,t) = \frac{L_{ts}(0,0)}{2\int_0^t \frac{\exp\left(-\frac{x^2}{2\sigma^2}\right)}{\sqrt{2\pi\sigma^2}}dx[\alpha-1]+1} \quad (15)$$

with $L_{ts}(0,0) = L_{ts}(p,0)$, since at time $t = 0$, no unimer would have time to add onto the surviving seeds.

Incorporating Equation (15) into Equation (7) leads to:

$$L_{mic}(p,t) = (p+1)\frac{L_{mic}(0,0)}{2\int_0^t \frac{\exp\left(-\frac{x^2}{2\sigma^2}\right)}{\sqrt{2\pi\sigma^2}}dx\left(\frac{L_{ts}(0,0)}{L_{c,f}}-1\right)+1} \quad (16)$$

Equation (16) can thus be used to fit $L_{mic}(p,t)/(1+p)$ versus time (Figure 5a). For this fit, we used the values of $\sigma = 2700$ min, $L_{mic}(0,0) = 65$ nm, $L_{ts}(0,0) = 41$ nm and $L_{c,f} = 32$ nm, which is the value expected from the superblob approach [36]. In a previous work, we have shown that in dilute solution seeds dissolve in a cooperative (explosive) process. Thus, seeds that survive dissolution would be expected to have a length similar to the original seeds, i.e., 43.5 nm, close to the value of $L_{ts}(0,0) = 41$ nm used to fit the data.

Fits of $L_{mic}(p,t)$ as a function of p and t are shown in Figure 5b,c. Despite the inherent uncertainty in the measurements of the micelle lengths, we could fit reasonably well the experimental data shown in Figure 5b,c using Equation (16).

3.3. Growth Kinetics at 75 °C

Equation (16) gives the evolution of the number average lengths of the micelles once they were fully regrown at room temperature. The lengths of these micelles were only dependent on the number of seeds present in solution, i.e., seed fragmentation. Growth kinetics could thus be ignored.

The situation is quite different for the seeds trapped at 75 °C, since the lengths of the micelles as a function of time depend on both micelle fragmentation and growth kinetics. Figure 2c,d, shows that the trapped seeds annealed for 2640 min were twice longer when $p = 4.9$ than in absence of unimer added to the solution ($p = 0$). The master curve obtained for $L_{mic}(p,t)/(p+1)$ versus time (Figure 5a), however, suggests that fragmentation did not depend on the amount of unimer added. From these two observations, we conclude that fragmentation and micelles growth were independent from each other, in our experimental conditions. Indeed, if fragmentation was related to micelle growth, then a larger amount of micelles would have fragmented when the micelles were longer, and $L_{mic}(p,t)/(p+1)$ would have varied with p. Therefore, fragmentation and growth can be seen as two independent time functions. We can thus write:

$$L_{ts}(p,t) = f(t)g(p,t) \quad (17)$$

where $f(t)$ is the fragmentation function given in Equation (12), while $g(p,t)$ describes micelle growth as a function of p in absence of fragmentation:

$$g(p,t) = \left[\frac{m_{uni}(p,0) - m_{uni}(p,t)}{m_{ts}(p,0)} + 1\right]L_{ts}(p,0) \quad (18)$$

with $L_{ts}(p,0) = L_{ts}(0,0)$, and $m_{ts}(p,0) = m_{ts}(0,0)$.

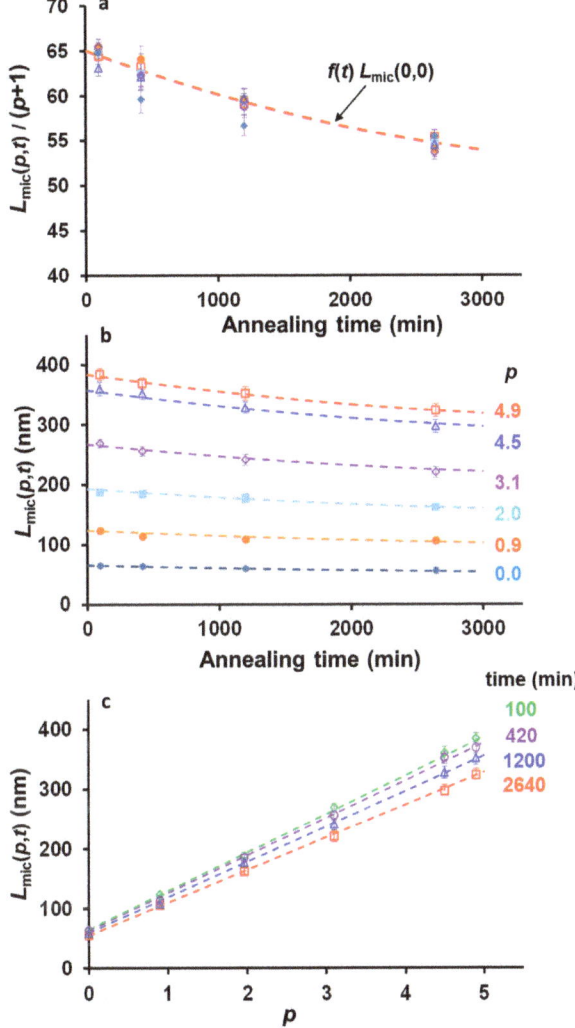

Figure 5. (a) Plot of $[L_{mic}(p,t)/(1+p)]$ as a function of annealing time (symbols). Fitting Equation (16) to these data (dashed line) led to the values of σ = 2700 min, $L_{mic}(0,0)$ = 65 nm, $L_{ts}(0,0)$ = 41 nm and $L_{c,f}$ = 32 nm. Plots of the length of the regrown micelles, $L_{mic}(p,t)$, fitted to Equation (16) (dashed lines) as a function of (**b**) time and (**c**) p. Error bars correspond to the s.e.m. of the length distributions.

As shown in Supporting Information, taking into account the effect of the molecular weight distribution of PFS$_{53}$-b-PI$_{637}$, leads to:

$$g(p,t) = L_{ts}(0,0)\left[1 + \left(\frac{L_{mic}(0,0)}{L_{ts}(0,0)}(1+p) - 1\right)\left(1 - e^{-(k^*t)^\beta}\right)\right] \quad (19)$$

where k^* is the growth rate constant and β, the stretching exponent.

A stretched exponential is a signature of a distribution of growth rates [45,46]. This kind of distribution could arise if the deposition rate of a unimer were highly sensitive either to the length of its PFS block or the block ratio of the BCP. Even though PFS$_{53}$-b-PI$_{637}$ has a narrow molecular weight distribution (Đ = 1.01), it is not monodispersed. The stretched exponential fit suggests that fractionation affects the micelle growth at 75 °C [47]. This result is consistent with the observation made in one of the rare studies of the fractionation of narrowly dispersed BCP by crystallization [48]. It is also in agreement with the recent work from Song et al., who studied the CDSA of crystalline-coil BCPs with corona-forming block of various molecular weight distributions [49].

Incorporating Equations (16) and (19) into Equation (17) finally gives the growth kinetics of the micelles at 75 °C in the presence of fragmentation.

$$L_{ts}(p,t) = \frac{L_{ts}(0,0)\left[1 + \left(\frac{L_{mic}(0,0)}{L_{ts}(0,0)}(1+p) - 1\right)\left(1 - e^{-(k^*t)^\beta}\right)\right]}{2\int_0^t \frac{\exp\left(-\frac{x^2}{2\sigma^2}\right)}{\sqrt{2\pi\sigma^2}} dx \left(\frac{L_{ts}(0,0)}{L_{c,f}} - 1\right) + 1} \quad (20)$$

Fits of $L_{ts}(p,t)$ as a function of p and t are shown in Figure 6. To fit the growth kinetics data (Figure 6a,b), we used $k^* = 1.7 \times 10^{-5}$ min^{-1} and $\beta = 0.51$. Interestingly, these values compared well with the data obtained by Boott et al. [11] for the growth of PFS$_{63}$-b-PDMS$_{513}$ in n-hexane at different temperatures. Indeed, extrapolation of the Eyring plot (ln(k'/T) versus 1/T) that they generated from kinetic data obtained at different temperatures led to a rate constant $k' = 4.4 \times 10^{-5}$ min^{-1} (or $k' = 7.3 \times 10^{-7}$ s^{-1}), which is in the same order of magnitude as the rate constant deduced from Equation (20).

3.4. Evaluation of the Growth Rate as a Function of Unimer Concentration

Equation (20) gives us the possibility to evaluate the growth rate of the core crystalline micelles as a function of unimer concentration:

$$G(p,t) = \frac{\partial L_{ts}(p,t)}{\partial t} \quad (21)$$

By itself, $G(p,t)$ is extremely complex. We are, however, interested in evaluating the variation of $G(p,t)$ as a function of p. For this reason, one simply needs to rewrite Equation (20) to isolate all the terms that depend on p:

$$L_{ts}(p,t) = \frac{L_{ts}(0,0)e^{-(k^*t)^\beta} + L_{mic}(0,0)\left(1 - e^{-(k^*t)^\beta}\right)}{f(t)} + p\frac{L_{mic}(0,0)\left(1 - e^{-(k^*t)^\beta}\right)}{f(t)} = x(t) + p\,y(t) \quad (22)$$

Therefore $G(p,t)$ is given by:

$$G(p,t) = \frac{dx(t)}{dt} + p\frac{dy(t)}{dt} \quad (23)$$

$G(p,t)$ thus increases linearly with p, and since p is proportional to the unimer concentration, c, we can write that $G \propto c^\gamma$, with $\gamma = 1$. This special case indicates that the micelle growth proceeded in the mononucleation regime, where one nucleus adds at a time on the growth face. The linear increase of $G(p,t)$ with p is in agreement with the results of Monte Carlo simulations reported by Hu et al. [50,51] for planar growth of 1D crystals.

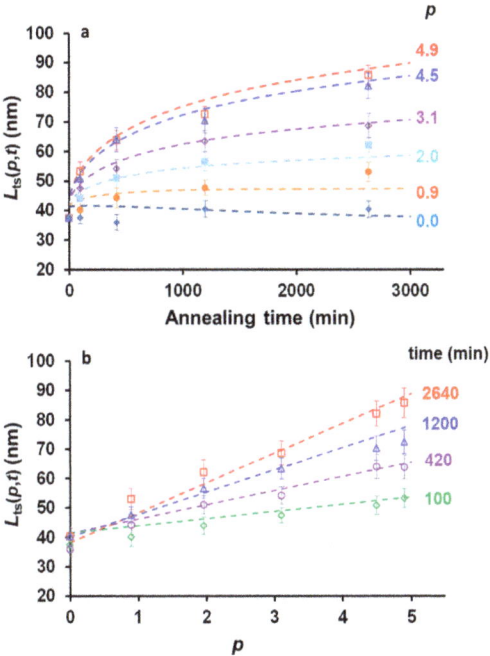

Figure 6. Plot of the of the length of the trapped crystallites, $L_{ts}(p,t)$ fitted to Equation (20) as a function of (**a**) time, and (**b**) p.

4. Conclusions

In summary, we have demonstrated that a seed-trapping protocol can be used to investigate the crystal growth kinetics of 1D micelles at elevated temperatures where both seed dissolution and fragmentation happened. Seed trapping proved particularly efficient in delineating the surviving seeds, allowing us to measure them after different annealing time. By comparing the length of the trapped seeds with the lengths of micelles that followed the exact same thermal history without being trapped, we could develop a kinetic model that accounted for seed fragmentation during crystallite growth. We considered that the probability that a seed fragment was decreasing with time, following a normal distribution function. The increase in number of seeds in solution could thus be described by a cumulative distribution function that is not dependent on the amount of unimer present in solution, and thus on the seeded growth kinetics. As previously reported, we observed that the seeded growth kinetics could be well modeled by a stretched exponential, which we believe is due to the fractionation by crystallization of the narrowly dispersed BCP [52]. Finally, we used our growth kinetics model to evaluate the variation of the growth rate as a function of the amount of unimer present in solution. We found that the 1D growth of the crystallites occurs in the mononucleation regime.

This study shows how seed trapping can be applied to study 1D micelle growth of other crystalline-coil BCPs, as well as more complex systems such as the parallel growth of two or three 1D crystals from a single BCP crystal face [53].

Supplementary Materials: The following are available online at https://www.mdpi.com/article/10.3390/polym13183122/s1. Supporting discussion Section I: Effect of lateral growth on $L_{mic}(p,t)$ during annealing. Supporting discussion Section II: Effect of fragmentation on $L_{mic}(p,t)$ during annealing. Supporting discussion Section III: Growth kinetics at 75 °C. Table S1: Values of $L_{mic}(p,t)$ of the PFS_{53}-b-PI_{637} crystallites heated for different annealing times at 75 °C and cooled to room temperature (control experiments), Table S2: Values of $L_{ts}(p,t)$ of the PFS_{53}-b-PI_{637} crystallites heated for different annealing time at 75 °C (seed-trapping experiments), Figure S1: Effect of PFS53-b-PI637 seed crystallite history on their stability at RT, Figure S2: Effect of PFS53-b-PI637 seed crystallite history on their stability against dissolution upon heating at 75 °C, Figure S3: Comparison of the histograms of the length distribution PFS53-b-PI637 seed crystallites that were trapped with PFS_{60}-b-$PDMS_{660}$ unimer after being heated at 75 °C, Figure S4: TEM micrographs of PFS_{53}-b-PI_{637} seeds trapped after ((a) 100 min, ((b) 420 min, ((c) 1200 min and (d) 2640 min of annealing at 75 °C without PFS_{53}-b-PI_{637} unimer added ($p = 0$), Figure S5: TEM micrographs of PFS_{53}-b-PI_{637} seeds trapped after (a) 100 min, (b) 420 min, (c) 1200 min and (d) 2640 min of annealing at 75 °C in the presence of an initial mass ratio of PFS_{53}-b-PI_{637} unimer added to PFS_{53}-b-PI_{637} seeds $p = 0.9$, Figure S6: TEM micrographs of PFS_{53}-b-PI_{637} seeds trapped after ((a) 100 min, ((b) 420 min, ((c) 1200 min and ((d) 2640 min of annealing at 75 °C in the presence of an initial mass ratio of PFS_{53}-b-PI_{637} unimer added to PFS_{53}-b-PI_{637} seeds $p = 2$, Figure S7: TEM micrographs of PFS_{53}-b-PI_{637} seeds trapped after ((a) 100 min, ((b) 420 min, ((c) 1200 min and (d() 2640 min of annealing at 75 °C in the presence of an initial mass ratio of PFS_{53}-b-PI_{637} unimer added to PFS_{53}-b-PI_{637} seeds $p = 3.1$, Figure S8: TEM micrographs of PFS_{53}-b-PI_{637} seeds trapped after ((a) 100 min, ((b) 420 min, (c) 1200 min and (d) 2640 min of annealing at 75 °C in the presence of an initial mass ratio of PFS_{53}-b-PI_{637} unimer added to PFS_{53}-b-PI_{637} seeds $p = 4.5$, Figure S9: TEM micrographs of PFS_{53}-b-PI_{637} seeds trapped after (a) 100 min, (b) 420 min, (c) 1200 min and (d) 2640 min of annealing at 75 °C in the presence of an initial mass ratio of PFS_{53}-b-PI_{637} unimer added to PFS_{53}-b-PI_{637} seeds $p = 4.9$, Figure S10: Histograms of the length distributions of PFS_{53}-b-PI_{637} trapped seeds after (a) 100 min, (b) 420 min, (c) 1200 min and (d) 2640 min of annealing at 75 °C without PFS_{53}-b-PI_{637} unimer added ($p = 0$), Figure S11: Histograms of the length distributions of PFS_{53}-b-PI_{637} trapped seeds after (a) 100 min, (b) 420 min, (c) 1200 min and (d) 2640 min of annealing at 75 °C in the presence of an initial mass ratio of PFS_{53}-b-PI_{637} unimer added to PFS_{53}-b-PI_{637} seeds $p = 1$, Figure S12: Histograms of the length distributions of PFS_{53}-b-PI_{637} trapped seeds after (a) 100 min, (b) 420 min, (c) 1200 min and (d) 2640 min of annealing at 75 °C in the presence of an initial mass ratio of PFS_{53}-b-PI_{637} unimer added to PFS_{53}-b-PI_{637} seeds $p = 2$, Figure S13: Histograms of the length distributions of PFS_{53}-b-PI_{637} trapped seeds after (a) 100 min, (b) 420 min, (c) 1200 min and (d) 2640 min of annealing at 75 °C in the presence of an initial mass ratio of PFS_{53}-b-PI_{637} unimer added to PFS_{53}-b-PI_{637} seeds $p = 3$, Figure S14: Histograms of the length distributions of PFS_{53}-b-PI_{637} trapped seeds after (a) 100 min, (b) 420 min, (c) 1200 min and (d) 2640 min of annealing at 75 °C in the presence of an initial mass ratio of PFS_{53}-b-PI_{637} unimer added to PFS_{53}-b-PI_{637} seeds $p = 4.5$, Figure S15: Histograms of the length distributions of PFS_{53}-b-PI_{637} trapped seeds after (a) 100 min, (b) 420 min, (c) 1200 min and (d) 2640 min of annealing at 75 °C in the presence of an initial mass ratio of PFS_{53}-b-PI_{637} unimer added to PFS_{53}-b-PI_{637} seeds $p = 4.8$, Figure S16: Histograms of the length of PFS_{53}-b-PI_{637} seed crystallites in decane that were trapped with added PFS_{60}-b-$PDMS_{660}$ unimer after being heated at 75 °C, then cooled to RT and stained with Karstedt's catalyst, Figure S17. Evolution of the number average lengths of the number average lengths of trapped seeds as a function of time for $p = 0$, 0.9, 2, 3.1, 4.5 and 4.9, Figure S18: Evolution of the number average lengths of PFS_{53}-b-PI_{637} seed crystallites annealed in decane at 75 °C, and cooled to 23 °C, as a function of the annealing time (control experiment), Figure S19: Plot of $L_{mic}(p,t)$ as a function of $(1 + p) L_{mic}(100,t)$.

Author Contributions: G.G. conceived the project, performed the experiments. P.A.R. synthesized the polymers. G.G. and M.A.W. prepared the manuscript. The whole project was supervised by M.A.W. All authors have read and agreed to the published version of the manuscript.

Funding: This research was funded by the Science and Technology Commission of Shanghai Municipality, China (21ZR1415400).

Institutional Review Board Statement: Not applicable.

Informed Consent Statement: Not applicable.

Acknowledgments: G.G. thanks Sepehr Mastour Tehrani for his helpful comments. G.G. and M.A.W. thank Anton Zilman for their valuable comments and suggestions.

Conflicts of Interest: The authors declare no competing financial interests.

References

1. Keller, A. A note on single crystals in polymers: Evidence for a folded chain configuration. *Philos. Mag.* **1957**, *2*, 1171–1175. [CrossRef]
2. Storks, K.H. An Electron Diffraction Examination of Some Linear High Polymers. *J. Am. Chem. Soc.* **1938**, *60*, 1753–1761. [CrossRef]
3. Bassett, D.C.; Keller, A. On the habits of polyethylene crystals. *Philos. Mag.* **1962**, *7*, 1553–1584. [CrossRef]
4. Lotz, B.; Kovacs, A.J.; Bassett, G.A.; Keller, A. Properties of copolymers composed of one poly-ethylene-oxide and one polystyrene block. *Kolloid-Z. Z. Polym.* **1966**, *209*, 115–128. [CrossRef]
5. Zhang, B.; Chen, J.; Baier, M.C.; Mecking, S.; Reiter, R.; Mülhaupt, R.; Reiter, G. Molecular-Weight-Dependent Changes in Morphology of Solution-Grown Polyethylene Single Crystals. *Macromol. Rapid Commun.* **2015**, *36*, 181–189. [CrossRef]
6. Xu, J.; Ma, Y.; Hu, W.; Rehahn, M.; Reiter, G. Cloning polymer single crystals through self-seeding. *Nat. Mater.* **2009**, *8*, 348–353. [CrossRef]
7. Zhang, B.; Chen, J.; Zhang, H.; Baier, M.C.; Mecking, S.; Reiter, R.; Mülhaupt, R.; Reiter, G. Annealing-induced periodic patterns in solution grown polymer single crystals. *RSC Adv.* **2015**, *5*, 12974–12980. [CrossRef]
8. Blundell, D.J.; Keller, A. The concentration dependence of the linear growth rate of polyethylene crystals from solution. *J. Polym. Sci. Part B* **1968**, *6*, 433–440. [CrossRef]
9. Dosiere, M.; Colet, M.-C.; Point, J.J. An isochronous decoration method for measuring linear growth rates in polymer crystals. *J. Polym. Sci. Part B* **1986**, *24*, 345–356. [CrossRef]
10. Tian, M.; Dosière, M.; Hocquet, S.; Lemstra, P.J.; Loos, J. Novel Aspects Related to Nucleation and Growth of Solution Grown Polyethylene Single Crystals. *Macromolecules* **2004**, *37*, 1333–1341. [CrossRef]
11. Boott, C.E.; Leitao, E.M.; Hayward, D.W.; Laine, R.F.; Mahou, P.; Guerin, G.; Winnik, M.A.; Richardson, R.M.; Kaminski, C.F.; Whittell, G.R.; et al. Probing the Growth Kinetics for the Formation of Uniform 1D Block Copolymer Nanoparticles by Living Crystallization-Driven Self-Assembly. *ACS Nano* **2018**, *12*, 8920–8933. [CrossRef] [PubMed]
12. Petzetakis, N.; Dove, A.P.; O'Reilly, R.K. Cylindrical micelles from the living crystallization-driven self-assembly of poly(lactide)-containing block copolymers. *Chem. Sci.* **2011**, *2*, 955–960. [CrossRef]
13. Schmelz, J.; Schedl, A.E.; Steinlein, C.; Manners, I.; Schmalz, H. Length Control and Block-Type Architectures in Worm-like Micelles with Polyethylene Cores. *J. Am. Chem. Soc.* **2012**, *134*, 14217–14225. [CrossRef]
14. Qian, J.; Li, X.; Lunn, D.; Gwyther, J.; Hudson, Z.; Kynaston, E.; Rupar, P.A.; Winnik, M.A.; Manners, I. Uniform, High Aspect Ratio Fiber-like Micelles and Block Co-micelles with a Crystalline π-Conjugated Polythiophene Core by Self-Seeding. *J. Am. Chem. Soc.* **2014**, *136*, 4121–4124. [CrossRef] [PubMed]
15. Mihut, A.M.; Drechsler, M.; Möller, M.; Ballauff, M. Sphere-to-Rod Transition of Micelles formed by the Semicrystalline Polybutadiene-block-Poly(ethylene oxide) Block Copolymer in a Selective Solvent. *Macromol. Rapid Commun.* **2010**, *31*, 449–453. [CrossRef] [PubMed]
16. Pitto-Barry, A.; Kirby, N.; Dove, A.; Oreilly, R. Expanding the scope of the crystallization-driven self-assembly of polylactide-containing polymers. *Polym. Chem.* **2014**, *5*, 1427–1436. [CrossRef]
17. Lazzari, M.; Scalarone, D.; Vazquez-Vazquez, C.; López-Quintela, M.A. Cylindrical Micelles from the Self-Assembly of Polyacrylonitrile-Based Diblock Copolymers in Nonpolar Selective Solvents. *Macromol. Rapid Commun.* **2008**, *29*, 352–357. [CrossRef]
18. Crassous, J.J.; Schurtenberger, P.; Ballauff, M.; Mihut, A.M. Design of block copolymer micelles via crystallization. *Polymer* **2015**, *62*, A1–A13. [CrossRef]
19. Wang, X.; Guerin, G.; Wang, H.; Wang, Y.; Manners, I.; Winnik, M.A. Cylindrical Block Copolymer Micelles and Co-Micelles of Controlled Length and Architecture. *Science* **2007**, *317*, 644–647. [CrossRef]
20. He, F.; Gädt, T.; Jones, M.; Scholes, G.D.; Manners, I.; Winnik, M.A. Synthesis and Self-Assembly of Fluorescent Micelles from Poly(ferrocenyldimethylsilane-b-2-vinylpyridine-b-2,5-di(2′-ethylhexyloxy)-1,4-phenylvinylene) Triblock Copolymer. *Macromolecules* **2009**, *42*, 7953–7960. [CrossRef]
21. Nazemi, A.; Boott, C.; Lunn, D.; Gwyther, J.; Hayward, D.W.; Richardson, R.; Winnik, M.A.; Manners, I. Monodisperse Cylindrical Micelles and Block Comicelles of Controlled Length in Aqueous Media. *J. Am. Chem. Soc.* **2016**, *138*, 4484–4493. [CrossRef] [PubMed]
22. Jia, L.; Zhao, G.; Shi, W.; Coombs, N.; Gourevich, I.; Walker, G.C.; Guerin, G.; Manners, I.; Winnik, M.A. A design strategy for the hierarchical fabrication of colloidal hybrid mesostructures. *Nat. Commun.* **2014**, *5*, 3882. [CrossRef] [PubMed]
23. Jia, L.; Tong, L.; Liang, Y.; Petretic, A.; Guerin, G.; Manners, I.; Winnik, M.A. Templated Fabrication of Fiber-Basket Polymersomes via Crystallization-Driven Block Copolymer Self-Assembly. *J. Am. Chem. Soc.* **2014**, *136*, 16676–16682. [CrossRef]
24. Rupar, P.A.; Chabanne, L.; Winnik, M.A.; Manners, I. Non-Centrosymmetric Cylindrical Micelles by Unidirectional Growth. *Science* **2012**, *337*, 559–562. [CrossRef]
25. Qiu, H.; Hudson, Z.M.; Winnik, M.A.; Manners, I. Multidimensional hierarchical self-assembly of amphiphilic cylindrical block comicelles. *Science* **2015**, *347*, 1329–1332. [CrossRef]
26. Hudson, Z.; Lunn, D.; Winnik, M.A.; Manners, I. Colour-tunable fluorescent multiblock micelles. *Nat. Commun.* **2014**, *5*, 3372. [CrossRef]

27. Guerin, G.; Cruz, M.; Yu, Q. Formation of 2D and 3D multi-tori mesostructures via crystallization-driven self-assembly. *Sci. Adv.* **2020**, *6*, eaaz7301. [CrossRef] [PubMed]
28. Jarrett-Wilkins, C.N.; Pearce, S.; Macfarlane, L.R.; Davis, S.A.; Faul, C.F.J.; Manners, I. Surface Patterning of Uniform 2D Platelet Block Comicelles via Coronal Chain Collapse. *ACS Macro Lett.* **2020**, *9*, 1514–1520. [CrossRef]
29. Armitstead, K.; Goldbeck-Wood, G.; Keller, A. Polymer crystallization theories. In *Macromolecules: Synthesis, Order and Advanced Properties*; Springer: Berlin/Heidelberg, Germany, 1992; pp. 219–312. ISBN 9783540544906.
30. Lauritzen, J.I.; Hoffman, J.D. Theory of formation of polymer crystals with folded chains in dilute solution. *J. Res. Natl. Bur. Stand. Sect. A* **1960**, *64*, 73–102. [CrossRef]
31. Hoffman, J.D.; Lauritzen, J.I.; Passaglia, E.; Ross, G.S.; Frolen, L.J.; Weeks, J.J. Kinetics of polymer crystallization from solution and the melt. *Kolloid-Zeitschrift Zeitschrift Polymere* **1969**, *231*, 564–592. [CrossRef]
32. Frank, F. Nucleation-controlled growth on a one-dimensional growth of finite length. *J. Cryst. Growth* **1974**, *22*, 233–236. [CrossRef]
33. Toda, A.; Kiho, H.; Miyaji, H.; Asai, K. A Kinetic Theory on the Growth Rate of Polymer Single Crystals. *J. Phys. Soc. Jpn.* **1985**, *54*, 1411–1422. [CrossRef]
34. Kundagrami, A.; Muthukumar, M. Continuum theory of polymer crystallization. *J. Chem. Phys.* **2007**, *126*, 144901. [CrossRef] [PubMed]
35. Keller, A.; Pedemonte, E. A study of growth rates of polyethylene single crystals. *J. Cryst. Growth* **1973**, *18*, 111–123. [CrossRef]
36. Guerin, G.; Rupar, P.A.; Manners, I.; Winnik, M.A. Explosive dissolution and trapping of block copolymer seed crystallites. *Nat. Commun.* **2018**, *9*, 1158. [CrossRef] [PubMed]
37. Guerin, G.; Molev, G.; Pichugin, D.; Rupar, P.A.; Qi, F.; Cruz, M.; Manners, I.; Winnik, M.A. Effect of Concentration on the Dissolution of One-Dimensional Polymer Crystals: A TEM and NMR Study. *Macromolecules* **2019**, *52*, 208–216. [CrossRef]
38. Guerin, G.; Molev, G.; Rupar, P.A.; Manners, I.; Winnik, M.A. Understanding the Dissolution and Regrowth of Core-Crystalline Block Copolymer Micelles: A Scaling Approach. *Macromolecules* **2020**, *53*, 10198–10211. [CrossRef]
39. Rupar, P.A.; Cambridge, G.; Winnik, M.A.; Manners, I. Reversible Cross-Linking of Polyisoprene Coronas in Micelles, Block Comicelles, and Hierarchical Micelle Architectures Using Pt(0)—Olefin Coordination. *J. Am. Chem. Soc.* **2011**, *133*, 16947–16957. [CrossRef]
40. Birshtein, T.; Borisov, O.; Zhulina, Y.; Khokhlov, A.; Yurasova, T. Conformations of comb-like macromolecules. *Polym. Sci. USSR* **1987**, *29*, 1293–1300. [CrossRef]
41. Panyukov, S.; Zhulina, E.; Sheiko, S.S.; Randall, G.; Brock, J.; Rubinstein, M. Tension Amplification in Molecular Brushes in Solutions and on Substrates. *J. Phys. Chem. B* **2009**, *113*, 3750–3768. [CrossRef]
42. Panyukov, S.V.; Sheiko, S.S.; Rubinstein, M. Amplification of Tension in Branched Macromolecules. *Phys. Rev. Lett.* **2009**, *102*, 148301. [CrossRef]
43. Park, I.; Nese, A.; Pietrasik, J.; Matyjaszewski, K.; Sheiko, S.S. Focusing bond tension in bottle-brush macromolecules during spreading. *J. Mater. Chem.* **2011**, *21*, 8448–8453. [CrossRef]
44. Qian, J.; Lu, Y.; Cambridge, G.; Guerin, G.; Manners, I.; Winnik, M.A. Polyferrocenylsilane Crystals in Nanoconfinement: Fragmentation, Dissolution, and Regrowth of Cylindrical Block Copolymer Micelles with a Crystalline Core. *Macromolecules* **2012**, *45*, 8363–8372. [CrossRef]
45. Johnston, D.C. Stretched exponential relaxation arising from a continuous sum of exponential decays. *Phys. Rev. B* **2006**, *74*, 184430. [CrossRef]
46. Choi, S.; Lodge, T.P.; Bates, F.S. Mechanism of Molecular Exchange in Diblock Copolymer Micelles: Hypersensitivity to Core Chain Length. *Phys. Rev. Lett.* **2010**, *104*, 047802. [CrossRef] [PubMed]
47. Point, J.J.; Colet, M.C.; Dosiere, M. Experimental criterion for the crystallization regime in polymer crystals grown from dilute solution: Possible limitation due to fractionation. *J. Polym. Sci. Part B* **1986**, *24*, 357–388. [CrossRef]
48. Lotz, B.; Kovacs, A.J. Propriétés Des Copolymères Biséquencés Polyoxyéthylène-Polystyrène. *Kolloid-Z. Z. Polym.* **1966**, *209*, 97–114. [CrossRef]
49. Song, S.; Zhou, H.; Manners, I.; Winnik, M.A. Block copolymer self-assembly: Polydisperse corona-forming blocks leading to uniform morphologies. *Chem* **2021**. [CrossRef]
50. Zhou, Y.; Hu, W. Kinetic Analysis of Quasi-One-Dimensional Growth of Polymer Lamellar Crystals in Dilute Solutions. *J. Phys. Chem. B* **2013**, *117*, 3047–3053. [CrossRef]
51. Shu, R.; Zha, L.; Eman, A.A.; Hu, W. Fibril Crystal Growth in Diblock Copolymer Solutions Studied by Dynamic Monte Carlo Simulations. *J. Phys. Chem. B* **2015**, *119*, 5926–5932. [CrossRef]
52. Pennings, A.J. Fractionation of polymers by crystallization from solutions. II. *J. Polym. Sci. Part C* **1962**, *16*, 1799–1812. [CrossRef]
53. Qiu, H.; Gao, Y.; Du, V.A.; Harniman, R.; Winnik, M.A.; Manners, I. Branched Micelles by Living Crystallization-Driven Block Copolymer Self-Assembly under Kinetic Control. *J. Am. Chem. Soc.* **2015**, *137*, 2375–2385. [CrossRef] [PubMed]

Article

Precise Tuning of Polymeric Fiber Dimensions to Enhance the Mechanical Properties of Alginate Hydrogel Matrices

Zehua Li [1,2], Amanda K. Pearce [1], Andrew P. Dove [1,*] and Rachel K. O'Reilly [1,*]

[1] School of Chemistry, University of Birmingham, Edgbaston, Birmingham B15 2TT, UK; ZXL798@student.bham.ac.uk (Z.L.); a.k.pearce@bham.ac.uk (A.K.P.)
[2] Department of Chemistry, University of Warwick, Gibbet Hill Road, Coventry CV4 7AL, UK
* Correspondence: a.dove@bham.ac.uk (A.P.D.); r.oreilly@bham.ac.uk (R.K.O.)

Abstract: Hydrogels based on biopolymers, such as alginate, are commonly used as scaffolds in tissue engineering applications as they mimic the features of the native extracellular matrix (ECM). However, in their native state, they suffer from drawbacks including poor mechanical performance and a lack of biological functionalities. Herein, we have exploited a crystallization-driven self-assembly (CDSA) methodology to prepare well-defined one-dimensional micellar structures with controlled lengths to act as a mimic of fibrillar collagen in native ECM and improve the mechanical strength of alginate-based hydrogels. Poly(ε-caprolactone)-b-poly(methyl methacrylate)-b-poly(N,N-dimethyl acrylamide) triblock copolymers were self-assembled into 1D cylindrical micelles with precise lengths using CDSA epitaxial growth and subsequently combined with calcium alginate hydrogel networks to obtain nanocomposites. Rheological characterization determined that the inclusion of the cylindrical structures within the hydrogel network increased the strength of the hydrogel under shear. Furthermore, the strain at flow point of the alginate-based hydrogel was found to increase with nanoparticle content, reaching an improvement of 37% when loaded with 500 nm cylindrical micelles. Overall, this study has demonstrated that one-dimensional cylindrical nanoparticles with controlled lengths formed through CDSA are promising fibrillar collagen mimics to build ECM scaffold models, allowing exploration of the relationship between collagen fiber size and matrix mechanical properties.

Keywords: crystallization-driven self-assembly; calcium alginate hydrogel; cylindrical micelles

Citation: Li, Z.; Pearce, A.K.; Dove, A.P.; O'Reilly, R.K. Precise Tuning of Polymeric Fiber Dimensions to Enhance the Mechanical Properties of Alginate Hydrogel Matrices. *Polymers* **2021**, *13*, 2202. https://doi.org/10.3390/polym13132202

Academic Editor: Holger Schmalz

Received: 27 May 2021
Accepted: 24 June 2021
Published: 2 July 2021

Publisher's Note: MDPI stays neutral with regard to jurisdictional claims in published maps and institutional affiliations.

Copyright: © 2021 by the authors. Licensee MDPI, Basel, Switzerland. This article is an open access article distributed under the terms and conditions of the Creative Commons Attribution (CC BY) license (https://creativecommons.org/licenses/by/4.0/).

1. Introduction

Fibrillar collagen is the basic building block for tissues and represents over 90% of the total abundant collagen family [1,2]. In vivo, most ECMs are formed by a hydrogel-like network of fibrous proteins comprised of fibrillar collagen, elastin, and a soft matrix combining proteoglycans and polysaccharides [3,4]. The physical properties and mechanical responses exhibited by ECMs are significantly influenced by their filamentous nature, which affects downstream cellular processes such as mechanotransduction [5,6]. During healthy aging, the fibrillar collagen content in many organs and tissues decreases, thus diminishing the integrity and strength of the ECM, which are fundamental properties for correct tissue and organ function [7,8]. For example, age-related skin wrinkling and stiffening of arteries are caused by a loss in integrity of fibrous proteins with age, which changes the mechanical properties of tissue [9]. In this regard, hydrogels are promising synthetic matrix materials to act as models to explore the effect of fibrillar collagen size on ECM mechanical properties. Indeed, not only can hydrogels reproduce ECM microenvironments, but also their mechanical properties [10]. Hence, by tailoring and designing the features of hydrogel-based systems, they can precisely satisfy specific biomedical applications [11,12].

Alginate is a widely used polysaccharide [13] and has been widely used as a matrix material in tissue engineering applications as a result of advantages such as low cytotoxicity, biocompatibility, non-immunogenicity, injectability, and low cost [14]. However, naturally-derived hydrogels suffer the drawback of being structurally weaker than synthetic hydrogels [15]. In addition, as alginate-derived hydrogels are hydrophilic and don't possess a fibrillar structure or nanofillers, they have weak mechanical properties (i.e., strength and stiffness) [16,17] as well as a lack of biological functionalities [16]. In an effort to overcome this, a range of nanocomposite hydrogels have been explored through the combination of nanoparticles and natural biopolymers [18–21], where the mechanical properties of the natural hydrogel network were shown to be enhanced by the interaction between the added nanostructures and the biopolymer chains. Such interactions were capable of tuning the stiffness, porosity, and nanostructure of the hydrogel, thus improving its overall performance as a biomaterial [10,11,22–24]. Despite some promising achievements to date—for example, exploring the effect of nanoparticle morphology on the mechanical properties of hydrogels [11,24]—precise control of nanocomposite dimensions and this influence on hydrogel mechanical properties is rarely reported and hence remains an attractive and challenging area of research.

Generally, an ideal hydrogel biomaterial should satisfy the criteria of biocompatibility, biodegradability, and appropriate mechanical properties [25–28] such as maintaining a certain mechanical strength and stiffness even in a swollen state [29]. Furthermore, the features of incorporated nanoparticulate filament structures are particularly important for providing directionality and guidance within the composite material [30–32]. For instance, filamentous extracellular protein networks show stiffening with increasing shear strain, likely caused by collective rearrangements of the filaments [33,34]. However, contrary to filamentous networks, many synthetic fibrillar hydrogels are shear-thinning, whereby the viscosity decreases in response to increased shear strain as the hydrogel network breaks. This property is often exploited in tissue engineering, where the hydrogel is able to be injected into the body with a syringe [35], followed by rapid recovery of the mechanical properties [10]. This particular shear-thinning behavior means the process of using polymer fibers to improve the mechanical properties of nanocomposite hydrogels is complicated. In order to understand this complexity, modeling the role of filament-like nanoparticles and their dimensions in the enhancement of shear strength of alginate-based hydrogels can provide crucial insights into the relationship between fibrillar collagen size and the mechanical properties of the native ECM [36].

In order to effectively mimic the fibrous architecture of different lengths, precise control over the formation of synthetic polymer fibers is vital. The ability to tailor the dimensions of nanostructures [24] enables polymer fibers to exhibit intriguing similarities to natural fibers, and therefore access to fibrillar collagen mimics. Polymer self-assembly through solution crystallization of block copolymers, termed crystallization-driven self-assembly (CDSA), with a semi-crystalline core-forming block has the remarkable ability to precisely control the dimensions of polymer fibers and to date has been demonstrated for a range of polymer blocks including polyferrocenylsilane [37], poly(ε-caprolactone) [38], polylactide [39], polypeptoids [40], and oligo(p-phenylenevinylene) [41]. Previous reports have used biocompatible and biodegradable poly(ε-caprolactone) (PCL)-based copolymers to form cylindrical assemblies with controlled dimensions by epitaxial growth, which led to the manufacture of biocompatible fibrillar hydrogels in situ [38]. The influence of the polymer nanoparticle shape on material properties was further explored through morphology control, which was achievable through CDSA. When calcium-alginate hydrogels were blended with nanostructures of different morphologies, including 0D spheres, 1D cylinders, and 2D platelets, the mechanical strength of the nanocomposite hydrogels was enhanced to different degrees. The 2D nanostructures exhibited a significant increase compared to the 0D or 1D nanoparticle counterparts [24]. Although this work demonstrated that PCL-based cylindrical micelles could form hydrogels and that nanoparticle morphologies could direct the mechanical strength of nanocomposite hydrogels, to date, utilizing PCL-based cylindrical micelles of controllable lengths to precisely tune the mechanical properties of nanocomposite hydrogels has yet to be achieved. The outcome of such advances could help to further understand the relationship between nanofiller dimensions and nanocomposite hydrogel strength.

Based on the above, in this study we aimed to introduce polymeric cylinders as a fibrillar structure within naturally-derived hydrogels and explore the influence of cylinder dimensions on material properties (Scheme 1). To achieve this, poly(ε-caprolactone)-*block*-poly(methyl methacrylate)-*block*-poly(N, N-dimethyl acrylamide) (PCL-*b*-PMMA-*b*-PDMA) triblock copolymers were synthesized (Scheme 2) and assembled into cylindrical micelles of controlled dimensions through CDSA. We demonstrated that the strength of nanocomposite hydrogels under shear could be improved when blended with 1D cylindrical nanoparticles, whereby the 500 nm cylindrical nanoparticles significantly increased the resistance of the nanocomposite hydrogels to flow under strain in comparison to their counterparts with other lengths. Overall, this work showed that nanoparticle cylinder length is one of the critical factors in precisely tuning the mechanical properties of alginate-based hydrogel networks, and that such nanocomposites show potential as collagen fiber mimics to explore the effect of dimensions on ECM mechanical properties.

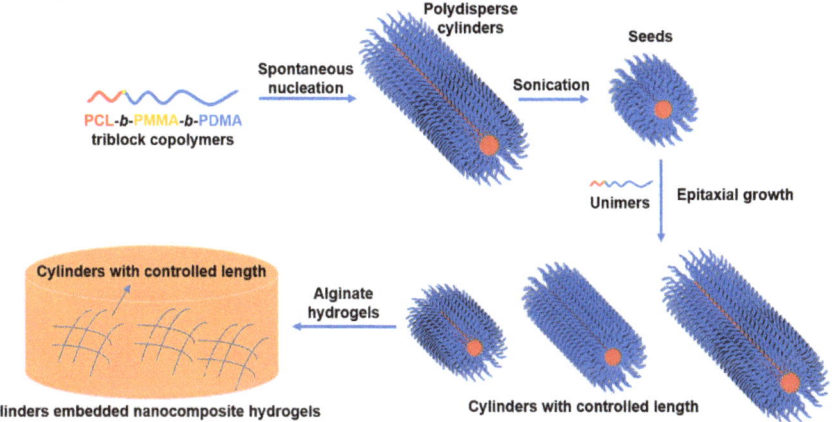

Scheme 1. Formation of PCL-based cylindrical micelles with various lengths by epitaxial growth, and the fabrication of cylinder embedded nanocomposite hydrogels.

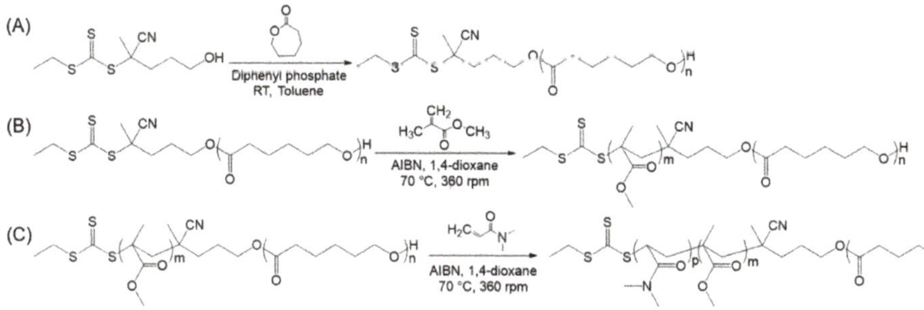

Scheme 2. (**A**) Synthesis of PCL macro-CTA by ROP with a dual-headed CTA initiator and a DPP catalyst. (**B**) Synthesis of PCL-*b*-PMMA diblock copolymer by RAFT polymerization using AIBN as initiator. (**C**) Synthesis of PCL-*b*-PMMA-*b*-PDMA triblock copolymer by RAFT polymerization using AIBN as initiator.

2. Materials and Methods

2.1. Materials

Chemicals and solvents were purchased from Sigma-Aldrich, Acros, Fluka, TCI, Fisher Chemical, Alfa Aesar or VWR. ε-Caprolactone (ε-CL) monomer was distilled over calcium hydride before being stored in a glove box under an inert atmosphere. Diphenyl phosphate (DPP) was recrystallized once from dried $CHCl_3$/Hexane (3:1) and dried over phosphorus pentoxide (P_2O_5) before use. (-)-Sparteine was dried over calcium hydride and distilled before use. 1,4-Dioxane, chloroform, methyl methacrylate (MMA), and N,N-dimethyl acrylamide (DMA) were purified by passing through basic alumina before use. 2,2'-Azobis(2-methylpropionitrile) (AIBN) was received from Molekula, recrystallized from methanol, and stored at 4 °C. Deuterated solvents were received from Apollo Scientific.

2.2. Typical Procedure for the ROP of PCL Homopolymers

In a nitrogen-filled glove box, solutions of DPP (49 mg, 0.20 mmol) in dry toluene (2 mL) and dual-headed CTA (52 mg, 0.21 mmol) in dry toluene (12.4 mL) were added to ε-CL (1.44 g, 12.62 mmol). After stirring for 11 h at room temperature, the solution was removed from the glove box, precipitated three times into ice-cold diethyl ether and collected by centrifugation. The resultant yellow polymer was dried under vacuum over phosphorus pentoxide for 2 days. The products were analyzed by SEC chromatograms and it was ensured there were no shoulders or tails in both sides of high or low molecular weight before proceeding with RAFT polymerizations and self-assembly. ^{1}H NMR (300 MHz, $CDCl_3$) δ/ppm: 4.06 (t, 100 H, CH_2OCO), 3.65 (t, 2 H, CH_2OH), 2.30 (t, 100 H, $OCOCH_2$), 1.73–1.33 (m, 330 H, $OCO(CH_2)_5OH$), M_n = 6.2 kg mol^{-1}, DP = 52. SEC chromatograms (DMF, PMMA standard): M_n = 13.1 kg mol^{-1}, M_w = 14.3 kg mol^{-1}, $Đ_M$ = 1.09.

2.3. Typical Procedure for the Synthesis of PCL-b-PMMA Diblock Copolymers

PCL$_{50}$ (500 mg, 0.08 mmol), MMA (420 mg, 4.20 mmol), and AIBN (1.38 mg, 8.39 × 10^{-3} mmol) were dissolved in 1,4-dioxane (1.40 mL) and placed in an ampoule. The solution was then freeze-pump-thawed three times and heated for 5 h at 70 °C. The reaction was quenched by immersion of the ampoule in the ice bath, and the polymer was precipitated in ice-cold methanol three times before being dried under vacuum and analyzed. ^{1}H NMR (400 MHz, $CDCl_3$) δ/ppm: 4.06 (t, 100 H, CH_2OCO), 3.59 (45 H, $COOCH_3$), 2.30 (t, 100 H, $OCOCH_2$), 1.91–1.81 (2 m, 8 H, CCH_2, PMMA), 1.72–1.33 (m, 305 H, $OCO(CH_2)_5OH$), 1.02–0.83 (m, 36 H, CH_3, PMMA), M_n = 8.2 kg mol^{-1}, DP = 20. SEC chromatograms (DMF, PMMA standard): M_n = 15.9 kg mol^{-1}, M_w = 17.4 kg mol^{-1}, $Đ_M$ = 1.10.

2.4. Typical Procedure for the Synthesis of PCL-b-PMMA-b-PDMA Triblock Copolymers

PCL$_{50}$-b-PMMA$_{20}$ (150 mg, 0.02 mmol), DMA (396 mg, 4.00 mmol), and AIBN (0.3 mg, 1.83 × 10^{-3} mmol) were dissolved in 1,4-dioxane (316 µL) and placed in an ampoule. The solution was then freeze-pump-thawed three times and heated for 1 h at 70 °C. The reaction was quenched by immersion of the ampoule in the ice bath, and the polymer was precipitated in ice-cold diethyl ether three times before being dried under vacuum and analyzed. ^{1}H NMR (400 MHz, $CDCl_3$) δ/ppm: 4.04 (t, 100 H, CH_2OH), 3.58 (45 H, $COOCH_3$), 3.10–2.44 (m, 889 H, $N(CH_3)_2$, $CHCH_2$, PDMA), 2.28 (t, 100 H, $OCOCH_2$), 1.79–1.23 (m, 674 H, $OCO(CH_2)_5OH$ (PCL), CCH_2 (PMMA), $CHCH_2$ (PDMA)), 1.02–0.83 (m, 48 H, CH_3, PMMA), M_n = 27.6 kg mol^{-1}, DP = 196. SEC chromatograms (DMF, PMMA standard): M_n = 34.2 kg mol^{-1}, M_w = 38.5 kg mol^{-1}, $Đ_M$ = 1.12.

2.5. Typical Crystallization-Driven Self-Assembly Method for the Self-Nucleation of PCL Block Copolymers

As a typical procedure of self-assembly conditions, PCL-b-PMMA-b-PDMA triblock copolymer (25 mg) was added to 5 mL of ethanol (5.0 mg mL^{-1}) in a vial. The samples were heated to 70 °C or 90 °C without stirring for 3 h before cooling to room temperature. Samples were imaged after 2 weeks of aging at room temperature.

2.6. Typical Gel Formation of Nanocomposite Calcium Alginates

Alginate gels were prepared at 1.5 wt. % sodium alginate. Before use, sodium alginate (19.9 mg, 0.1 mmol) was heated in the water to 70 °C for 1 h to aid dissolution and cooled to room temperature. Micelles were dispersed in the water for 2 h before stirring with calcium carbonate (5.0 mg, 0.05 mmol), followed by addition to the sodium alginate solution and vortexing for 1 min. After the addition of D-glucono-δ-lactone (GDL) (17.8 mg, 0.1 mmol), the gel was again vortexed for one minute before incubating at room temperature for two days.

3. Results and Discussion

3.1. Triblock Copolymer Synthesis and Characterization

To investigate the effect of polymer fiber dimensions on their ability to enhance the mechanical properties of hydrogels, cylindrical micelles with controlled lengths were prepared. According to previous literature, PCL block copolymers are able to produce cylindrical nanostructures by crystallization-driven self-assembly methodologies [42–44], where the cylinder length is controlled by epitaxial growth [24]. To overcome the disassembly of cylindrical micelles in water, which was attributed to the swelling of the corona block and subsequent fracture by the stress induced to the crystalline core, a glassy, highly hydrophobic polymer block was used to protect the PCL core. Therefore, following this strategy, poly(ε-caprolactone)-*block*-poly(methyl methacrylate)-*block*-poly(N, N-dimethyl acrylamide) (PCL-*b*-PMMA-*b*-PDMA) triblock copolymers were prepared.

The PCL-based triblock copolymers were synthesized using a combination of ring-opening polymerization (ROP) of ε-caprolactone (ε-CL) and reversible addition-fragmentation chain transfer (RAFT) polymerization of methyl methacrylate (MMA) and N, N-dimethyl acrylamide (DMA), respectively. First, the ROP of ε-CL was performed in a nitrogen-filled glove box at room temperature (RT) using a dual-headed initiator/chain transfer agent (CTA) and diphenyl phosphate (DPP) catalyst in dry toluene (Scheme 2A). The successful synthesis of PCL was indicated by proton nuclear magnetic resonance (^{1}H NMR) spectroscopy, where the methylene resonances of the repeat units were observed at $\delta = 4.06$ ppm and $\delta = 2.30$ ppm, while the end group signal was at $\delta = 3.65$ ppm (Figure S1). The degree of polymerization (DP) was calculated by end-group analysis comparing the integrations of characteristic signals, giving a DP of 55 and thus an M_n of 6.5 kg mol^{-1} (Table S1). Analysis of the PCL homopolymer by size exclusion chromatography (SEC) in tetrahydrofuran (THF) eluent revealed a narrow dispersity ($Đ_M = 1.06$, PMMA standards), which indicated a well-controlled polymerization, as well as no transesterification as detected by MALDI ToF mass spectrum (Figure S5). Importantly, the SEC chromatogram from the UV-Vis detector at $\lambda = 309$ nm confirmed the retention of the trithiocarbonate end group on the PCL homopolymer.

In order to prevent disassembly of the cylindrical micelles in water and subsequent rapid polymer precipitation, MMA was chosen as a hydrophobic RAFT compatible monomer to serve as an interfacial block before incorporation of the DMA hydrophilic block. The RAFT polymerization of MMA was carried out in 1,4-dioxane at 70 °C using the PCL macro-CTA as a RAFT agent (Scheme 2B). The reaction reached 40% monomer conversion after 5 h, and then purification was performed by precipitation into ice-cold methanol to obtain the PCL-*b*-PMMA diblock copolymer. The ^{1}H NMR spectroscopic analysis revealed a DP of 19 and thus an M_n of 8.4 kg mol^{-1}, using the methyl resonances of the PMMA repeat units at $\delta = 3.59$ ppm and $\delta = 1.02$ ppm (Figure S2). The subsequent RAFT polymerization of DMA was finally undertaken using the trithiocarbonate end group of the PCL-*b*-PMMA macro-CTA in 1,4-dioxane at 70 °C (Scheme 2C). The reaction reached 71% monomer conversion as determined by ^{1}H NMR spectroscopy after 24 h, followed by quenching the polymerization and precipitation in ice-cold diethyl ether. The successful synthesis of the PCL-*b*-PMMA-*b*-DMA triblock copolymer was confirmed by ^{1}H NMR spectroscopy (Figure S3), where the methyl resonances of the DMA repeat unit were observed at $\delta = 3.10$–2.44 ppm, giving a calculated DP of 196 and an M_n of 27.6 kg mol^{-1}. SEC analysis with DMF as eluent showed a narrow molecular weight distribution with a clear shift in molecular weight after each block and no observable evidence of low molecular weight species ($Đ_M$ = 1.12, PMMA standards, Figures S4 and S5B,C).

3.2. Crystallization-Driven Self-Assembly for the Production of Cylindrical Micelles

To investigate the effect of nanoparticle dimensions on the mechanical strength of nanocomposite hydrogels, cylindrical micelles with different lengths were prepared and subsequently blended into alginate hydrogels to mimic the native ECM with a fibrillar structure. Initially, polydisperse 1D cylindrical micelles were prepared from the poly(ε-caprolactone)-based block copolymers using the spontaneous nucleation method as previously reported [38] (Scheme 1, Figure 1Aa). The crystalline nature of the PCL-based cylindrical micelles was confirmed using Wide Angle X-ray Scattering (WAXS) (Figure S6), where 2theta (2θ) peaks of crystalline PCL were observed at ca. 21° and 24° [45,46]. In order to transform the polydisperse cylinders into micelles of controlled dimensions, a living CDSA method was employed. In this process, crystalline seed micelles of uniform size are obtained through probe sonication of the polydisperse cylinders, which can then serve as initiation sites for further seeded growth of free polymer (unimer) in solution. With the controlled addition of polymer unimers, cylindrical micelles with low dispersity and precise length can be obtained, analogous to a living polymerization.

Spontaneously nucleated polydisperse micrometer-long cylinders (Figure 1Aa) were diluted in solution to 0.5 mg mL^{-1} and subjected to probe sonication to prepare seed micelles. In order to restrain undesired crystallization caused by local high temperatures, the sonication was performed under a controlled temperature of 0 °C, and the total time of 20 min was divided into 10 rounds of 2 min, with the micelle solution left cooling in the ice bath for at least 10 min between sonication rounds. Finally, uniform crystalline seeds were obtained with a number average length of 65 nm (Figure 1Ab,B).

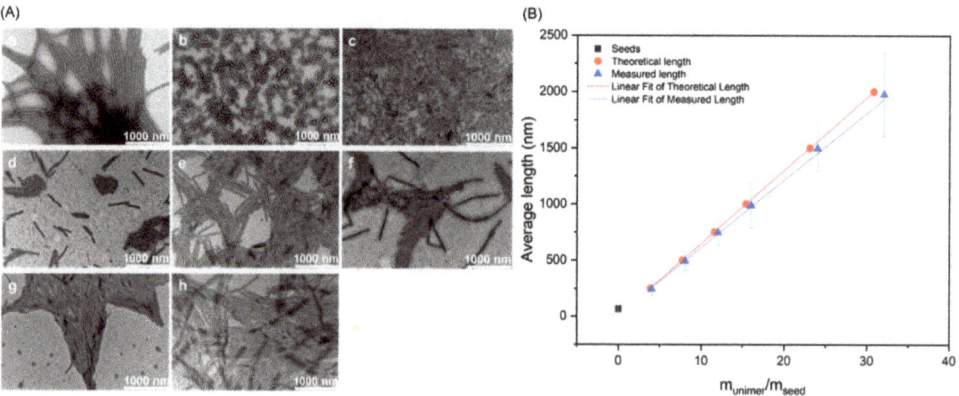

Figure 1. (**A**) TEM micrographs of (**a**) polydisperse cylindrical and (**b**) monodisperse seed micelles. Epitaxial growth of cylinders targeting (**c**) 250 nm, (**d**) 500 nm, (**e**) 750 nm, (**f**) 1000 nm, (**g**) 1500 nm, and (**h**) 2000 nm length values. 1% uranyl acetate was used as a negative stain. Scale bar = 1 μm. (**B**) Linear epitaxial growth of PCL$_{52}$-b-PMMA$_{20}$-b-PDMA$_{196}$ cylinders with narrow length dispersity (blue triangles, error bars represent the standard deviation (σ) of the length distribution) in comparison to the theoretical length (red circles). The cylindrical micelles were grown from ca. 65 nm seeds.

For the epitaxial growth step, a unimer solution was prepared by dissolving the PCL$_{50}$-b-PMMA$_{20}$-b-PDMA$_{200}$ triblock copolymers in tetrahydrofuran (THF) to form a 100 mg mL^{-1} solution. THF was chosen for this process as it is a good solvent for both blocks and also miscible with ethanol, which was the chosen corona-selective solvent. Different length cylindrical micelles were targeted by employing different unimer to seed ratios, whereby the unimer THF solution was added at the desired ratio into a

0.5 mg mL^{-1} seed micelle solution, followed by solvent evaporation at room temperature (RT) to obtain stable structures in ethanol. Controlled linear epitaxial growth was visually observed, with the longer cylinder solutions being more turbid, and further confirmed by TEM imaging. The analysis showed that monodisperse cylindrical micelles with precise micrometer lengths had been achieved, and that these were in agreement with predicted lengths (Figure 1Ac–h,B, Table 1). Indeed, the length displayed by the cylinders was proportional to the amount of unimer added (Table 1), which confirmed the controlled nature of the epitaxial growth process (Figure S7). The monodisperse cylindrical micelles were subsequently transferred into a pure aqueous phase by dialysis against water for 72 h. Importantly, the nanoparticles were confirmed to still exist as stable cylindrical nanostructures without disassembly or degradation after the dialysis process (Figure S8), therefore giving unprecedented access to collagen fiber mimics.

Table 1. Length dispersity of cylinders formed by epitaxial growth of PCL$_{50}$-b-PMMA$_{20}$-b-PDMA$_{200}$ cylindrical micelles.

Target (nm)	L_w [a] (nm)	L_n [b] (nm)	L_w/L_n
Seeds	68	65	1.05
250 nm	262	246	1.07
500 nm	506	491	1.03
750 nm	762	746	1.02
1000 nm	1029	988	1.04
1500 nm	1561	1532	1.02
2000 nm	1984	1911	1.04

Imaged by TEM; one hundred cylinders per sample were counted to measure the length of cylinders by ImageJ. [a] L_w = weight average length. [b] L_n = number average length.

We next sought to investigate the relationship between the length of polymer fibers and mechanical strength of the nanocomposite hydrogels. Alginate was chosen as the hydrogel model matrix material in this work due to its high biocompatibility, low cost, facile processing technology, and ubiquity in translational medicine. Alginate is an anionic polysaccharide which forms a hydrogel structure through crosslinking with cationic substances, typically calcium [47,48]. In order to exclude any influences of surface charge on hydrogel formation, we first measured the ζ-potential of the PCL$_{50}$-b-PMMA$_{20}$-b-PDMA$_{200}$ triblock copolymer cylinders. This was found to be ca. −4.34 mV, and thus could be considered approximately neutral [49] and as such avoid any additional effects of micelle surface charge on crosslinking.

Calcium-crosslinked alginate hydrogels (without cylinders) were used as a control group and were prepared by the combination of sodium alginate (1.5 wt %), calcium carbonate (CaCO$_3$, 0.5 eq.), and D-glucono-δ-lactone (GDL, 1.0 eq.). The low solubility of CaCO$_3$ relative to other crosslinking agents such as CaCl$_2$ or CaSO$_4$·2H$_2$O allowed for a slower gelation process. In turn, this CaCO$_3$-GDL system ensured uniform dispersion of calcium throughout the gel system before crosslinking occurred, preventing a heterogeneous structure which can lead to weak mechanical properties with low reproducibility [48]. Following establishment of the protocol for control calcium-crosslinked alginate hydrogels, the precise length polymer fibers were then incorporated during the crosslinking process, and TEM imaging was used to confirm that the cylinders remained intact during the mixing and vortexing process (Figure S9). The calcium concentration was kept constant, and different equivalents of monodisperse PCL$_{50}$-b-PMMA$_{20}$-b-PDMA$_{200}$ cylindrical micelles were added at RT.

The mechanical strength of the resultant nanocomposite hydrogels was characterized through oscillatory rheology measurements. A control was performed in each dataset to ensure that the trends were not affected by slight variations in sample preparation such as lab temperature, which could cause a small shift in the magnitude of the theoretical moduli between data sets. For all samples, a gel-like behavior was confirmed by comparing the storage and loss modulus (G' and G''), whereby G' was higher than G'' for the entire range of frequency sweep (Figure S10). Amplitude sweep tests also provided strain-dependent information. Specifically, the broad linear-viscoelastic (LVE) region and the hydrogel network breakdown were observed as the strain increased. In the LVE region, no substantial change in G' was observed when increasing the nanoparticle content, which indicated the low embrittlement of the nanocomposite hydrogels as a consequence of the addition of fillers. At high strain, the strain at the flow point (τ_f) was obtained (i.e., intersection of the curves for G' and G''), with the value of shear stress being determined at the crossover point ($G' = G''$, Figure S11).

The strain-dependent response of the control alginate hydrogel without nanoparticles showed a low strain value (*ca.* 20%) at the flow point. After the addition of 0.04, 0.06, 0.08, 0.10, and 0.12 wt % cylindrical micelles, the flow strain increased in comparison to the control group (Figure 2, Table S2). Hence, polymer cylindrical micelles with different lengths could enhance the mechanical strength of the nanocomposite calcium-crosslinked alginate hydrogels. In particular, the strain at the flow point of the nanocomposite hydrogel embedded with 500 nm PCL_{50}-*b*-$PMMA_{20}$-*b*-$PDMA_{200}$ cylindrical micelles distinctly increased up to ca. 37% (at 0.10 wt %). We hypothesized that this phenomenon was related to the pore size of the calcium-crosslinked alginate hydrogels, with the 500 nm cylinders better able to pack into the pores of the hydrogels. The strain values at the flow point were found to increase and then decrease with the cylinder wt % for each cylinder length—although the maximum value was obtained at different wt % for different lengths of cylinder. Notably, this behavior was not observed with cylinders of 1000 nm and 1500 nm lengths, for which no significant enhancement was observed. Furthermore, when the cylindrical micelles' nanoparticle content reached 0.12 wt %, the hydrogel strain value decreased for all systems, which was ascribed to steric hindrance [24,50]. In this case, we postulate that surplus cylinders were unable to pack inside the pores of the hydrogel network, thus reducing homogeneity and, consequently, disrupting hydrogel formation and yielding poorer mechanical features. Finally, polydisperse cylinders were prepared by mixing equal masses of 250 nm, 500 nm, 750 nm, 1000 nm, and 1500 nm cylinders to investigate the extent to which mixed micelles could affect the mechanical properties of hydrogels. The polydisperse cylinders showed a robust enhancement of the hydrogel strain properties, which was second only to the 500 nm samples.

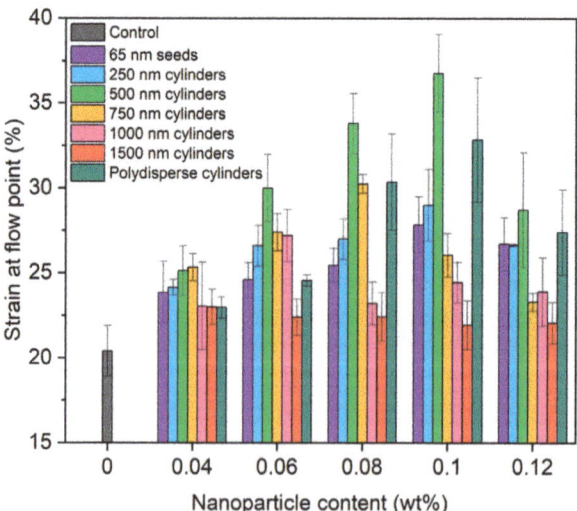

Figure 2. Histogram comparing the strain values at the flow point for nanocomposite hydrogels incorporating cylindrical micelles at different wt %. Error bars represent the standard deviation of the data.

These results showed that the dimensions of added fibrillar nanoparticles play an essential role in tuning the mechanical strength of nanocomposite hydrogels. The addition of cylindrical nanoparticles less than 750 nm in length was able to enhance the strain at flow point for the alginate hydrogels; however, further increasing cylinder length to 1000 nm and 1500 nm resulted in no improvement in mechanical properties. We hypothesize that the short cylinders of 65 nm, 250 nm, 500 nm, and 750 nm lengths could provide an efficient contribution to the mechanical resistance of hydrogels under shear due to being readily distributed throughout the hydrogel network. However, when the cylinders were overly short (65 nm or 250 nm in length), the cylinder ends were not able to effectively connect the hydrogel network in the pores, thus limiting such improvement. In contrast, 1000 nm and 1500 nm cylinders hindered the enhancement, likely because of the steric hindrance of long cylinders. Importantly, our results showed that when cylinders of 500 nm in length were incorporated, the resistance of the hydrogels to shear could be substantially enhanced through uniform embedding of the polymer fibers, and therefore that cylindrical micelles that can fit best in the pores of hydrogels should be preferentially considered as additives.

4. Conclusions

In this work, the relationship between the length of polymer fibers and the strain at the flow point of nanocomposite calcium-crosslinked alginate hydrogels was studied. Monodisperse PCL_{50}-b-$PMMA_{20}$-b-$PDMA_{200}$ cylindrical micelles were achieved through CDSA to give neutrally charged cylinders of 65 nm, 250 nm, 500 nm, 750 nm, 1000 nm, and 1500 nm in length, which were subsequently embedded within calcium-alginate hydrogel matrices. Oscillatory rheology measurements showed that cylinders of 750 nm in length and below, as well as a sample of polydisperse cylindrical micelles, were able to enhance the strain at flow point of the alginate hydrogels. When cylinders of 500 nm length were incorporated, the nanocomposite hydrogels showed an enhancement of strain at flow point by up to ca. 37%. In comparison, no significant improvement was observed for 1000 nm and 1500 nm samples. Overall, this work demonstrates that hydrogel strength can be enhanced through uniform embedding of polymer fibers in a size-dependent manner, and therefore that nanocomposite hydrogels show potential to mimic the native ECM as scaffolds for a wide range of biomedical applications in vivo, such as manufacturing artificial skin or tissues.

Supplementary Materials: The following are available online at https://www.mdpi.com/article/10.3390/polym13132202/s1. Figure S1. ^{1}H NMR spectra (300 MHz, $CDCl_3$) of PCL_{52} homopolymer. Figure S2. ^{1}H NMR spectra (300 MHz, $CDCl_3$) of PCL_{52}-b-$PMMA_{20}$ diblock copolymer. Figure S3. ^{1}H NMR spectra (300 MHz, $CDCl_3$) of PCL_{52}-b-$PMMA_{20}$-b-$PDMA_{196}$ triblock copolymer. Figure S4. Overlaid (A) RI and (B) UV (λ = 309 nm) SEC chromatograms of PCL macro-CTA, PCL-b-PMMA diblock copolymer, and PCL-b-PMMA-b-PDMA triblock copolymer using DMF with 5 mM NH_4BF_4 as the eluent and PMMA standards. Figure S5. (A) MALDI ToF mass spectra of PCL_{57} homopolymers, which showed a m/z difference of 114.14 equivalent to a PCL repeat unit and, therefore, minimal transesterification. (B) DOSY NMR spectra (500 MHz, $CDCl_3$) of PCL_{52}-b-$PMMA_{20}$ diblock copolymers. (C) DOSY NMR spectra (500 MHz, $CDCl_3$) of PCL_{52}-b-$PMMA_{20}$-b-$PDMA_{196}$ triblock copolymers. Figure S6. WAXS spectra obtained of nanoparticles prepared from PCL_{52}-based triblock copolymers, exhibiting the 2θ peaks at ca. 21° and 24°. Figure S7. Diameter distribution of (a) polydisperse cylinders, (b) seeds, (c) 250 nm cylinders, (d) 500 nm cylinders, (e) 750 nm cylinders, (f) 1000 nm cylinders, (g) 1500 nm cylinders, and (h) 2000 nm cylinders. Figure S8. TEM micrographs of (a) polydisperse cylinders, (b) seeds, (c) 250 nm cylinders, (d) 500 nm cylinders, (e) 750 nm cylinders, (f) 1000 nm cylinders, (g) 1500 nm cylinders, and (h) 2000 nm cylinders in water. 1% uranyl acetate was used as a negative stain. Scale bar = 1 μm. Figure S9. Dynamic oscillatory frequency sweeps at 0.5% strain of calcium-crosslinked alginate hydrogels at 0.50 eq. calcium with 0 wt% (control), 0.04 wt%, 0.06 wt%, 0.08 wt%, 0.10 wt%, and 0.12 wt% PCL_{50}-b-$PMMA_{20}$-b-$PDMA_{200}$ cylindrical micelles with different lengths, which included (a) 65 nm, (b) 250 nm, (c) 500 nm, (d) 750 nm, (e) 1000 nm, (f) 1500 nm, and (g) polydisperse cylinders. Figure S10. Strain-dependent oscillatory rheology measurements at 10 rad s^{-1} angular frequency (ω) of calcium-crosslinked alginate hydrogels at 0.50 eq. calcium with 0 wt% (control), 0.04 wt%, 0.06 wt%, 0.08 wt%, 0.10 wt%, and 0.12 wt% PCL_{50}-b-$PMMA_{20}$-b-$PDMA_{200}$ cylindrical micelles with different lengths, which included (a) 65 nm, (b) 250 nm, (c) 500 nm, (d) 750 nm, (e) 1000 nm, (f) 1500 nm, and (g) polydisperse cylinders. Table S1. Polymer characterization data for PCL_{52}-b-$PMMA_{20}$-b-$PDMA_{196}$ block copolymers.Table S2 Strain values at the flow point for nanocomposite alginate hydrogels enriched with different concentrations of PCL_{50}-b-$PMMA_{20}$-b-$PDMA_{200}$ cylinders with different lengths. Data are presented as average ± standard deviation.

Author Contributions: A.P.D. and R.K.O. conceived the project. Z.L., A.K.P., A.P.D. and R.K.O. designed the experiments, and Z.L. conducted the studies and performed all characterization and analyses. All authors contributed to writing and editing the manuscript have read and agreed to the published version of the manuscript.

Funding: This research was funded through a PhD studentship to Z.L. from the China Scholarship Council.

Institutional Review Board Statement: Not applicable.

Informed Consent Statement: Not applicable.

Data Availability Statement: The data presented in this study are available from the corresponding author upon reasonable request.

Acknowledgments: The China Scholarship Council and the University of Warwick are acknowledged for a Joint Scholarship to Z.L. Carl Reynolds (School of Metallurgy and Materials, the University of Birmingham) and Maria Pérez-Madrigal (School of Chemistry, the University of Birmingham) are thanked for their assistance and scientific discussion.

Conflicts of Interest: The authors declare no conflict of interest.

References

1. Gelse, K.; Pöschl, E.; Aigner, T. Collagens—structure, function, and biosynthesis. *Adv. Drug Deliv. Rev.* **2003**, *55*, 1531–1546. [CrossRef]
2. Bierbaum, S.; Hintze, V.; Scharnweber, D. 2.8 Artificial extracellular matrices to functionalize biomaterial surfaces. In *Comprehensive Biomaterials II*; Ducheyne, P., Ed.; Elsevier: Oxford, UK, 2017; pp. 147–178.
3. Theocharis, A.D.; Skandalis, S.S.; Gialeli, C.; Karamanos, N.K. Extracellular matrix structure. *Adv. Drug Deliv. Rev.* **2016**, *97*, 4–27. [CrossRef]
4. Bosman, F.T.; Stamenkovic, I. Functional structure and composition of the extracellular matrix. *J. Pathol.* **2003**, *200*, 423–428. [CrossRef]
5. Wozniak, M.A.; Chen, C.S. Mechanotransduction in development: A growing role for contractility. *Nat. Rev. Mol. Cell Biol.* **2009**, *10*, 34–43. [CrossRef]
6. DuFort, C.C.; Paszek, M.J.; Weaver, V.M. Balancing forces: Architectural control of mechanotransduction. *Nat. Rev. Mol. Cell Biol.* **2011**, *12*, 308–319. [CrossRef] [PubMed]
7. Dworatzek, E.; Baczko, I.; Kararigas, G. Effects of aging on cardiac extracellular matrix in men and women. *Proteom. Clin. Appl.* **2016**, *10*, 84–91. [CrossRef]
8. Swynghedauw, B. Molecular mechanisms of myocardial remodeling. *Physiol. Rev.* **1999**, *79*, 215–262. [CrossRef]
9. Baldwin, A.K.; Simpson, A.; Steer, R.; Cain, S.A.; Kielty, C.M. Elastic fibres in health and disease. *Expert Rev. Mol. Med.* **2013**, *15*, e8. [CrossRef] [PubMed]
10. Prince, E.; Kumacheva, E. Design and applications of man-made biomimetic fibrillar hydrogels. *Nat. Rev. Mater.* **2019**, *4*, 99–115. [CrossRef]
11. Zhan, H.; Löwik, D.W.P.M. A Hybrid Peptide Amphiphile Fiber PEG Hydrogel Matrix for 3D Cell Culture. *Adv. Funct. Mater.* **2019**, *29*, 1808505. [CrossRef]
12. Nowak, A.P.; Breedveld, V.; Pakstis, L.; Ozbas, B.; Pine, D.J.; Pochan, D.; Deming, T.J. Rapidly recovering hydrogel scaffolds from self-assembling diblock copolypeptide amphiphiles. *Nature* **2002**, *417*, 424–428. [CrossRef] [PubMed]
13. Nicolas, J.; Magli, S.; Rabbachin, L.; Sampaolesi, S.; Nicotra, F.; Russo, L. 3D Extracellular Matrix Mimics: Fundamental Concepts and Role of Materials Chemistry to Influence Stem Cell Fate. *Biomacromolecules* **2020**, *21*, 1968–1994. [CrossRef]
14. Cohen, D.L.; Malone, E.; Lipson, H.O.D.; Bonassar, L.J. Direct freeform fabrication of seeded hydrogels in arbitrary geometries. *Tissue Eng.* **2006**, *12*, 1325–1335. [CrossRef] [PubMed]
15. Hinderer, S.; Layland, S.L.; Schenke-Layland, K. ECM and ECM-like materials—Biomaterials for applications in regenerative medicine and cancer therapy. *Adv. Drug Deliv. Rev.* **2016**, *97*, 260–269. [CrossRef] [PubMed]
16. Lee, K.Y.; Mooney, D.J. Hydrogels for tissue engineering. *Chem. Rev.* **2001**, *101*, 1869–1880. [CrossRef]

17. Kim, M.S.; Kim, G. Three-dimensional electrospun polycaprolactone (PCL)/alginate hybrid composite scaffolds. *Carbohydr. Polym.* **2014**, *114*, 213–221. [CrossRef] [PubMed]
18. Xiao, L.; Zhu, J.; Londono, J.D.; Pochan, D.J.; Jia, X. Mechano-responsive hydrogels crosslinked by block copolymer micelles. *Soft Matter* **2012**, *8*, 10233–10237. [CrossRef]
19. Xiao, L.; Liu, C.; Zhu, J.; Pochan, D.J.; Jia, X. Hybrid, elastomeric hydrogels crosslinked by multifunctional block copolymer micelles. *Soft Matter* **2010**, *6*, 5293–5297. [CrossRef] [PubMed]
20. Li, L.; Jiang, R.; Chen, J.; Wang, M.; Ge, X. In situ synthesis and self-reinforcement of polymeric composite hydrogel based on particulate macro-RAFT agents. *RSC Adv.* **2017**, *7*, 1513–1519. [CrossRef]
21. Tan, M.; Zhao, T.; Huang, H.; Guo, M. Highly stretchable and resilient hydrogels from the copolymerization of acrylamide and a polymerizable macromolecular surfactant. *Polym. Chem.* **2013**, *4*, 5570–5576. [CrossRef]
22. Pek, Y.S.; Wan, A.C.; Shekaran, A.; Zhuo, L.; Ying, J.Y. A thixotropic nanocomposite gel for three-dimensional cell culture. *Nat. Nanotechnol.* **2008**, *3*, 671–675. [CrossRef]
23. Merino, S.; Martin, C.; Kostarelos, K.; Prato, M.; Vazquez, E. Nanocomposite Hydrogels: 3D Polymer–Nanoparticle Synergies for On-Demand Drug Delivery. *ACS Nano* **2015**, *9*, 4686–4697. [CrossRef] [PubMed]
24. Arno, M.C.; Inam, M.; Weems, A.C.; Li, Z.; Binch, A.L.; Platt, C.I.; Richardson, S.M.; Hoyland, J.A.; Dove, A.P.; O'Reilly, R.K. Exploiting the role of nanoparticle shape in enhancing hydrogel adhesive and mechanical properties. *Nat. Commun.* **2020**, *11*, 1420. [CrossRef] [PubMed]
25. Tibbitt, M.W.; Rodell, C.B.; Burdick, J.A.; Anseth, K.S. Progress in material design for biomedical applications. *Proc. Natl. Acad. Sci. USA* **2015**, *112*, 14444–14451. [CrossRef] [PubMed]
26. Kesireddy, V.; Kasper, F.K. Approaches for building bioactive elements into synthetic scaffolds for bone tissue engineering. *J. Mater. Chem. B* **2016**, *4*, 6773–6786. [CrossRef]
27. Kharkar, P.M.; Kiick, K.L.; Kloxin, A.M. Designing degradable hydrogels for orthogonal control of cell microenvironments. *Chem. Soc. Rev.* **2013**, *42*, 7335–7372. [CrossRef]
28. Vedadghavami, A.; Minooei, F.; Mohammadi, M.H.; Khetani, S.; Kolahchi, A.R.; Mashayekhan, S.; Sanati-Nezhad, A. Manufacturing of hydrogel biomaterials with controlled mechanical properties for tissue engineering applications. *Acta Biomater.* **2017**, *62*, 42–63. [CrossRef] [PubMed]
29. Caló, E.; Khutoryanskiy, V.V. Biomedical applications of hydrogels: A review of patents and commercial products. *Eur. Polym. J.* **2015**, *65*, 252–267. [CrossRef]
30. Sano, K.; Ishida, Y.; Aida, T. Synthesis of anisotropic hydrogels and their applications. *Angew. Chem. Int. Ed.* **2018**, *57*, 2532–2543. [CrossRef]
31. Rose, J.C.; Gehlen, D.B.; Haraszti, T.; Köhler, J.; Licht, C.J.; De Laporte, L. Biofunctionalized aligned microgels provide 3D cell guidance to mimic complex tissue matrices. *Biomaterials* **2018**, *163*, 128–141. [CrossRef] [PubMed]
32. Lutolf, M.P.; Hubbell, J.A. Synthetic biomaterials as instructive extracellular microenvironments for morphogenesis in tissue engineering. *Nat. Biotechnol.* **2005**, *23*, 47–55. [CrossRef]
33. Münster, S.; Jawerth, L.M.; Leslie, B.A.; Weitz, J.I.; Fabry, B.; Weitz, D.A. Strain history dependence of the nonlinear stress response of fibrin and collagen networks. *Proc. Natl. Acad. Sci. USA* **2013**, *110*, 12197–12202. [CrossRef]
34. Brown, A.E.X.; Litvinov, R.I.; Discher, D.E.; Purohit, P.K.; Weisel, J.W. Multiscale Mechanics of Fibrin Polymer: Gel Stretching with Protein Unfolding and Loss of Water. *Science* **2009**, *325*, 741–744. [CrossRef]
35. Guvendiren, M.; Lu, H.D.; Burdick, J.A. Shear-thinning hydrogels for biomedical applications. *Soft Matter* **2012**, *8*, 260–272. [CrossRef]
36. Black, L.D.; Allen, P.G.; Morris, S.M.; Stone, P.J.; Suki, B. Mechanical and Failure Properties of Extracellular Matrix Sheets as a Function of Structural Protein Composition. *Biophys. J.* **2008**, *94*, 1916–1929. [CrossRef]
37. Wang, X.; Guerin, G.; Wang, H.; Wang, Y.; Manners, I.; Winnik, M.A. Cylindrical block copolymer micelles and co-micelles of controlled length and architecture. *Science* **2007**, *317*, 644–647. [CrossRef]
38. Arno, M.C.; Inam, M.; Coe, Z.; Cambridge, G.; Macdougall, L.J.; Keogh, R.; Dove, A.P.; O'Reilly, R.K. Precision Epitaxy for Aqueous 1D and 2D Poly(ε-caprolactone) Assemblies. *J. Am. Chem. Soc.* **2017**, *139*, 16980–16985. [CrossRef] [PubMed]
39. Sun, L.; Petzetakis, N.; Pitto-Barry, A.; Schiller, T.L.; Kirby, N.; Keddie, D.J.; Boyd, B.J.; O'Reilly, R.K.; Dove, A.P. Tuning the size of cylindrical micelles from poly (L-lactide)-b-poly (acrylic acid) diblock copolymers based on crystallization-driven self-assembly. *Macromolecules* **2013**, *46*, 9074–9082. [CrossRef]
40. Shi, Z.; Wei, Y.; Zhu, C.; Sun, J.; Li, Z. Crystallization-Driven Two-Dimensional Nanosheet from Hierarchical Self-Assembly of Polypeptoid-Based Diblock Copolymers. *Macromolecules* **2018**, *51*, 6344–6351. [CrossRef]
41. Tao, D.; Feng, C.; Cui, Y.; Yang, X.; Manners, I.; Winnik, M.A.; Huang, X. Monodisperse Fiber-like Micelles of Controlled Length and Composition with an Oligo(p-phenylenevinylene) Core via "Living" Crystallization-Driven Self-Assembly. *J. Am. Chem. Soc.* **2017**, *139*, 7136–7139. [CrossRef] [PubMed]
42. Zhang, J.; Wang, L.Q.; Wang, H.; Tu, K. Micellization Phenomena of Amphiphilic Block Copolymers Based on Methoxy Poly(ethylene glycol) and Either Crystalline or Amorphous Poly(caprolactone-b-lactide). *Biomacromolecules* **2006**, *7*, 2492–2500. [CrossRef] [PubMed]
43. Du, Z.X.; Xu, J.T.; Fan, Z.Q. Regulation of Micellar Morphology of PCL-b-PEO Block Copolymers by Crystallization Temperature. *Macromol. Rapid Commun.* **2008**, *29*, 467–471. [CrossRef]

44. Chan, S.-C.; Kuo, S.W.; Lu, C.H.; Lee, H.F.; Chang, F.C. Syntheses and characterizations of the multiple morphologies formed by the self-assembly of the semicrystalline P4VP-b-PCL diblock copolymers. *Polymer* **2007**, *48*, 5059–5068. [CrossRef]
45. Muñoz-Bonilla, A.; Cerrada, M.L.; Fernández-García, M.; Kubacka, A.; Ferrer, M.; Fernández-García, M. Biodegradable Polycaprolactone-Titania Nanocomposites: Preparation, Characterization and Antimicrobial Properties. *Int. J. Mol. Sci.* **2013**, *14*, 9249. [CrossRef] [PubMed]
46. Zhang, Y.; Huo, H.; Li, J.; Shang, Y.; Chen, Y.; Funari, S.S.; Jiang, S. Crystallization behavior of poly(ε-caprolactone) and poly(ε-caprolactone)/LiClO4 complexes from the melt. *CrystEngComm* **2012**, *14*, 7972–7980. [CrossRef]
47. Lee, K.Y.; Mooney, D.J. Alginate: Properties and biomedical applications. *Prog. Polym. Sci.* **2012**, *37*, 106–126. [CrossRef]
48. Kuo, C.K.; Ma, P.X. Ionically crosslinked alginate hydrogels as scaffolds for tissue engineering: Part 1. Structure, gelation rate and mechanical properties. *Biomaterials* **2001**, *22*, 511–521. [CrossRef]
49. Clogston, J.D.; Patri, A.K. Zeta Potential Measurement. In *Characterization of Nanoparticles Intended for Drug Delivery*; McNeil, S.E., Ed.; Humana Press: Totowa, NJ, USA, 2011; pp. 63–70.
50. Cheng, K.-C.; Huang, C.F.; Wei, Y.; Hsu, S.H. Novel chitosan–cellulose nanofiber self-healing hydrogels to correlate self-healing properties of hydrogels with neural regeneration effects. *NPG Asia Mater.* **2019**, *11*, 25. [CrossRef]

Article

Membrane Separation of Gaseous Hydrocarbons by Semicrystalline Multiblock Copolymers: Role of Cohesive Energy Density and Crystallites of the Polyether Block

Md. Mushfequr Rahman

Helmholtz-Zentrum Hereon, Institute of Membrane Research, Max-Planck-Straße 1, 21502 Geesthacht, Germany; mushfequr.rahman@hereon.de; Tel.: +49-415-287-2224

Abstract: The energy-efficient separation of hydrocarbons is critically important for petrochemical industries. As polymeric membranes are ideal candidates for such separation, it is essential to explore the fundamental relationships between the hydrocarbon permeation mechanism and the physical properties of the polymers. In this study, the permeation mechanisms of methane, ethane, ethene, propane, propene and n-butane through three commercial multiblock copolymers PEBAX 2533, PolyActive1500PEGT77PBT23 and PolyActive4000PEGT77PBT23 are thoroughly investigated at 33 °C. This study aims to investigate the influence of cohesive energy density and crystallites of the polyether block of multiblock copolymers on hydrocarbon separation. The hydrocarbon separation behavior of the polymers is explained based on the solution–diffusion model, which is commonly accepted for gas permeation through nonporous polymeric membrane materials.

Keywords: copolymer; membrane; hydrocarbon; cohesive energy density; gas separation; semicrystalline polymer

1. Introduction

Multiblock copolymers, which have alternating series of polyether-based soft blocks and a glassy or semicrystalline hard block (e.g., polyamide [1,2], polyester [3,4], polyimide [5] and polyurethane [6]), have earned a reputation as ideal membrane materials for gas separation. The structure–property relationships of this class of polymers have been an intriguing field of research in the last decade. A good microphase separation between the hard and soft blocks is highly desirable to enhance the gas separation performance of these polymers [7,8]. Gas permeation occurs through the phase composed of the soft polyether blocks while the hard blocks provide mechanical strength and a film-forming ability [4,9]. Since the gas permeation behavior of such polymers can be tuned by choosing the type, content and length of the hard and soft blocks, these multiblock copolymers are often regarded as a versatile tool in the design of gas separation membranes [7,10–16]. These types of multiblock copolymers have been extensively explored for the separation of carbon dioxide from light gases, e.g., nitrogen and hydrogen [17–20]. The quadrupolar moment of carbon dioxide has a distinct affinity towards the polar ether oxygens, which makes them ideal membrane materials through which carbon dioxide permeates faster than the light gases. Polymers with several polyethers (e.g., poly(ethylene oxide) (PEO) [15,21], poly(propylene oxide) [22–24] and poly(tetramethylene oxide) (PTMO) [25–27]) as soft blocks have been explored as gas separation membrane materials. Multiblock copolymers with PEO as a soft block have higher carbon dioxide selectivity than others due to higher ether oxygen content. For the separation of carbon dioxide from light gases, the PEO blocks are expected to be in an amorphous state under these operating conditions because the presence of PEO crystallites reduces the gas permeability substantially. Gas permeation through non-porous polymers occurs due to the presence of fractional free volume (i.e., the volume unoccupied by the polymer chains). Gases cannot permeate through the perfectly packed chain-folded crystallites because they are too dense. In other words, the PEO crystallites do not have the free-volume to allow permeation of the gases, which lowers the gas permeability. The crystallinity of the multiblock copolymers has a relatively small influence on selectivity. The overall selectivity of the polymers is mostly dictated by the amorphous phase. However, in some studies, it has been reported that the crystallites contribute to the polymer's size-sieving ability and, thereby, have a negative impact on the selectivity of carbon dioxide over light gases [15]. Unlike carbon dioxide separation, the hydrocarbon separation mechanism of these multiblock copolymers is relatively less explored. The separation of hydrocarbons is crucial in the petrochemical industry because it is related to obtaining high-quality fuel and raw materials for bulk chemical production (e.g., ethene for polyethylene) [28]. The conventional separation techniques (e.g., cryogenic distillation) have a large energy penalty compared to membrane separation technology. Thus, the fundamental knowledge regarding the correlation between the separation mechanism of hydrocarbons with the physical properties of the polymeric membrane materials is crucial. In this work, we have investigated the permeability, diffusivity and solubility of methane, ethane, ethene, propane, propene and n-butane through three multiblock copolymers. The three commercially available multiblock copolymers, PEBAX 2533, PolyActive1500PEGT77PBT23 and PolyActive4000PEGT77PBT23, are denoted as P2533, P1500 and P4000, respectively. The chemical structure of these polymers is provided in Figure 1. P2533 is composed of 80 wt% PTMO and 20 wt% polyamide 12 [1,2,18]. P1500 and P4000 are composed of 77 wt% poly(ethylene glycol)terephthalate and 23% poly(butylene terephthalate) [3,17]. P1500 and P4000 contain PEO segments of 1500 and 4000 g/mol, respectively. A thorough investigation of these three polymers revealed important information regarding the role of cohesive energy density and the crystallites of the polyether blocks on the gas permeation mechanism.

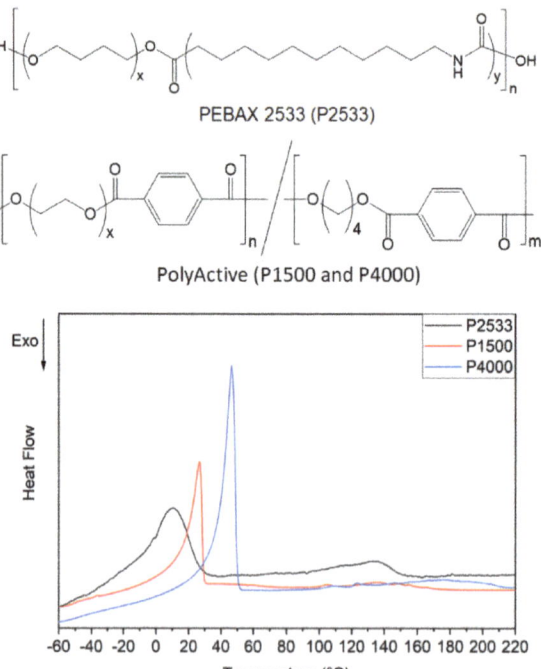

Figure 1. Chemical structures and DSC thermograms (second heating trace) of P2533, P1500 and P4000.

2. Materials and Methods

PolyActive was purchased from PolyVation. PEBAX 2533 was purchased from ARKEMA. The solvents dichloromethane (purity 99.0%) and n-butanol (purity 99.5%) were purchased from Merck KGaA and Scharlau Chemie S.A., respectively. The chemicals were used as received without any further purification.

Dense films were prepared in Teflon molds via solution casting. P2533 was dissolved in n-butanol at 70 °C for 2 h. P1500 and P4000 were dissolved in dichloromethane at room temperature. The obtained homogeneous solutions were poured in a Teflon mold. The solution of P1500 and P4000 was evaporated at room temperature, while the solution of P2533 was evaporated at 40 °C. The films were dried under a vacuum overnight at 30 °C. Membrane thickness was measured by a digital micrometer, and they varied from 100 to 300 μm.

Differential scanning calorimetry (DSC) was used to study the melting transitions of P2533, P1500 and P4000, ranging from −100 °C to 250 °C. All DSC runs were performed in a DSC 1 (Star system) from Mettler Toledo using a nitrogen purge gas stream (60 mL/min) at a scan rate of 10 K/min. Heating and cooling scans were performed by initially heating the sample to 100 °C to erase the effects of residual solvent, and then the sample was cooled to −100 °C. Finally, a second heating scan was performed up to 250 °C. The DSC thermograms presented in Figure 1 correspond to the second heating cycle.

Single gas permeability of the prepared dense membranes was determined by the constant volume and variable pressure ("time-lag"). Permeability (P), diffusivity (D) and solubility (S) are determined at 33 °C from the pressure increase curves obtained during the "time-lag" experiments using the following equations:

$$P = D.S = \frac{V_p . l}{A.R.T.\Delta t} \ln \frac{p_f - p_{p1}}{p_f - p_{p2}} \tag{1}$$

$$D = \frac{l^2}{\theta} \tag{2}$$

where V_p is the permeate volume, l is the membrane thickness, A is the membrane area, R is the gas constant, p_f is the feed pressure considered constant in the time range Δt, p_{p1} and p_{p2} are permeate pressures at times 1 and 2, Δt is the time difference between two points (1 and 2) on the pressure curve and θ is the time lag.

The ideal selectivity of the membranes is determined according to the following equation:

$$\alpha_{A/B} = \frac{P_A}{P_B} = \frac{D_A}{D_B} \cdot \frac{S_A}{S_B} \tag{3}$$

where $\alpha_{A/B}$ is the ideal selectivity, and P_A and P_B are single gas permeabilities of the two gases A and B, respectively.

3. Results and Discussion

Figure 1 shows the second heating traces of the DSC thermograms of P2533, P1500 and P4000. The microphase separated multiblock copolymer P2533 has two distinct melting endotherms for the PTMO and polyamide 12 block, respectively. Two separate melting endotherms are also visible for P1500 and P4000 for the PEO and poly(butylene terephthalate), respectively [17]. The onset and endset of the melting endotherms of the PEO block of P1500 and P4000 are significantly different from each other. For P4000, the melting of the PEO block starts at 40 °C and ends at 49 °C. The gas permeation properties through the polymers are investigated at 33 °C. From the DSC thermograms, it is evident that at 33 °C, the polyether blocks of P2533 and P1500 are completely amorphous while that of P4000 is semicrystalline. Since the content of the polyether blocks P2533, P1500 and P4000 are rather similar, this study sheds light on the permeation mechanism of the hydrocarbons through the amorphous PTMO block, amorphous PEO block and semicrystalline PEO block, respectively. The permeabilities of all the hydrocarbons through P2533 > P1500 >

P4000 are presented in Figure 2. The semicrystalline nature of the PEO blocks is responsible for the low permeability of the gases through P4000. The densely packed crystallites are impermeable for hydrocarbons. Gases permeate through the amorphous part of the polymer only. The substantially high gas permeability through P2533 compared to P1500, despite the similar amorphous polyether content, stems from the difference in the cohesive energy density of the polyether blocks of these two polymers. For all of these polymers, permeabilities of paraffinic hydrocarbons (i.e., methane, ethane, propane and butane) increase with the molecular size (Figure 2). Despite the smaller size, the permeabilities of ethene are slightly higher than that of ethane. Similarly, the permeabilities of propene are higher than propane. To get a clear understanding of the gas permeation behavior of these polymers, it is important to examine the influence of the cohesive energy density and crystallinity of the polyether blocks on the solubilities and diffusivities of the gases.

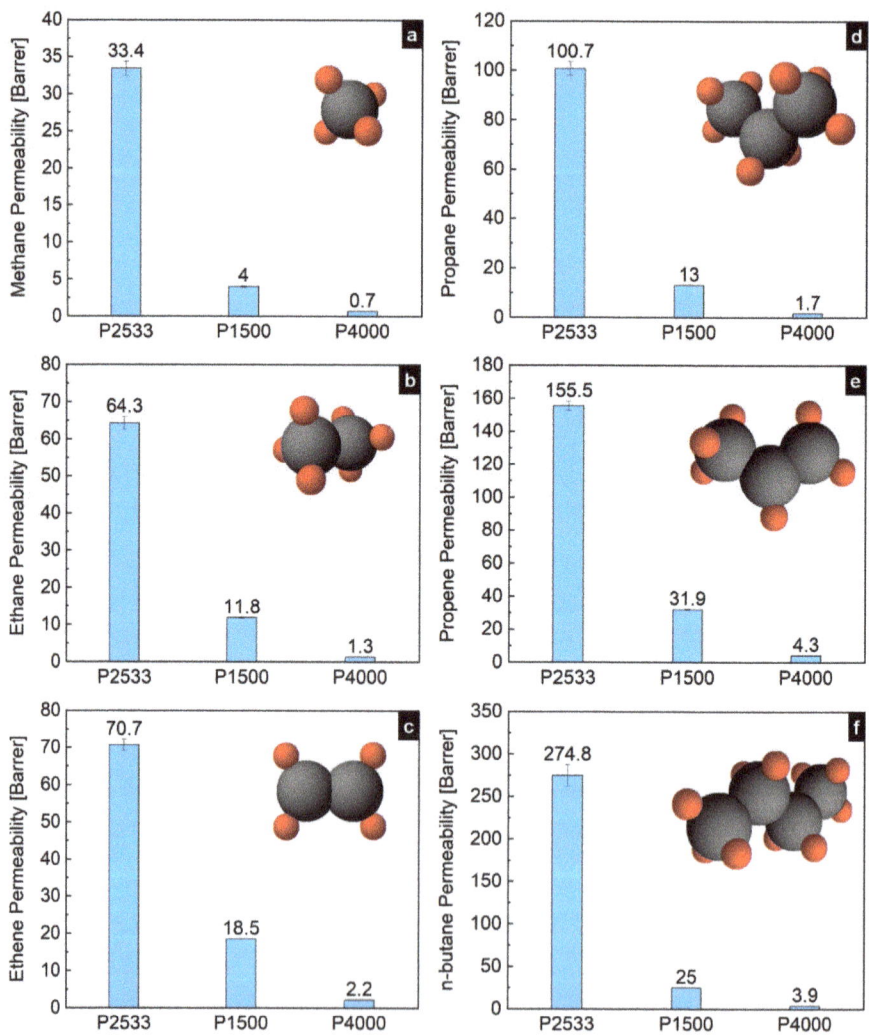

Figure 2. Permeabilities of (**a**) methane, (**b**) ethane, (**c**) ethene, (**d**) propane, (**e**) propene and (**f**) n-butane through P2533, P1500 and P4000.

By definition, cohesive energy density is the internal energy of a substance per unit volume, measuring the interaction energy between the molecules of the substance at a fixed temperature: the stronger the interaction between the molecules, the higher the cohesive energy density. The square root of the cohesive energy density is the Hildebrand solubility parameter. For small molecules, cohesive energy density can be determined experimentally from the enthalpy of vaporization [29]. As the polymers degrade before vaporization, the cohesive energy density and the Hildebrand solubility parameter for polymers are usually theoretically determined by a group contribution method [30]. Efforts have been made to extract the cohesive energy density from other parameters, e.g., surface tension and thermal expansivity [31]. The cohesive energy density of PEO and PTMO are widely reported in the literature. While there are inherent limitations in determining the absolute values, it is common knowledge that PEO has a higher cohesive energy than PTMO [31,32]. The attractive forces between the polyether segments stem from polar ether oxygen. The higher ether oxygen content of PEO leads to a higher cohesive energy density compared to that of PTMO.

A combination of two thermodynamic processes is involved in the dissolution of a gas molecule in a polymeric membrane—the condensation of the gas molecules at the surface of the membrane and the formation of a molecular scale gap in the polymer to accommodate the gas molecule [33]. The first process is dictated by the inherent condensability of a gas molecule. The critical temperature of the gas molecule is widely used as a measure of condensability. Therefore, in Figure 3, the solubilities of the hydrocarbons in P2533, P1500 and P4000 are plotted against their critical temperature. The second process of gas dissolution is related to the cohesive energy density of the polymer. The higher the cohesive energy density, the higher the energy demand to open up a molecular scale gap to accommodate the gas molecule at the surface of the membrane. Since the cohesive energy density of the PTMO segments is significantly lower than the PEO segments, the PTMO containing P2533 can accommodate the condensed gas molecules more easily than the PEO containing P1500 and P4000. Therefore, the solubilities of the gases are higher in P2533 compared to P1500 and P4000. From Figure 3, it is clear that the cohesive energy densities of the polymer segments have a stronger influence on the solubility of a gas with low condensability, e.g., methane. For a gas with low condensability, the solubility is mostly determined by the energy required to form a molecular-scale gap in the polymer for a gas molecule. As the condensability of the gas increases, this process starts to become less dominant when determining the solubility. In P2533, the solubilities of the gases increase systematically with the critical temperature, i.e., the condensability of the gases. However, in P1500 and P4000, the solubilities of the hydrocarbons are not dependent only on the condensability of the gases. While the solubility of paraffinic hydrocarbons increases linearly, the olefinic hydrocarbons (i.e., ethene and propene) have higher solubility than expected from their condensability. Considering the condensability of the gases, ethene is expected to have a slightly lower solubility in the polymers than ethane, and propene is expected to have a slightly lower solubility than propane. Figure 3 shows that in P1500 and P4000, the solubility of ethene is slightly higher than ethane while that of propene is significantly higher than propane. Hence, the olefinic hydrocarbons have a specific affinity towards the polymers, which the paraffinic hydrocarbons do not have. The specific affinity originates from the polar ether oxygen of the polymers and the double bonded carbons of the gases [34]. Since both PTMO and PEO contain ether oxygen, this observation leads to a conjecture that the specific affinity between polyether and olefinic gases impacts the solubility selectivity of the gases only when the cohesive energy between the polyether segments is high enough.

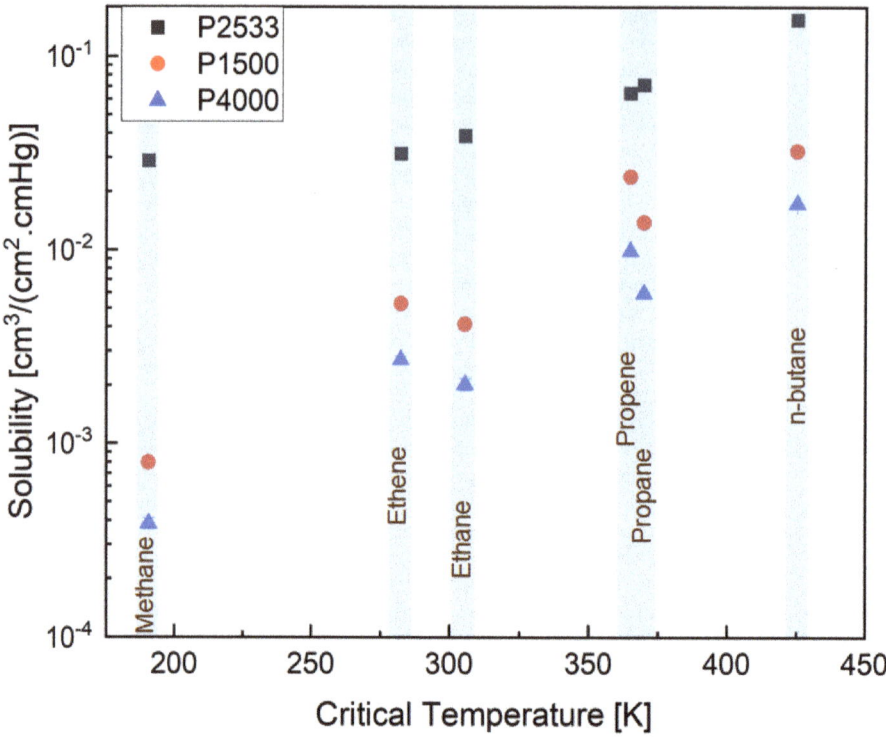

Figure 3. Solubility of the gases in P2533, P1500 and P4000 vs. critical temperature of the gases.

The diffusion of a gas molecule through the polymeric membranes consists of a series of diffusive jumps. In a rubbery polymer, the diffusive jumps are facilitated by the formation of transient free volumes. The segments of the rubbery polymers have sufficient energy for chain rotation, translational motion and vibrational motion, which creates transient free volumes and allows the gas molecules to jump from one site to another. Hence, it is intuitive that cohesive energy density, combined with other factors, influences the formation of the transient free volume in rubbery polymers. Due to the influence of other factors (e.g., the chemical structure of the polymer), a high cohesive energy density does not necessarily translate in low fractional free volume [29]. To examine diffusion, it is equally important to consider the properties of gas molecules because the diffusive jumps are a function of the size of the gases. There are several scaling parameters for the size of a gas molecule, e.g., kinetic diameter and critical volume. Since most of the hydrocarbons are non-spherical, critical volume is a more accurate scaling parameter than kinetic diameter to compare the diffusion of the hydrocarbons [35–37]. For this reason, in Figure 4, the diffusivities are plotted against the critical volumes of the hydrocarbons (the critical volume values reported by Li et al. [38] are used). The diffusivities of the hydrocarbons decrease with the increasing critical volume of the gases in P1500 and P4000, which is not the case in P2533. The diffusion of the gases through the PTMO containing P2533 does not vary significantly upon the change of the hydrocarbons' size. Hence, the higher cohesive energy density of the PEO segments leads to the size-sieving ability of P1500 and P4000.

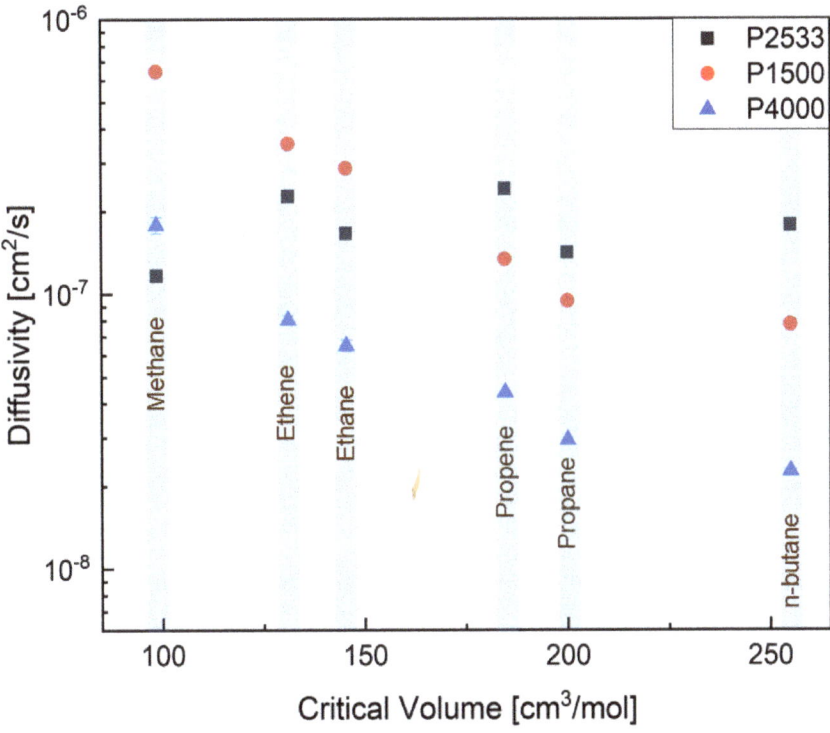

Figure 4. Diffusivity of the gases in P2533, P1500 and P4000 vs. critical volume of the gases.

The influence of the crystallites' presence on gas permeation behavior is often explained using a two-phase model originally proposed by Michaels et al. [39–41] for polyethylene. The model assumes that due to the presence of impermeable crystallites, the gas molecules have to follow a rather tortuous path. Moreover, the crystallites also reduce the mobility of the neighboring amorphous segments. Thus, along with the content of the amorphous fraction, it is necessary to consider a tortuosity factor, τ, and a chain immobilization factor, β, to explain the gas permeation through semicrystalline polymers. According to this model

$$P_s = \frac{P_a}{\tau\beta}\varnothing_a \qquad (4)$$

$$S_s = S_a\varnothing_a \qquad (5)$$

$$D_s = \frac{D_a}{\tau\beta} \qquad (6)$$

where ϕ_a is the volume fraction of the amorphous phase in the polymer, P_s is the permeability of a gas through a semicrystalline polymer, P_a is the permeability of a gas through a pure amorphous polymer, S_s is the solubility coefficient of a gas in a semicrystalline polymer, S_a is the solubility coefficient of a gas in a pure amorphous polymer, D_s is the diffusion coefficient of a gas through a semicrystalline polymer, and D_a is the diffusion coefficient of the gas through a completely amorphous polymer. P4000 has lower solubilities due to the lower amorphous PEO content than P1500 (Figure 3). The diffusivities of the hydrocarbons through P4000 are also lower (Figure 4) than those through P1500. However, both polymers show a similar trend of change in the diffusivities and solubilities of the hydrocarbons as a function of critical volume and temperature, respectively. An accurate analysis of the impact of cohesive energy density and PEO crystallites is possible by comparing the permselectivities, diffusion selectivities and solubility selectivities of the gas pairs.

Figures 5–7 show the contribution of the diffusion and solubility selectivities to the permselectivities of paraffinic hydrocarbon pairs. All three polymers selectively permeate the larger paraffinic hydrocarbons of the gas pairs due to the dominant solubility selectivity over diffusion selectivity. For P2533, the diffusion selectivity of the paraffinic gas pairs is in the range of 0.9–1.5. Since the diffusion selectivities are almost equal to one, P2533 has almost no size-sieving ability, and the permselectivity is merely determined by the solubility selectivity. However, that is not the case for P1500 and P4000. Both of these polymers have substantially stronger solubility selectivities for the paraffinic hydrocarbon pairs than those of P2533. For the gas pairs with significant differences, e.g., n-butane/methane (Figure 6c) and propane/methane (Figure 6f), the solubility selectivities of P1500 and P4000 are 6–8 times higher than those of P2533. However, owing to the cohesive energy density of the PEO blocks, since P1500 and P4000 have size-sieving abilities (i.e., smaller gases diffuse faster), the diffusion selectivity values are far below one. The counteracting influence of the faster diffusivities of smaller gases and the higher solubilities of larger gases in P1500 and P4000 reduces the permselectivities of the paraffinic hydrocarbon pairs compared to those of P2533. The permselectivities of n-butane over ethane (Figure 5d) and methane (Figure 6a) are slightly higher for P4000 than those for P1500, which implies that the presence of PEO crystallites leads to slightly higher permselectivities of these two gas pairs, originating from solubility selectivities. The difference in the permselectivities and solubility selectivities of P1500 and P4000 for the other paraffinic and olefinic/paraffinic hydrocarbon pairs are relatively insignificant. The PEO crystallites significantly reduce the available surface area for the dissolution of the gases. The significantly higher condensability of n-butane over ethane and methane translates into a slight increase in the solubility selectivity in the presence of the PEO crystallites. However, the trend of solubilities (Figure 3) and the difference in solubility selectivities of all hydrocarbon pairs (Figures 5–8) in P1500 and P4000 imply that the characteristic thermodynamic properties of the amorphous PEO of these two polymers are similar. The PEO crystallites of P4000 do not impose sufficient stress to alter the thermodynamic property of the amorphous PEO. The similar trend of diffusivities (Figure 4) and the similar diffusion selectivities of all hydrocarbon pairs (Figures 5–8) in P1500 and P4000 prove that the PEO crystallites do not alter the size-sieving characteristics of the amorphous PEO part of P4000. From an energy consideration, the chain immobilization factor, β (Equation (6)), is a function of the size of permeating gases. The similar size-sieving ability of P1500 and P4000 implies that the chain immobilization factor, β (Equation (6)), does not play a significant role in the diffusion of hydrocarbons through P4000. The intercrystalline spaces of P4000 are sufficiently larger than the critical volume of n-butane (i.e., the largest hydrocarbon used in this study), which makes β relatively insignificant for the gases' diffusion. The crystallites of PEO contribute to the tortuosity factor, τ (Equation (6)), only. As the hydrocarbons have to follow a long, tortuous path due to the presence of the PEO crystallites, the diffusivities through P4000 are lower than those through P1500.

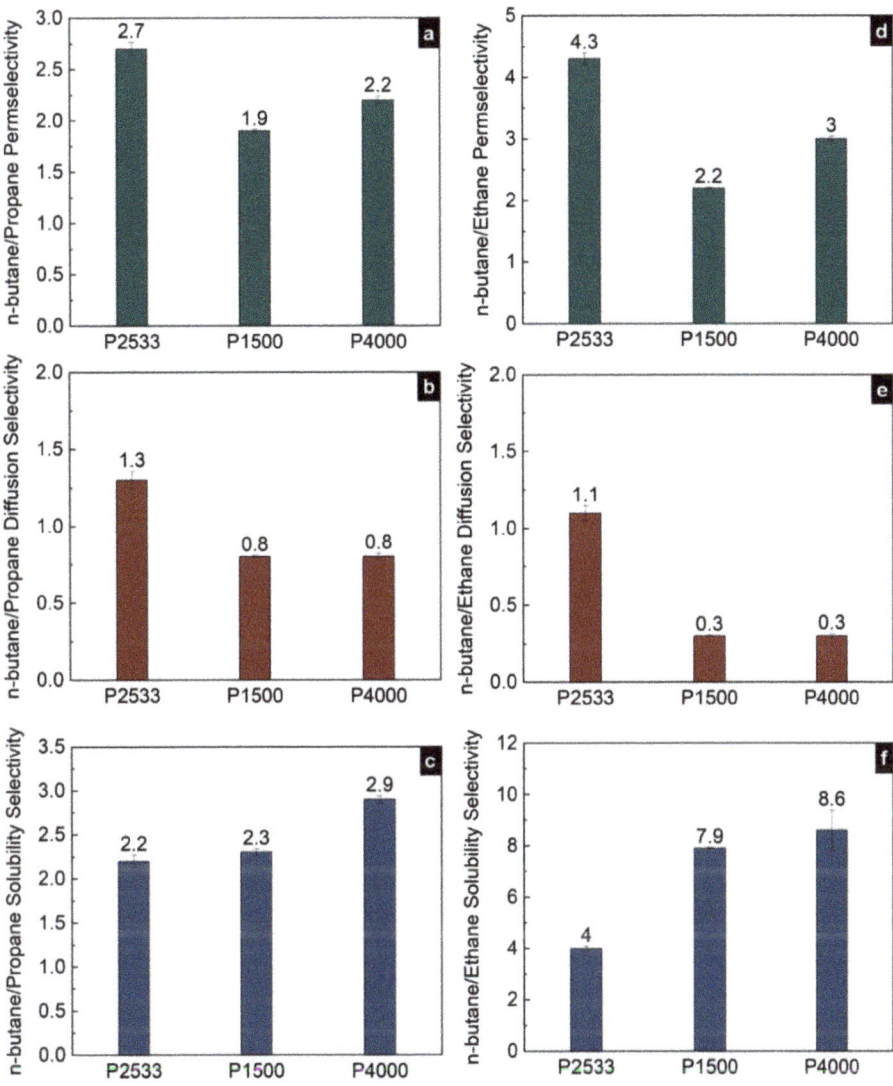

Figure 5. Permselectivities (**a**,**d**), diffusion selectivities (**b**,**e**) and solubility selectivities (**c**,**f**) of n-butane/propane and n-butane/ethane gas pairs in P2533, P1500 and P4000.

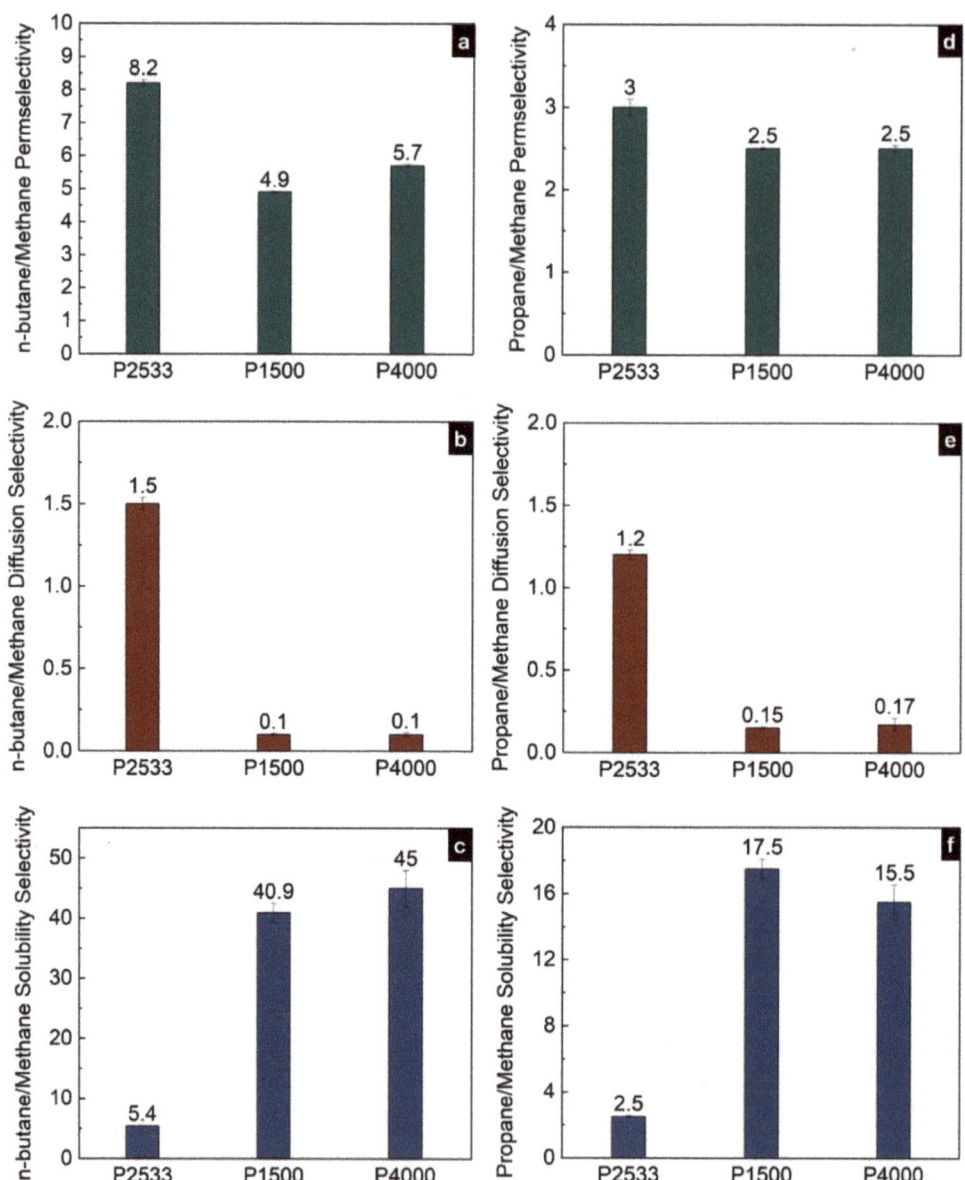

Figure 6. Permselectivities (**a,d**), diffusion selectivities (**b,e**) and solubility selectivities (**c,f**) of n-butane/methane and propane/methane gas pairs in P2533, P1500 and P4000.

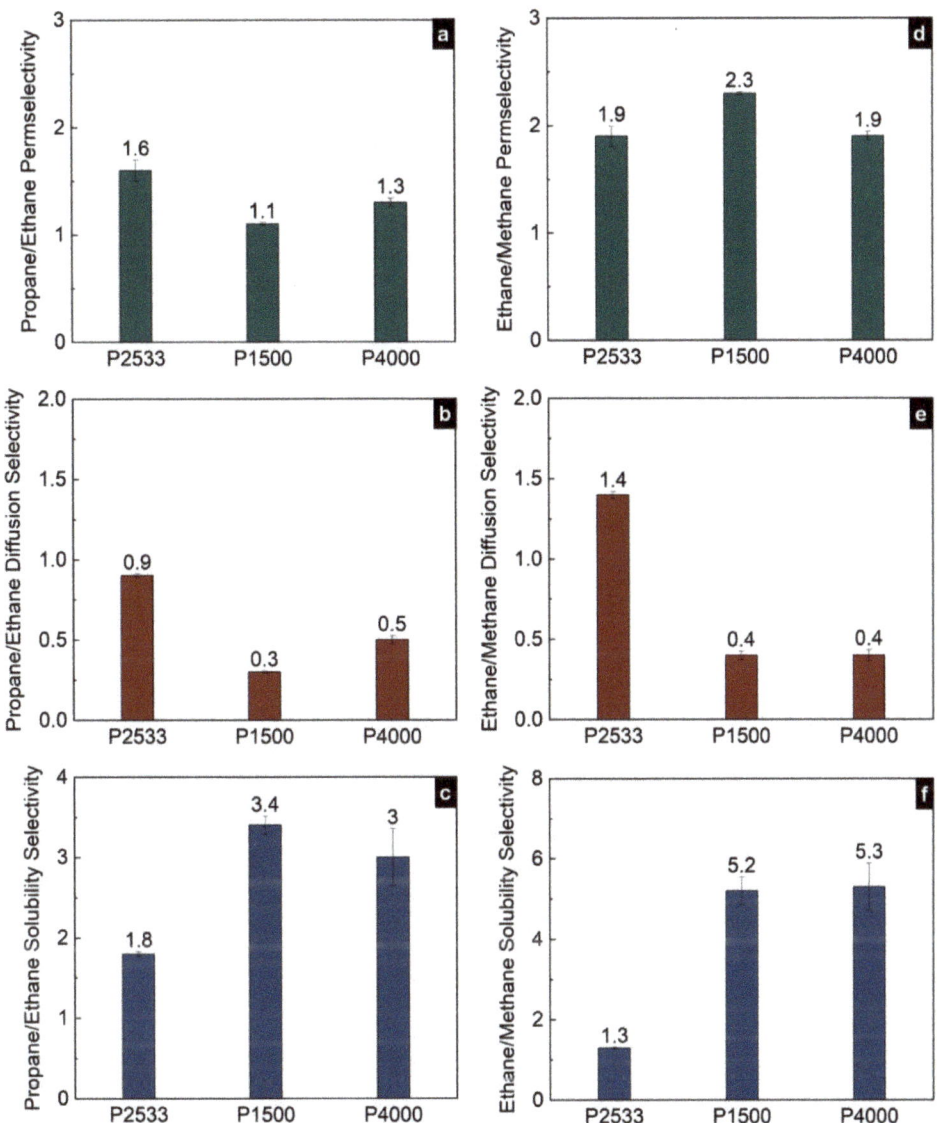

Figure 7. Permselectivities (**a**,**d**), diffusion selectivities (**b**,**e**) and solubility selectivities (**c**,**f**) of propane/ethane and ethane/methane gas pairs in P2533, P1500 and P4000.

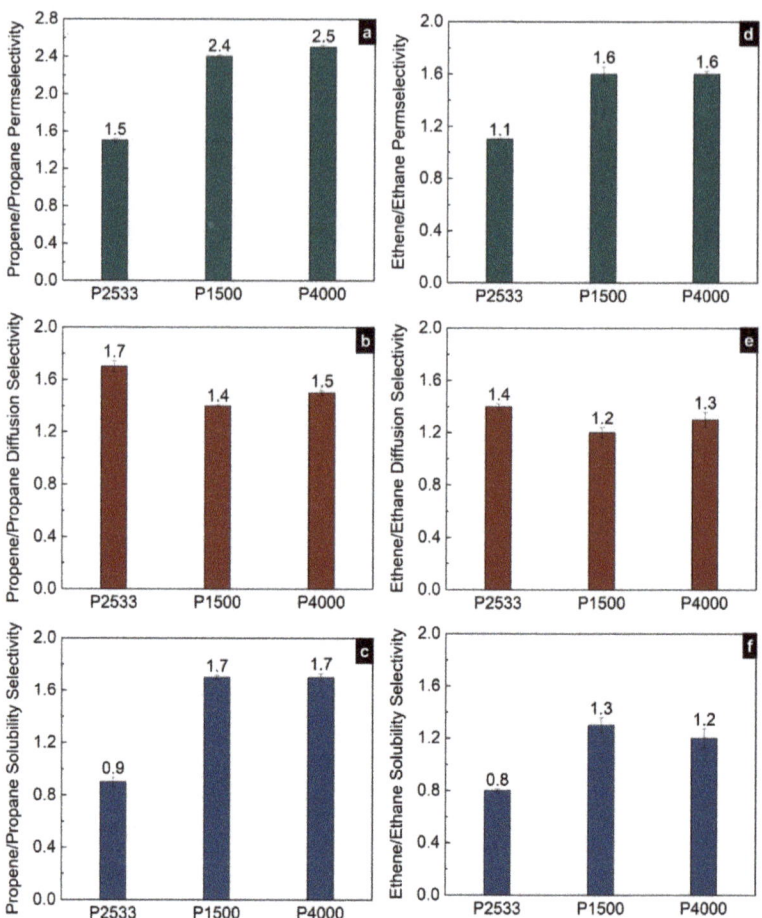

Figure 8. Permselectivities (**a,d**), diffusion selectivities (**b,e**) and solubility selectivities (**c,f**) of propene/propane and ethene/ethane gas pairs in P2533, P1500 and P4000.

4. Conclusions

The detailed investigation of gas permeation through the three commercial multiblock copolymers P2533, P1500 and P4000 demonstrates that for the separation of paraffinic hydrocarbons, the multiblock copolymers with PTMO segments are better membrane materials compared to those containing PEO segments. The lower cohesive energy density of the PTMO containing polymer leads to higher permeability and permselectivity of the paraffinic hydrocarbons. Due to the lower cohesive energy density, the PTMO containing polymer facilitates the dissolution of the hydrocarbons, while the diffusion of the hydrocarbons through the polymer is almost independent of the size of the gases. Under these circumstances, the solubility selectivity determines the overall permselectivity for the PTMO containing polymer. The higher cohesive energy density not only lowers the permeability of hydrocarbons but also imparts a size-sieving property in the PEO containing polymers. The size-sieving property counteracts the solubility selectivity and lowers the permselectivity of the PEO-containing polymers. However, the PEO containing polymers are an ideal choice for olefin/paraffin separation. The PEO-containing polymers allow the selective permeation of olefins over paraffins due to higher solubility, which stems from the specific affinity of the polar ether oxygen towards the double bonds of olefins. The polymer's size-sieving ability remains unchanged in the presence of PEO crystallites because the intercrystalline space is large enough to allow permeation of the hydrocarbons. The presence of PEO crystallites causes a slight permselectivity improvement for the n-butane/methane and the n-butane/ethane gas pairs due to higher solubility selectivity. However, considering the loss of permeability, the presence of impermeable PEO crystallites in the membrane is undesirable for hydrocarbon separation.

Funding: This research received no external funding.

Institutional Review Board Statement: Not applicable.

Informed Consent Statement: Not applicable.

Data Availability Statement: The data presented in this study are available on request from the corresponding author.

Acknowledgments: The author acknowledges the technical support of Silvio Neumann for DSC measurements.

Conflicts of Interest: The author declares no conflict of interest.

References

1. Bondar, V.I.; Freeman, B.D.; Pinnau, I. Gas sorption and characterization of poly(ether-b-amide) segmented block copolymers. *J. Polym. Sci. Part B Polym. Phys.* **1999**, *37*, 2463–2475. [CrossRef]
2. Bondar, V.I.; Freeman, B.D.; Pinnau, I. Gas transport properties of poly(ether-b-amide) segmented block copolymers. *J. Polym. Sci. Part B Polym. Phys.* **2000**, *38*, 2051–2062. [CrossRef]
3. Rahman, M.M.; Abetz, C.; Shishatskiy, S.; Martin, J.; Müller, A.J.; Abetz, V. CO_2 Selective PolyActive Membrane: Thermal Transitions and Gas Permeance as a Function of Thickness. *ACS Appl. Mater. Interfaces* **2018**, *10*, 26733–26744. [CrossRef] [PubMed]
4. Metz, S.J.; Mulder, M.H.V.; Wessling, M. Gas-Permeation Properties of Poly(ethylene oxide) Poly(butylene terephthalate) Block Copolymers. *Macromolecules* **2004**, *37*, 4590–4597. [CrossRef]
5. Okamoto, K.-i.; Fuji, M.; Okamyo, S.; Suzuki, H.; Tanaka, K.; Kita, H. Gas permeation properties of poly(ether imide) segmented copolymers. *Macromolecules* **1995**, *28*, 6950–6956. [CrossRef]
6. Semsarzadeh, M.A.; Sadeghi, M.; Barikani, M. Effect of Chain Extender Length on Gas Permeation Properties of Polyurethane Membranes. *Iran. Polym. J.* **2008**, *17*, 431–440.
7. Reijerkerk, S.R.; Arun, A.; Gaymans, R.J.; Nijmeijer, K.; Wessling, M. Tuning of mass transport properties of multi-block copolymers for CO_2 capture applications. *J. Membr. Sci.* **2010**, *359*, 54–63. [CrossRef]
8. Chen, H.; Xiao, Y.; Chung, T.-S. Synthesis and characterization of poly (ethylene oxide) containing copolyimides for hydrogen purification. *Polymer* **2010**, *51*, 4077–4086. [CrossRef]
9. Sahin, O.; Magonov, S.; Su, C.; Quate, C.F.; Solgaard, O. An atomic force microscope tip designed to measure time-varying nanomechanical forces. *Nat. Nano* **2007**, *2*, 507–514. [CrossRef]
10. Husken, D.; Feijen, J.; Gaymans, R.J. Hydrophilic segmented block copolymers based on poly(ethylene oxide) and monodisperse amide segments. *J. Polym. Sci. Part A Polym. Chem.* **2007**, *45*, 4522–4535. [CrossRef]
11. Husken, D.; Krijgsman, J.; Gaymans, R.J. Segmented blockcopolymers with uniform amide segments. *Polymer* **2004**, *45*, 4837–4843. [CrossRef]
12. Gaymans, R.J. Segmented copolymers with monodisperse crystallizable hard segments: Novel semi-crystalline materials. *Prog. Polym. Sci.* **2011**, *36*, 713–748. [CrossRef]
13. Husken, D.; Feijen, J.; Gaymans, R.J. Segmented Block Copolymers with Terephthalic-Extended Poly(ethylene oxide) Segments. *Macromol. Chem. Phys.* **2008**, *209*, 525–534. [CrossRef]
14. Krijgsman, J.; Gaymans, R.J. Tensile and elastic properties of thermoplastic elastomers based on PTMO and tetra-amide units. *Polymer* **2004**, *45*, 437–446. [CrossRef]
15. Husken, D.; Visser, T.; Wessling, M.; Gaymans, R.J. CO_2 permeation properties of poly(ethylene oxide)-based segmented block copolymers. *J. Membr. Sci.* **2010**, *346*, 194–201. [CrossRef]
16. Reijerkerk, S.R.; Ijzer, A.C.; Nijmeijer, K.; Arun, A.; Gaymans, R.J.; Wessling, M. Subambient Temperature CO_2 and Light Gas Permeation through Segmented Block Copolymers with Tailored Soft Phase. *ACS Appl. Mater. Interfaces* **2010**, *2*, 551–560. [CrossRef] [PubMed]
17. Rahman, M.M.; Filiz, V.; Shishatskiy, S.; Abetz, C.; Georgopanos, P.; Khan, M.M.; Neumann, S.; Abetz, V. Influence of Poly(ethylene glycol) Segment Length on CO_2 Permeation and Stability of PolyActive Membranes and Their Nanocomposites with PEG POSS. *ACS Appl. Mater. Interfaces* **2015**, *7*, 12289–12298. [CrossRef]
18. Rahman, M.M.; Filiz, V.; Shishatskiy, S.; Abetz, C.; Neumann, S.; Bolmer, S.; Khan, M.M.; Abetz, V. PEBAX® with PEG Functionalized POSS as Nanocomposite Membranes for CO_2 Separation. *J. Membr. Sci.* **2013**, *437*, 286–297. [CrossRef]

19. Rahman, M.M.; Shishatskiy, S.; Abetz, C.; Georgopanos, P.; Neumann, S.; Khan, M.M.; Filiz, V.; Abetz, V. Influence of Temperature upon Properties of Tailor-Made PEBAX® MH 1657 Nanocomposite Membranes for Post-Combustion CO_2 Capture. *J. Membr. Sci.* **2014**, *469*, 344–354. [CrossRef]
20. Rahman, M.M.; Filiz, V.; Khan, M.M.; Gacal, B.N.; Abetz, V. Functionalization of POSS nanoparticles and fabrication of block copolymer nanocomposite membranes for CO_2 separation. *React. Funct. Polym.* **2015**, *86*, 125–133. [CrossRef]
21. Yave, W.; Szymczyk, A.; Yave, N.; Roslaniec, Z. Design, synthesis, characterization and optimization of PTT-b-PEO copolymers: A new membrane material for CO_2 separation. *J. Membr. Sci.* **2010**, *362*, 407–416. [CrossRef]
22. Park, H.B.; Kim, C.K.; Lee, Y.M. Gas separation properties of polysiloxane/polyether mixed soft segment urethane urea membranes. *J. Membr. Sci.* **2002**, *204*, 257–269. [CrossRef]
23. Xiao, H.; Ping, Z.H.; Xie, J.W.; Yu, T.Y. Permeation of CO_2 through polyurethane. *J. Appl. Polym. Sci.* **1990**, *40*, 1131–1139. [CrossRef]
24. Huang, S.-L.; Lai, J.-Y. On the gas permeability of hydroxyl terminated polybutadiene based polyurethane membranes. *J. Membr. Sci.* **1995**, *105*, 137–145. [CrossRef]
25. Galland, G.; Lam, T.M. Permeability and diffusion of gases in segmented polyurethanes: Structure–properties relations. *J. Appl. Polym. Sci.* **1993**, *50*, 1041–1058. [CrossRef]
26. Hsieh, K.H.; Tsai, C.C.; Tseng, S.M. Vapor and gas permeability of polyurethane membranes. Part I. Structure-property relationship. *J. Membr. Sci.* **1990**, *49*, 341–350. [CrossRef]
27. Hsieh, K.H.; Tsai, C.C.; Chang, D.M. Vapor and gas permeability of polyurethane membranes. Part II. Effect of functional group. *J. Membr. Sci.* **1991**, *56*, 279–287. [CrossRef]
28. Yang, L.; Qian, S.; Wang, X.; Cui, X.; Chen, B.; Xing, H. Energy-efficient separation alternatives: Metal–organic frameworks and membranes for hydrocarbon separation. *Chem. Soc. Rev.* **2020**, *49*, 5359–5406. [CrossRef]
29. White, R.P.; Lipson, J.E.G. Free Volume, Cohesive Energy Density, and Internal Pressure as Predictors of Polymer Miscibility. *Macromolecules* **2014**, *47*, 3959–3968. [CrossRef]
30. Kubica, P.; Wolinska-Grabczyk, A. Correlation between Cohesive Energy Density, Fractional Free Volume, and Gas Transport Properties of Poly(ethylene-*co*-vinyl acetate) Materials. *Int. J. Polym. Sci.* **2015**, *2015*, 861979. [CrossRef]
31. Dee, G.T.; Sauer, B.B. The cohesive energy density of polymers and its relationship to surface tension, bulk thermodynamic properties, and chain structure. *J. Appl. Polym. Sci.* **2017**, *134*, 44431. [CrossRef]
32. Galin, M. Thermodynamic studies on polyether—Solvent systems by gas—liquid chromatography. *Polymer* **1995**, *36*, 3533–3539. [CrossRef]
33. Rahman, M.M.; Lillepärg, J.; Neumann, S.; Shishatskiy, S.; Abetz, V. A thermodynamic study of CO_2 sorption and thermal transition of PolyActive™ under elevated pressure. *Polymer* **2016**, *93*, 132–141. [CrossRef]
34. Lin, H.; Freeman, B.D. Gas Solubility, Diffusivity and Permeability in Poly(ethylene oxide). *J. Membr. Sci.* **2004**, *239*, 105–117. [CrossRef]
35. Teplyakov, V.; Meares, P. Correlation aspects of the selective gas permeabilities of polymeric materials and membranes. *Gas Sep. Purif.* **1990**, *4*, 66–74. [CrossRef]
36. Yampolskii, Y. 2—Fundamental science of gas and vapour separation in polymeric membranes. In *Advanced Membrane Science and Technology for Sustainable Energy and Environmental Applications*; Woodhead Publishing: Sawston, UK, 2011; pp. 22–55.
37. Matteucci, S.; Yampolskii, Y.; Freeman, B.D.; Pinnau, I. Transport of Gases and Vapors in Glassy and Rubbery Polymers. In *Materials Science of Membranes for Gas and Vapor Separationl*; John Wiley & Sons, Ltd.: Hoboken, NJ, USA, 2006; pp. 1–47.
38. Li, J.-R.; Kuppler, R.J.; Zhou, H.-C. Selective gas adsorption and separation in metal–organic frameworks. *Chem. Soc. Rev.* **2009**, *38*, 1477–1504. [CrossRef] [PubMed]
39. Michaels, A.S.; Parker, R.B. Sorption and flow of gases in polyethylene. *J. Polym. Sci.* **1959**, *41*, 53–71. [CrossRef]
40. Michaels, A.S.; Bixler, H.J. Solubility of gases in polyethylene. *J. Polym. Sci.* **1961**, *50*, 393–412. [CrossRef]
41. Michaels, A.S.; Bixler, H.J. Flow of gases through polyethylene. *J. Polym. Sci.* **1961**, *50*, 413–439. [CrossRef]

Article

Crystallization and Morphology of Triple Crystalline Polyethylene-*b*-poly(ethylene oxide)-*b*-poly(ε-caprolactone) PE-*b*-PEO-*b*-PCL Triblock Terpolymers

Eider Matxinandiarena [1], Agurtzane Múgica [1], Manuela Zubitur [2], Viko Ladelta [3], George Zapsas [3], Dario Cavallo [4], Nikos Hadjichristidis [3,*] and Alejandro J. Müller [1,5,*]

[1] POLYMAT and Department of Polymers and Advanced Materials: Physics, Chemistry and Technology, University of the Basque Country UPV/EHU, Paseo Manuel Lardizábal 3, 20018 Donostia-San Sebastián, Spain; eider.matxinandiarena@ehu.eus (E.M.); agurtzane.mugica@ehu.eus (A.M.)

[2] Department of Chemical and Environmental Engineering, University of the Basque Country UPV/EHU, Plaza Europa 1, 20018 Donostia-San Sebastián, Spain; manuela.zubitur@ehu.eus

[3] Polymer Synthesis Laboratory, KAUST Catalysis Center, Physical Sciences and Engineering Division, King Abdullah University of Science and Technology (KAUST), Thuwal 23955, Saudi Arabia; viko.ladelta@kaust.edu.sa (V.L.); georgios.zapsas@kaust.edu.sa (G.Z.)

[4] Department of Chemistry and Industrial Chemistry, University of Genova, Via Dodecaneso 31, 16146 Genova, Italy; dario.cavallo@unige.it

[5] Ikerbasque, Basque Foundation for Science, Plaza Euskadi 5, 48009 Bilbao, Spain

* Correspondence: Nikolaos.Hadjichristidis@kaust.edu.sa (N.H.); alejandrojesus.muller@ehu.es (A.J.M.)

Citation: Matxinandiarena, E.; Múgica, A.; Zubitur, M.; Ladelta, V.; Zapsas, G.; Cavallo, D.; Hadjichristidis, N.; Müller, A.J. Crystallization and Morphology of Triple Crystalline Polyethylene-*b*-poly(ethylene oxide)-*b*-poly(ε-caprolactone) PE-*b*-PEO-*b*-PCL Triblock Terpolymers. *Polymers* 2021, 13, 3133. https://doi.org/10.3390/polym13183133

Academic Editors: Volker Abetz and Holger Schmalz

Received: 2 September 2021
Accepted: 13 September 2021
Published: 16 September 2021

Publisher's Note: MDPI stays neutral with regard to jurisdictional claims in published maps and institutional affiliations.

Copyright: © 2021 by the authors. Licensee MDPI, Basel, Switzerland. This article is an open access article distributed under the terms and conditions of the Creative Commons Attribution (CC BY) license (https://creativecommons.org/licenses/by/4.0/).

Abstract: The morphology and crystallization behavior of two triblock terpolymers of polymethylene, equivalent to polyethylene (PE), poly (ethylene oxide) (PEO), and poly (ε-caprolactone) (PCL) are studied: $PE_{22}{}^{7.1}$-*b*-$PEO_{46}{}^{15.1}$-*b*-$PCL_{32}{}^{10.4}$ (T1) and $PE_{37}{}^{9.5}$-*b*-$PEO_{34}{}^{8.8}$-*b*-$PCL_{29}{}^{7.6}$ (T2) (superscripts give number average molecular weights in kg/mol and subscripts composition in wt %). The three blocks are potentially crystallizable, and the triple crystalline nature of the samples is investigated. Polyhomologation (C1 polymerization), ring-opening polymerization, and catalyst-switch strategies were combined to synthesize the triblock terpolymers. In addition, the corresponding PE-*b*-PEO diblock copolymers and PE homopolymers were also analyzed. The crystallization sequence of the blocks was determined via three independent but complementary techniques: differential scanning calorimetry (DSC), in situ SAXS/WAXS (small angle X-ray scattering/wide angle X-ray scattering), and polarized light optical microscopy (PLOM). The two terpolymers (T1 and T2) are weakly phase segregated in the melt according to SAXS. DSC and WAXS results demonstrate that in both triblock terpolymers the crystallization process starts with the PE block, continues with the PCL block, and ends with the PEO block. Hence triple crystalline materials are obtained. The crystallization of the PCL and the PEO block is coincident (i.e., it overlaps); however, WAXS and PLOM experiments can identify both transitions. In addition, PLOM shows a spherulitic morphology for the PE homopolymer and the T1 precursor diblock copolymer, while the other systems appear as non-spherulitic or microspherulitic at the last stage of the crystallization process. The complicated crystallization of tricrystalline triblock terpolymers can only be fully grasped when DSC, WAXS, and PLOM experiments are combined. This knowledge is fundamental to tailor the properties of these complex but fascinating materials.

Keywords: triblock terpolymers; polyethylene (PE); poly(ethylene oxide) (PEO); poly(ε-caprolactone) (PCL); tricrystalline spherulites

1. Introduction

Crystallization in block copolymers is a subject widely studied in the past decades [1–11]. It is vital to understand the morphology upon crystallization since it is directly related to the final properties of a material. Many applications can take advantage of these materials due to the different chemical nature of the segments that form a block copolymer [9,12–14]. In addition, many other factors such as composition, molecular weight, crystallization protocol, segregation strength, and block miscibility affect the crystallization behavior. As different morphologies can be developed, the final performance of the materials can be tuned by varying these factors [2,4,7,9,11,15–18].

AB-type diblock copolymers with one or two crystallizable blocks have been studied in the past few decades. Among medium or strongly segregated systems, the diblock copolymer PE-b-PLLA [18–24] is a well-known system. Müller et al. [19–21] reported strong segregation strength for these diblock copolymers and a lamellar morphology for compositions close to 50/50. Therefore, they did not see any spherulitic-type morphology as expected. When the content of PLLA in the diblock is between 89 and 96%, then spherulitic morphologies have been reported in the literature, as PLLA conforms the matrix phase [25]. The overall crystallization rate of both PLLA block and PE block in the diblock copolymers [19–21] was slower than that of the corresponding PLLA and PE homopolymers. In addition, coincident crystallization occurs, since the crystallization transitions of the PE block and the PLLA block overlap employing cooling rates higher than 2 °C/min.

Other double crystalline diblock copolymers show miscible or weakly segregated behaviors, and several studies have been reported about the crystallization process of these systems [7,26–33], although the most relevant ones are: PEO-b-PCL [7,34–51], PEO-b-PLLA [52–65], and PCL-b-PLA [29,30,32,66–74], because of their possible applications in the biomedical field due to the biodegradable and biocompatible nature of the blocks [14,75–78]. Additionally, some ABA-type systems have also been analyzed, such as PBT-b-PEO-b-PBT [79], PEO-b-PEB-b-PEO [80], or PLLA-b-PVDF-b-PLLA [81], for instance. The addition of a third potentially crystallizable block to diblock copolymers results in a more complex analysis of the crystallization behavior. Few studies have been published about tricrystalline triblock terpolymers, such as ABC-type triblock terpolymers and ABCBA pentablock terpolymers, including the apolar PE block, and the polar PEO, PCL, and PLLA blocks [17,24,37,42,82–96].

Palacios et al. studied the crystallization and morphology of ABC triblock terpolymers with three crystallizable blocks: PEO, PCL, and PLLA [92–95]. They [92] highlighted the triple crystalline nature of the PEO-b-PCL-b-PLLA triblock terpolymer, with the three different blocks crystallizing independently upon cooling from the melt. Even when changing the PLLA content, crystallization of the blocks follows this sequence: the PLLA block first, the PCL block second, and finally the PEO block. Melt miscibility of the three blocks was confirmed by SAXS. In addition, PLOM experiments showed that the first crystallized PLLA block determines the final morphology since the PCL block and the PEO block crystallized within the interlamellar regions of the PLLA templated spherulites, maintaining the superstructure determined by the PLLA block and forming triple crystalline spherulites. The crystallization of the PCL and the PEO blocks was evidenced by a change in the birefringence. There are several examples of confined crystallization of one block within the lamellae of another previously crystallized block [1,7,97,98].

Furthermore, by SAXS and AFM experiments, Palacios et al. [94] were able to identify a trilamellar self-assembly with lamellae of the three blocks at room temperature. Based on extensive observations and SAXS simulations, they proposed an alternation of single lamellae of PEO or PCL in between two PLLA lamellae. Very few reports have been published about the crystalline morphology in AB diblock copolymers and ABC triblock terpolymers from the melt by in situ AFM, and only two blocks crystallized in those samples [99,100]. Palacios et al. [17] analyzed by in situ hot-stage AFM the evolution of the trilamellar morphology upon melting of the PEO-b-PCL-b-PLLA triblock terpolymer. Three different lamellar populations were detected at different temperatures; the melting of each of the populations gives information about the corresponding block: the thinnest lamellae corresponded to the PEO block (the first block to melt at 45 °C), the medium size lamellae to the PCL block (melted at 60 °C), and the thickest lamellae to the highest melting temperature block, i.e., PLLA.

Still, few works have been published using the apolar PE block as one of the crystallizable blocks in triblock terpolymers. Müller et al. [96] analyzed the crystalline behavior and morphology of PE-b-PEO-b-PLLA and PE-b-PCL-b-PLLA triblock terpolymers employing different cooling rates. DSC, WAXS, and PLOM techniques were used to confirm the triple crystalline character of the copolymers. They concluded that there is no change in the sequential crystallization for the $PE_{21}{}^{2.6}$-b-$PEO_{32}{}^{4.0}$-b-$PLLA_{47}{}^{5.9}$ triblock terpolymer using 1 or 20 °C/min, since the sequence remains the following: the PE block crystallizes first, then the PLLA block, and finally the PEO block. However, the crystallization sequence changed in the $PE_{21}{}^{7.1}$-b-$PCL_{12}{}^{4.2}$-b-$PLLA_{67}{}^{23.0}$ triblock terpolymer, since when using 20 °C/min as cooling rate, the crystallization begins with the PE block. In contrast, at 1 °C/min the PLLA is the first block to crystallize. PLOM experiments showed that this variation in the crystallization sequence affects the final morphology, so the cooling rate is a factor that can be used to tune the final properties.

In the present work, the triple crystalline nature of PE-b-PEO-b-PCL triblock terpolymers is analyzed, varying molecular weight and block content. The corresponding PE-b-PEO diblock copolymers and PE homopolymers are also investigated. Samples were synthesized by combining polyhomologation and catalyst-switch strategies. We study the influence of molecular weight and block composition on the crystallization of these triblock terpolymers, consisting in an apolar (PE) and two polar blocks, PEO (biocompatible) and PCL (biodegradable). The study employs differential scanning calorimetry (DSC), in situ small-angle and wide-angle X-ray scattering (SAXS/WAXS), and polarized light optical microscopy (PLOM). Understanding the crystalline behavior and the analysis of the morphology is essential to tune crystallinity and obtain novel materials with enhanced properties.

2. Materials and Methods

2.1. Materials

All reagents used for the synthesis of the triblock terpolymers were purchased from Merck KGaA (Darmstadt, Germany). Two different "catalyst switch" strategies were used in the synthesis of the tricrystalline terpolymers poly (ethylene)-b-poly (ethylene oxide)-b-poly (ε-caprolactone) (PE-b-PEO-b-PCL). First, the polyhomologation of dimethylsulfoxonium methylide was performed to synthesize a hydroxyl-terminated polyethylene (PE-OH) macroinitiator [101]. Then, the strong phosphazene base t-BuP$_4$ was employed as the catalyst to promote the ring-opening polymerization (ROP) of ethylene oxide (EO) to obtain PE-b-PEO, followed by the addition of diphenyl phosphate (DPP) to neutralize t-BuP$_4$. For the ROP of ε-caprolactone (CL), two different catalysts were used, Sn(Oct)$_2$ for T1 (organic/metal catalyst-switch), and phosphazene base t-BuP$_2$ for T2 (organic/organic catalyst-switch). These catalyst switch strategies were applied to avoid as many possible side-reactions during the ROP of CL in toluene at 80 °C (Scheme 1) [102].

Table 1 shows the molecular weights of each of the blocks of the synthesized triblock terpolymers. The subscript numbers represent composition in wt %, and superscripts indicate M_n values of each block in kg/mol. The polyethylene block precursors are not 100% linear because of possible side reactions and monomer purity issues. NMR measurements indicate that the PE block of T1 (see Table 1) contains 0.32% propyl side groups and 3% methyl groups, and that of T2 contains 0.45% propyl side groups and 2% methyl groups. Different melting points are obtained because of this variation in microstructure, since the T_m value of $PE^{7.1}$ is 129.7 °C, while that of $PE^{9.5}$ is 117 °C (see Table S3), as the latter contains a higher amount of short-chain branches.

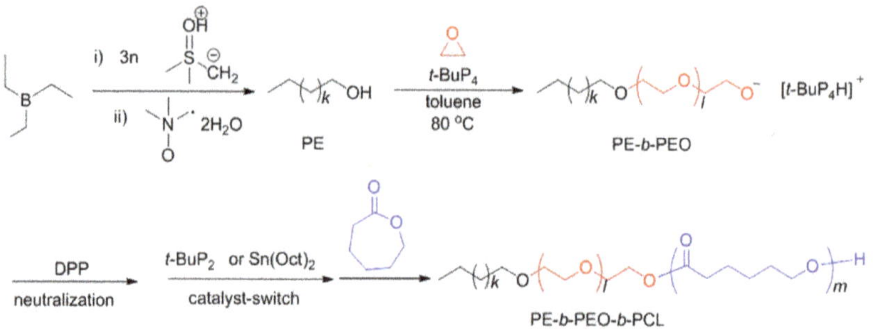

Scheme 1. Synthesis of tricrystalline terpolymer PE-b-PEO-b-PCL by a combination of polyhomologation and two catalyst-switch strategies [102].

Table 1. Number-average molecular weight (M_n) and polydispersity of homopolymers, diblock copolymers, and triblock terpolymers. Subscripts represent composition in wt %, and superscripts indicate M_n values of each block in kg/mol.

Sample	M_n PE (g/mol)	M_n PEO (g/mol)	M_n PCL (g/mol)	Đ [a]
PE$^{7.1}$	7100	-	-	1.32
PE$_{32}$$^{7.1}$-b-PEO$_{68}$$^{15.1}$	7100	15,100	-	<1.3
PE$_{22}$$^{7.1}$-b-PEO$_{46}$$^{15.1}$-b-PCL$_{32}$$^{10.4}$ (**T1**)	7100	15,100	10,400	<1.3
PE$^{9.5}$	9500	-	-	1.28
PE$_{52}$$^{9.5}$-b-PEO$_{48}$$^{8.8}$	9500	8800	-	<1.3
PE$_{37}$$^{9.5}$-b-PEO$_{34}$$^{8.8}$-b-PCL$_{29}$$^{7.6}$ (**T2**)	9500	8800	7600	<1.3

[a] In the case of PEO-containing polymers, the polydispersity values are not correct since PEO is adsorbed by the columns used in high-temperature GPC. Using a soluble in THF terpolymer (low molecular weight PE) we were able to prove that Đ of the terpolymers is below 1.3 (for more details see our Ref. [103]).

The formation of double crystalline copolymers and triple crystalline terpolymers was confirmed by differential scanning calorimetry (DSC), polarized light optical microscopy (PLOM), and X-ray diffraction (SAXS/WAXS).

2.2. Methods

2.2.1. Differential Scanning Calorimetry (DSC)

Non-isothermal DSC experiments were carried out with a Perkin Elmer DSC Pyris 1 (Perkin Elmer, Norwalk, USA) equipped with a refrigerated cooling system (Intracooler 2P). Indium and tin standards were used for the calibration of the equipment. Aluminum pans with about 3 mg of sample were tested using ultra-high quality nitrogen atmosphere.

A temperature range between 0 and 160 °C and 20 °C/min as cooling and heating rates were employed in non-isothermal DSC experiments. The samples are kept for 3 min 30 °C above the peak melting temperature of the block showing the highest melting temperature to erase the thermal history of the samples. They are then cooled down at 20 °C/min keeping them 1 min at low temperatures, and finally heating up also at 20 °C/min until the block at the highest temperature melts.

2.2.2. Small-Angle and Wide-Angle X-ray Scattering (SAXS/WAXS)

Simultaneous in situ small-angle X-ray scattering (SAXS) and wide-angle X-ray scattering (WAXS) experiments were performed at the ALBA Synchrotron facility in Barcelona (Barcelona, Spain), beamline BL11-NCD. A Linkam THMS600 (Linkam, Surrey, UK) hot stage coupled to a liquid nitrogen cooling system was used to cool and heat the samples, which were previously placed into glassy capillaries. The same thermal protocol adopted in the non-isothermal DSC experiments was used to get the SAXS/WAXS patterns, in which crystallization and melting of the samples are followed, thus obtaining comparable results by the two different techniques.

The X-ray energy source was 12.4 keV (λ = 1.03 Å). For the SAXS setup, the distance between the sample and the detector (ADSC Q315r detector, Poway, CA, USA, with a resolution of 3070 × 3070 pixels, pixel size of 102 µm^2) was 6463 mm with a tilt angle of 0°. Calibration was performed with silver behenate. Regarding WAXS configuration, a distance of 132.6 mm was used between the sample and the detector, with a tilt angle of 21.2°. Chromium (III) oxide (Rayonix LX255-HS detector, Evanston, IL, USA, with a resolution of 1920 × 5760 pixels, pixel size of 44 µm^2) was employed for calibration. Scattering intensity as a function of scattering vector, $q = 4\pi \sin\theta \lambda^{-1}$ data are obtained, where λ is the X-ray wavelength, and 2θ is the scattering angle.

2.2.3. Polarized Light Optical Microscopy (PLOM)

An Olympus BX51 polarized light optical microscope (Olympus, Tokyo, Japan) was used to follow the morphological changes occurring within the samples while cooled and heated at a constant rate of 20 °C/min. For accurate temperature control, a Linkam THMS600 (Linkam, Surrey, UK) hot stage with liquid nitrogen was used. Micrographs were recorded by an Olympus SC50 camera (Olympus, Tokyo, Japan). First, a glass slide in which samples are melted is used, with a glass coverslip, and then, 20 °C/min as cooling and heating rates are employed. Morphological variations that occur during the application of this constant rate are recorded as micrographs in which crystallization and melting of each of the blocks can be followed.

Furthermore, the software ImageJ [103] was used to analyze the micrographs by measuring transmitted light intensities. The increase in light intensity detected refers to the increase in crystal content of a certain sample since crystallization of one component has started. Crystallization of this component can be followed by the increase in intensity by decreasing temperature, and the temperature range at which crystallization of a specific block occurs can be determined. In order to detect intensity changes, the whole micrographs are considered as "region of interest". Thus, all superstructures that can be formed during the cooling scans contribute to this analysis. So, the entire crystallization process is followed by analyzing intensity changes as a function of temperature, and the crystallization temperature of a particular block of the diblock copolymers and triblock terpolymers can be determined.

3. Results and Discussion

3.1. Small-Angle X-ray Scattering (SAXS)

SAXS measurements are useful to study not only the phase segregation in the melt but also if the phase segregation is kept when the block components crystallize or if crystallization destroys it by breaking out the phase structure of the melt. Figure 1 shows the SAXS patterns of the homopolymer PE$^{7.1}$, the diblock copolymer PE$_{32}$$^{7.1}$-b-PEO$_{68}$$^{15.1}$, and the triblock terpolymer PE$_{22}$$^{7.1}$-b-PEO$_{46}$$^{15.1}$-b-PCL$_{32}$$^{10.4}$ (T1) upon cooling from the melt.

For the homopolymer PE$^{7.1}$ and the diblock copolymer PE$_{32}$$^{7.1}$-b-PEO$_{68}$$^{15.1}$ (Figure 1a,b), there is no phase segregation in the melt, as evidenced by the lack of scattering peaks in the molten state. The broad peak that appears at lower temperatures corresponds to the diffraction from crystalline lamellar stacks in the formed superstructures (i.e., spherulites or axialites).

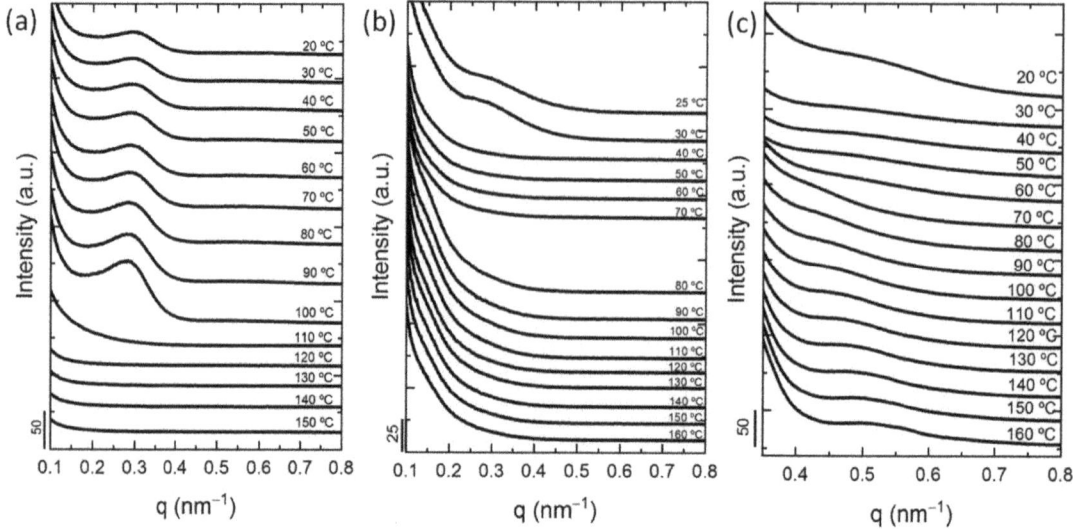

Figure 1. SAXS ramp down patterns at 20 °C/min for (a) PE$^{7.1}$, (b) PE$_{32}$$^{7.1}$-b-PEO$_{68}$$^{15.1}$, and (c) PE$_{22}$$^{7.1}$-b-PEO$_{46}$$^{15.1}$-b-PCL$_{32}$$^{10.4}$ (T1) at the indicated temperatures.

However, there is weak phase segregation for the PE$_{22}$$^{7.1}$-b-PEO$_{46}$$^{15.1}$-b-PCL$_{32}$$^{10.4}$ (T1) triblock terpolymer (Figure 1c) since there is a broad scattering peak in the melt, which disappears as crystallization breaks out when the first block upon cooling from the melt starts to crystallize (i.e., the PE block). This behavior is evidenced by the shift in q values between the reflection in the melt and the weaker reflection at room temperature, which appears at lower q values. The broad peak at room temperature corresponds to the average long period of the lamellae formed during the crystallization process because the phase structure established by phase segregation in the melt was destroyed by the break-out.

Figure 2 shows SAXS patterns of the homopolymer PE$^{9.5}$, the diblock copolymer PE$_{52}$$^{9.5}$-b-PEO$_{48}$$^{8.81}$, and the triblock terpolymer PE$_{37}$$^{9.5}$-b-PEO$_{34}$$^{8.8}$-b-PCL$_{29}$$^{7.6}$ (T2) at the indicated temperatures reached upon cooling. In this case, the behavior of the homopolymer PE$^{9.5}$ (Figure 2a) is the same as for the homopolymer PE$^{7.1}$ (Figure 1a) explained above, not showing any phase segregation in the melt, as expected for a homopolymer.

The diblock copolymer PE$_{52}$$^{9.5}$-b-PEO$_{48}$$^{8.81}$ (Figure 2b) and the triblock terpolymer PE$_{37}$$^{9.5}$-b-PEO$_{34}$$^{8.8}$-b-PCL$_{29}$$^{7.6}$ (T2) (Figure 2c) are phase segregated in the melt, with possible lamellar and interpenetrated morphologies, respectively, although more detailed analysis of the scattering curves would be needed to ascertain the exact melt morphology. The clear scattering peaks in the molten state in these two materials corroborate the phase segregation behavior; however, their phase segregation is weak, since when the first block crystallizes upon cooling, i.e., the PE block at 100 °C, the phase structure is destroyed, the one generated by phase segregation in the melt, as deduced by the change in q values and intensities of the scattering peaks.

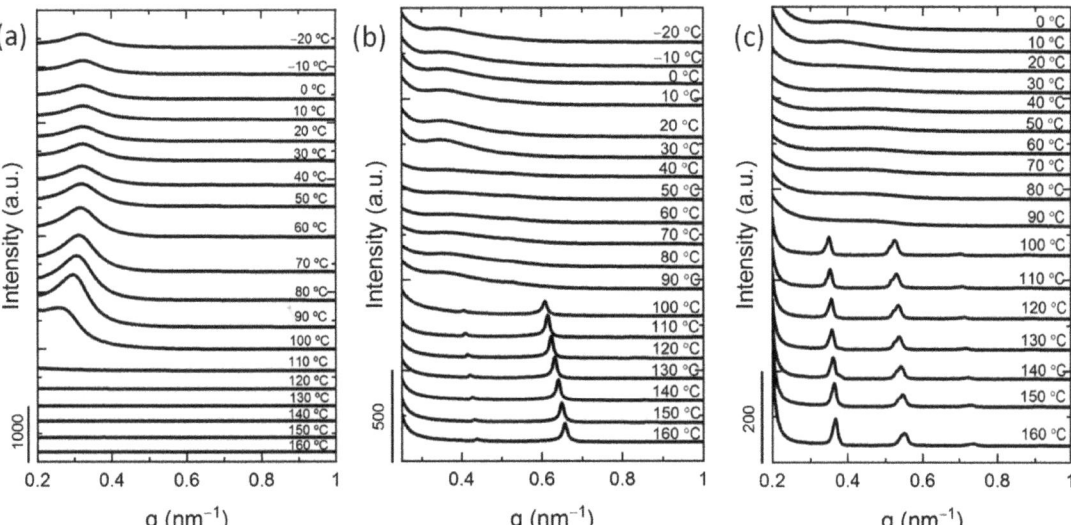

Figure 2. SAXS ramp down patterns at 20 °C/min for (a) PE$^{9.5}$, (b) PE$_{52}$$^{9.5}$-b-PEO$_{48}$$^{8.8}$, and (c) PE$_{37}$$^{9.5}$-b-PEO$_{34}$$^{8.8}$-b-PCL$_{29}$$^{7.6}$ (T2) at the indicated temperatures.

One way to predict the segregation strength in linear diblock copolymers is by multiplying the Flory-Huggins interaction parameter (χ) (evaluated at the interest temperature in the melt) by N (the total degree of polymerization). The estimation becomes more difficult in the case of triblock terpolymers. Different behaviors can be predicted depending on the segregation strength values. Values equal or lower to 10 indicates miscibility in the melt, between 10 and 30 weak phase segregation, between 30 and 50 intermediate segregation, and if values are higher than 50, the systems are strongly segregated. A rough approximation for each pair of blocks is reported in Table S1 (see Supporting Information), using the solubility parameters of PE, PEO, and PCL from the literature [60,104]. In this case, the predicted values suggest that at least the diblock copolymers should be strongly segregated, but the experimental SAXS findings indicate miscibility for PE$_{32}$$^{7.1}$-b-PEO$_{68}$$^{15.1}$ and weak segregation for the PE$_{52}$$^{9.5}$-b-PEO$_{48}$$^{8.8}$.

As the dominant behavior during crystallization is that of break out, the final morphology is that of crystalline lamellae arranged in superstructures like axialites or spherulites. Therefore, we will not explore in detail the morphology of the materials in the melt, as it is destroyed upon crystallization.

3.2. Non-Isothermal Crystallization by DSC

DSC cooling and heating scans of the homopolymers, diblock copolymers, and triblock terpolymers of the two systems (Table 1) are discussed in this section. In addition, all data obtained are collected in Tables S2–S4 (Supporting Information).

Figure 3 shows the cooling (A) and heating (B) DSC scans for the PE$^{7.1}$ homopolymer, PE$_{32}$$^{7.1}$-b-PEO$_{68}$$^{15.1}$ diblock copolymer, and PE$_{22}$$^{7.1}$-b-PEO$_{46}$$^{15.1}$-b-PCL$_{32}$$^{10.4}$ (T1) triblock terpolymer. The crystallization peak of each block (T_c) has been assigned using WAXS data collected under identical conditions at the synchrotron (shown and described below). The same color code is used throughout this work to highlight the crystallization and melting of the different blocks (blue for PCL, red for PEO, and violet for PE). The sharp exotherm (Figure 3A(a)) and subsequent endotherm (Figure 3B(a)) of the neat PE$^{7.1}$ precursor is a consequence of its linear character (synthesized by polyhomologation).

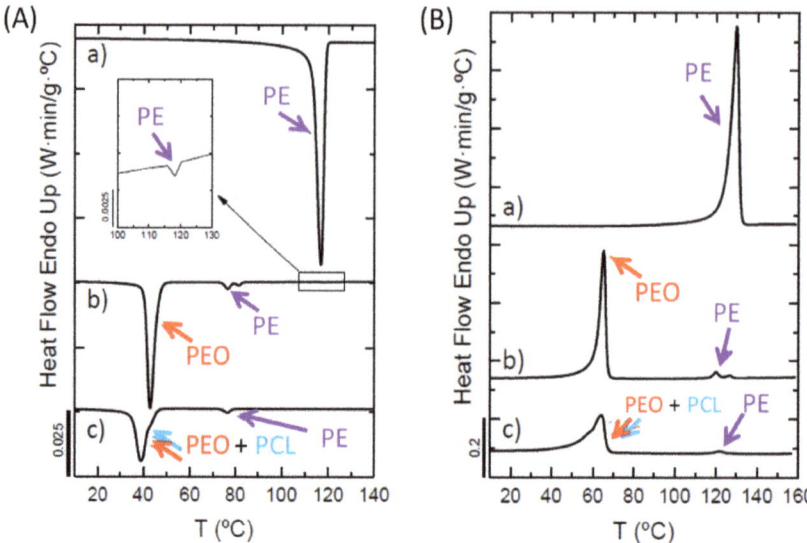

Figure 3. DSC scans at 20 °C/min for (a) $PE^{7.1}$, (b) $PE_{32}^{7.1}$-b-$PEO_{68}^{15.1}$, and (c) $PE_{22}^{7.1}$-b-$PEO_{46}^{15.1}$-b-$PCL_{32}^{10.4}$ (T1) of (**A**) cooling from the melt, with a close-up to notice the very first crystallization exotherm of the PE block and (**B**) subsequent heating with arrows indicating transitions for each block.

In the $PE_{32}^{7.1}$-b-$PEO_{68}^{15.1}$ diblock copolymer, PE (violet arrow) is the first block crystallizing upon cooling from the melt, and then the crystallization of the PEO block (red arrow) occurs (Figure 3Ab). The crystallization of the PE block does not occur in a unique step since three exothermic crystallization peaks appear for the PE block crystallization: at 118 °C, 82 °C, and 79 °C. This evidences that the PE block crystallizes in a fractionated way, which means that several crystallization exotherms appear at lower temperatures instead of a single crystallization exotherm corresponding to the PE block's bulk crystallization temperature. Note that as shown in Figure 1b, this diblock copolymer shows miscibility in the melt, and as crystallization occurs from a homogeneous melt, as well as only having 32 wt % of PE block content and a relatively low molecular weight, the crystallization of the PE block is somehow hindered, as evidenced by its crystallization enthalpy value of 22 J/g (Table S2). However, the sharp crystallization exotherm of the PEO block and the high block content (68 wt %) suggest its high crystallization ability, as the enthalpy for the PEO is 177 J/g (Table S2).

The crystallization in the $PE_{22}^{7.1}$-b-$PEO_{46}^{15.1}$-b-$PCL_{32}^{10.4}$ (T1) triblock terpolymer (Figure 3A(c)) starts with the PE block (violet arrow). In this case, the PE block content is low (22 wt %), and a very small crystallization exotherm is observed in the cooling scan (14 J/g) (Table S2). Crystallization continues with the PCL block (blue arrow) and the PEO block (red arrow). Although the crystallization peaks of the PEO and the PCL blocks are overlapped, WAXS results below demonstrate that the PCL block crystallizes some degrees above the PEO block (Figure 5c). As we are not able to distinguish between both transitions, an estimation of the crystallization enthalpies is reported in Table S2 by employing block content for the calculations.

Figure 3B shows the subsequent heating scans with the endothermic melting peaks (T_m) for each sample; data are collected in Table S3. The homopolymer $PE^{7.1}$ (Figure 3B(a)) shows a crystallinity value of 75% (Table S4), as expected, observing the sharp melting transition. For the diblock copolymer (Figure 3B(b)), melting starts with the PEO block (red) with a crystallinity value of 85%; and it continues with the PE block melting (violet), with a crystallinity value of only 7% (Table S4), because as previously mentioned, small block content and cooling from a homogenous melt are not the best scenarios to enhance crystallization. The overlapped melting peak at the lowest temperature for the triblock terpolymer (Figure 3B(c)) corresponds to the PEO (red) and the PCL (blue) blocks (an estimation of the crystallinity values is provided in Table S4), whereas the melting at the highest temperatures occurs for the PE block crystals, although its crystallinity degree is only 5% (Table S4) of its 32 wt % block content in the terpolymer.

Figure 4 shows the cooling and heating scans of the $PE^{9.5}$ homopolymer, the $PE_{52}{}^{9.5}$-b-$PEO_{48}{}^{8.8}$ diblock copolymer, and the $PE_{37}{}^{9.5}$-b-$PEO_{34}{}^{8.8}$-b-$PCL_{29}{}^{7.6}$ (T2) triblock terpolymer. The crystallization and melting transitions of the blocks in these samples (Figure 4A(c)–B(c)) follow the same trend described before in Figure 3, but with some differences due to the phase behavior of the materials.

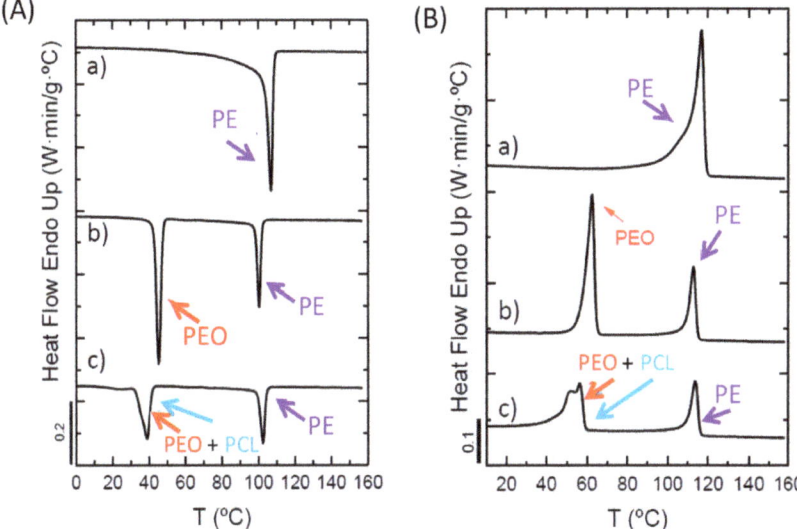

Figure 4. DSC scans at 20 °C/min for (a) $PE^{9.5}$, (b) $PE_{52}{}^{9.5}$-b-$PEO_{48}{}^{8.8}$, and (c) $PE_{37}{}^{9.5}$-b-$PEO_{34}{}^{8.8}$-b-$PCL_{29}{}^{7.6}$ (T2) of (**A**) cooling from the melt and (**B**) subsequent heating with arrows indicating transitions for each block.

The crystallization of $PE^{9.5}$ homopolymer (Figure 4A(a)) occurs in a single and sharp transition. For the $PE_{52}{}^{9.5}$-b-$PEO_{48}{}^{8.8}$ diblock copolymer (Figure 4A(b)), the crystallization of the PE block (violet) occurs at high temperatures, followed by the crystallization of the PEO block (red) at lower temperatures. Note that the PE block crystallizes in a unique crystallization step in this diblock copolymer, not in a fractionated way as in the previous diblock copolymer discussed before (Figure 3A(b)). The difference remains in the phase behavior in the melt, on the one hand, since this diblock copolymer shows weak phase segregation (as evidenced by SAXS experiments shown in Figure 2b), and the fact of being segregated in the melt enhances the crystallization ability of the PE block. In addition, the PE block content is higher in this copolymer (52 wt %) with a higher molecular weight (9500 vs. 7100 g/mol). So, higher PE content and cooling from a segregated melt, do not largely hinder its crystallization, showing a crystallization enthalpy of 81 J/g (Table S2).

The crystallization sequence in the $PE_{37}{}^{9.5}$-b-$PEO_{34}{}^{8.8}$-b-$PCL_{29}{}^{7.6}$ (T2) triblock terpolymer is the same as the one explained in the previous triblock terpolymer (T1) (Figure 3A(c)): first the PE block (violet), and then the PCL (blue) and PEO (red) blocks. Although, also in this case, there is an overlap of the crystallization peaks of the PCL and PEO blocks, WAXS measurements show () that the PEO block crystallizes a few degrees lower than the PCL block; and estimations of the enthalpies are provided in Table S2.

The subsequent heating scans are shown in Figure 4B. The homopolymer $PE^{9.5}$ in Figure 4B(a) shows a clear melting transition and a crystallinity value of 55% (Table S4). In the case of the $PE_{52}{}^{9.5}$ -b- $PEO_{48}{}^{8.8}$ diblock copolymer (Figure 4B(b)), the melting starts with the PEO block (red) and ends with the PE block (violet). As previously mentioned, segregation in the melt and higher PE content enhance its crystallization, and thus, a clear and sharp melting transition with a crystallinity value of 27% is obtained (Table S4). Finally, the $PE_{37}{}^{9.5}$-b-$PEO_{34}{}^{8.8}$-b-$PCL_{29}{}^{7.6}$ (T2) triblock terpolymer follows the same trend as in the triblock terpolymer T1 (Figure 3B(c)): melting of the PEO (red) and PCL (blue) blocks occur with a difference of some degrees, although not enough to distinguish between both DSC melting transitions (demonstrated by WAXS experiments in Figures S3(c)–S4(c)); and melting of the PE block showing a higher crystallinity degree (44%) (Table S3).

3.3. In Situ Wide Angle X-ray Scattering (WAXS) Real-Time Synchrotron Results

The crystallization of each block in the WAXS patterns is identified by analyzing the crystal planes indexing for the PE, PCL, and PEO blocks reported in Table S5 [29,32,38,60,64,66,92,105,106]. In addition, normalized intensity measurements as a function of temperature upon cooling from the melt (at 20 °C/min) are provided, confirming the samples' double and triple crystalline nature.

As shown in Figure 5, all blocks are able to crystallize, as demonstrated by the presence of their characteristic scattering peaks at certain q values, pointed out with the colors we are employing throughout the whole work.

The $PE^{7.1}$ homopolymer crystallization starts at 118 °C (Figure 5a), as its characteristic scattering peak at 15.4 nm^{-1} (violet arrow) corresponding to the (110) crystallographic plane appears at this temperature. Cooling down the sample, at 16.9 nm^{-1}, the other scattering peak of the (200) plane confirms PE crystallization. In addition, the normalized WAXS intensity calculation as a function of temperature for the PE_{110} (15.4 nm^{-1}) reflection in Figure 6a confirms the crystallization of the PE block by the sharp increase of the intensity.

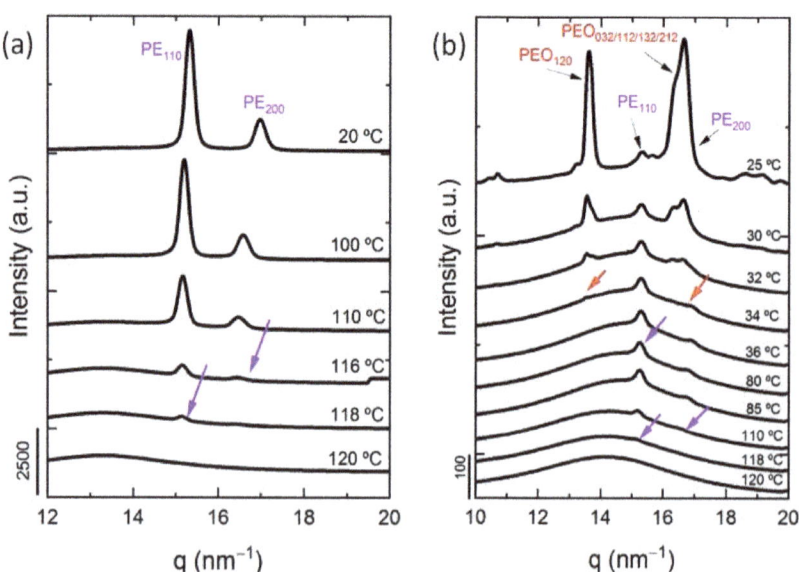

Figure 5. Cont.

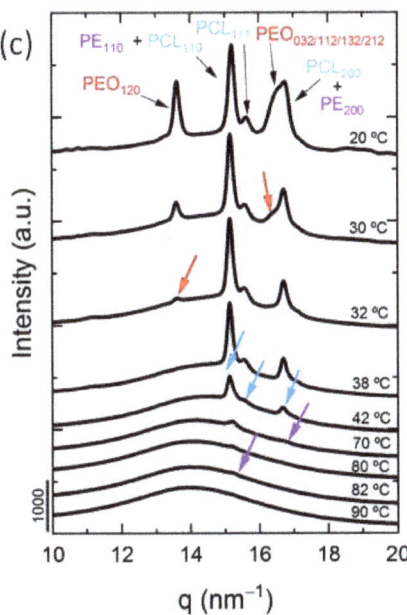

Figure 5. WAXS patterns upon cooling from the melt at 20 °C/min for (a) $PE^{7.1}$, (b) $PE_{32}^{7.1}$-b-$PEO_{68}^{15.1}$, and (c) $PE_{22}^{7.1}$-b-$PEO_{46}^{15.1}$-b-$PCL_{32}^{10.4}$ (T1) at different temperatures with colored arrows indicating crystallization of each block and the (hkl) planes.

Figure 5b shows that the first block to crystallize, during cooling from the melt, in the $PE_{32}^{7.1}$-b-$PEO_{68}^{15.1}$ diblock copolymer is PE at 118 °C (violet arrows) with its scattering peaks at 15.4 and 16.9 nm^{-1} (reflections (110) and (200), respectively). At lower temperatures, 34 °C, the PEO block (red arrows) starts to crystallize with its (120) and (032)/(112)/(132)/(212) reflections at 13.8 and 16.4 nm^{-1}, respectively. Although the crystallization of these two blocks is clear, the normalized WAXS intensities calculated in Figure 6b, show this sequential crystallization by analyzing separately the unique scattering peaks of the PEO_{120} (13.8 nm^{-1}) and the PE_{110} (15.4 nm^{-1}). At high temperatures, the intensity starts to increase at 118 °C due to PE crystallization, and the second increase at 82 °C also corresponds to PE, because as reported in Figure 3A(b), PE crystallizes in two steps.

Figure 5c corresponds to the $PE_{22}^{7.1}$-b-$PEO_{46}^{15.1}$-b-$PCL_{32}^{10.4}$ (T1) triblock terpolymer. In this case, the crystallization sequence starts with the PE crystallization (violet arrows), as evidenced by the PE_{110} reflection at 82 °C and the PE_{200} reflection at 70 °C. One may find this crystallization temperature low for the PE block, but as discussed previously in Figure 3A(c), the PE content is low (22 wt %) and the crystallization enthalpy is 14 J/g. The next block that crystallizes is the PCL block (blue arrows). At 42 °C, the PCL_{110} (15.5 nm^{-1}), PCL_{111} (15.6 nm^{-1}), and PCL_{200} (16.7 nm^{-1}) reflections prove the presence of PCL block crystals. The last block to crystallize upon cooling from the melt is the PEO block (red arrows). The presence of its scattering peak at 13.8 nm^{-1} corresponding to the (120) crystallographic plane at 32 °C confirms the crystallization. At lower temperatures, the other characteristic peak of PEO (16.4 nm^{-1}) appears at 30 °C corresponding to the (032/112/132/212) plane (Figure 5c).

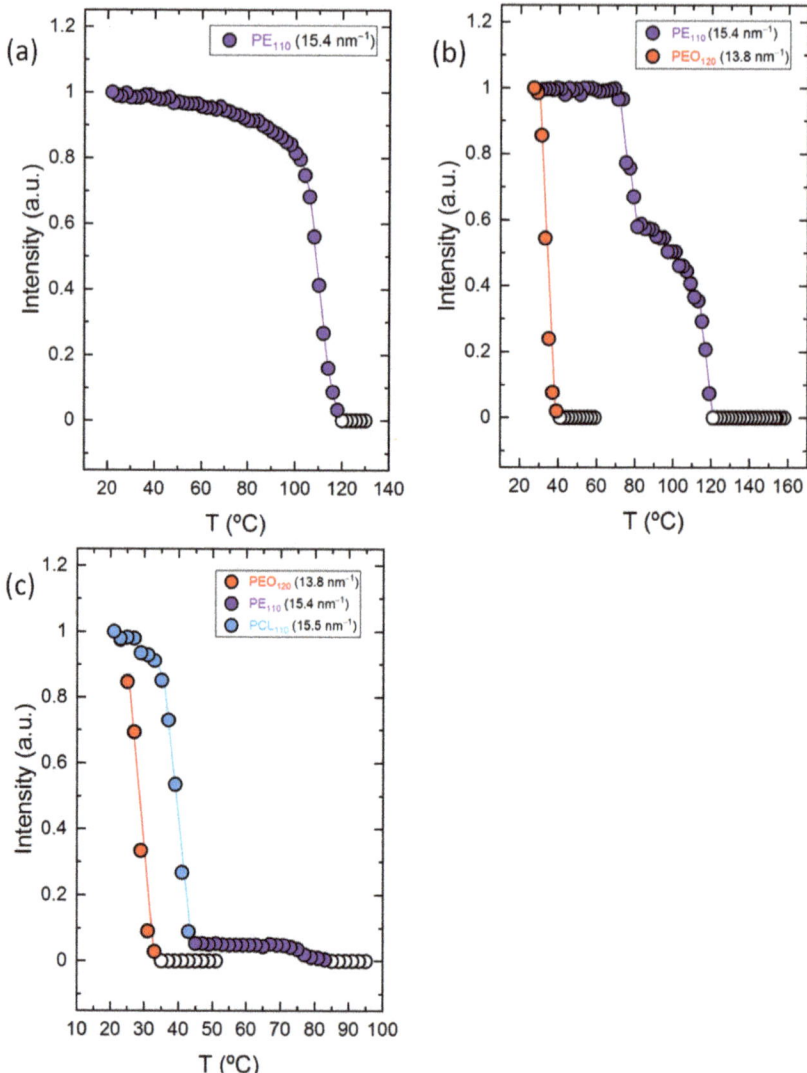

Figure 6. Normalized WAXS intensities as a function of temperature calculated from WAXS data represented in Figure 5 for (**a**) PE$^{7.1}$ (PE$_{110}$ (15.4 nm^{-1})), (**b**) PE$_{32}^{7.1}$-b-PEO$_{68}^{15.1}$ (PE$_{110}$ (15.4 nm^{-1}) and PEO$_{120}$ (13.8 nm^{-1})), and (**c**) PE$_{22}^{7.1}$-b-PEO$_{46}^{15.1}$-b-PCL$_{32}^{10.4}$ (PEO$_{120}$ (13.8 nm^{-1}), PE$_{110}$ (15.4 nm^{-1}) and PCL$_{110}$ (15.5 nm^{-1})) with colored data points and lines indicating crystallization of the corresponding blocks. Empty data points correspond to the molten state.

The normalized intensities are analyzed to detect the exact temperature at which each of the blocks crystallizes (Figure 6c). The joint reflections of PE$_{110}$ (15.4 nm^{-1}) and PCL$_{110}$ (15.5 nm^{-1}) are used to determine their crystallization temperature ranges. The first slight change in intensity at 82 °C confirms PE crystallization (violet), barely noticeable due to the low content of the PE block in the terpolymer (22 wt %). Then, the sharp increase at 42 °C indicates the crystallization of the PCL block (blue). The single PEO$_{120}$ (13.8 nm^{-1}) reflection (along with the other PE and PCL reflections) confirms its crystallization by a sharp increase in intensity.

Similarly, in Figures 7 and 8, WAXS patterns upon cooling the melt (at 20 °C/min) and the normalized intensity measurements confirm crystallization of all blocks in the other set of samples listed in Table 1: the homopolymer $PE^{9.5}$, the diblock copolymer $PE_{52}{}^{9.5}\text{-}b\text{-}PEO_{48}{}^{8.8}$, and the triblock terpolymer $PE_{37}{}^{9.5}\text{-}b\text{-}PEO_{34}{}^{8.8}\text{-}b\text{-}PCL_{29}{}^{7.6}$ (T2).

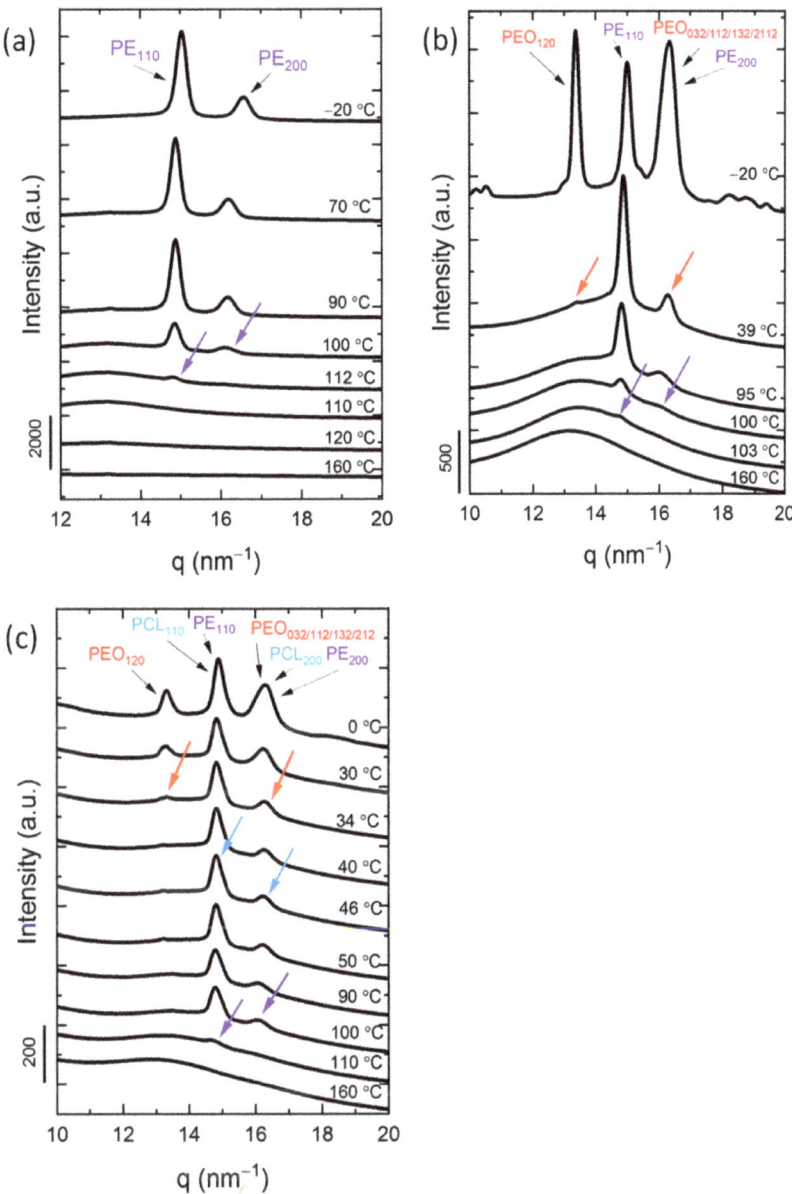

Figure 7. WAXS patterns upon cooling from the melt at 20 °C/min for (**a**) $PE^{9.5}$, (**b**) $PE_{52}{}^{9.5}\text{-}b\text{-}PEO_{48}{}^{8.8}$, and (**c**) $PE_{37}{}^{9.5}\text{-}b\text{-}PEO_{34}{}^{8.8}\text{-}b\text{-}PCL_{29}{}^{7.6}$ (T2) at different temperatures with colored arrows indicating crystallization of each block and (hkl) planes.

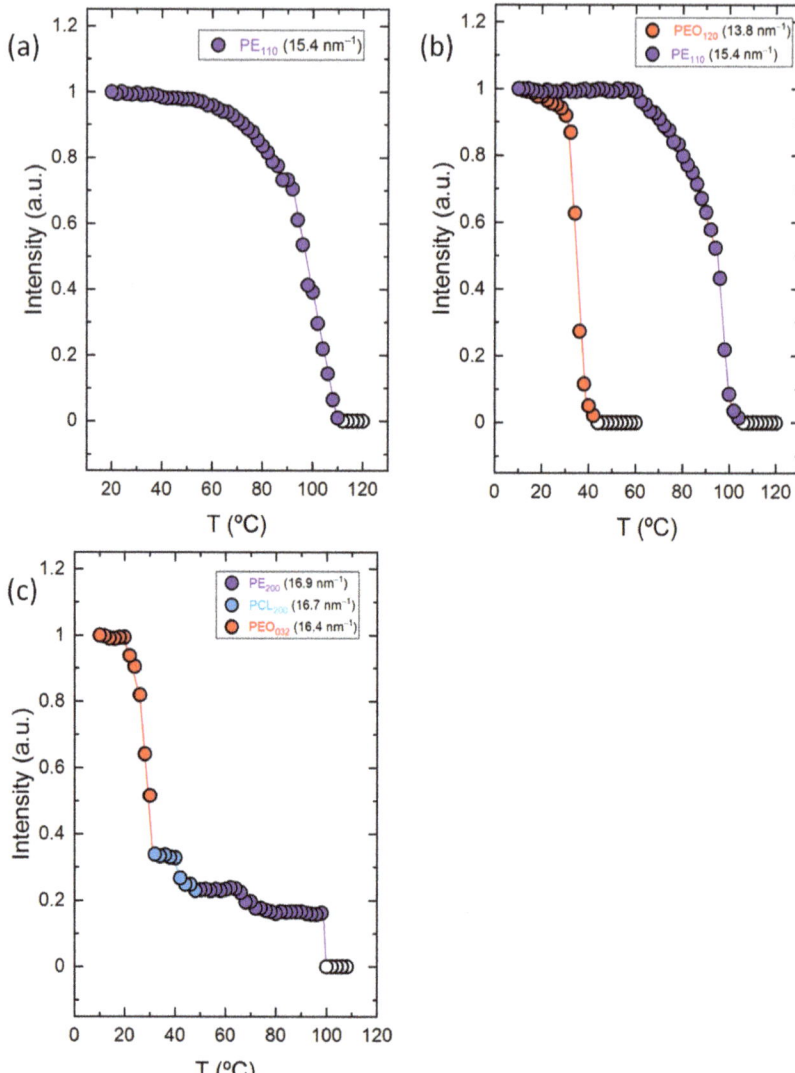

Figure 8. Normalized WAXS intensities as a function of temperature calculated from WAXS data represented in Figure 7 for (**a**) PE$^{9.5}$ (PE$_{110}$ (15.4 nm^{-1})), (**b**) PE$_{52}$$^{9.5}$-$b$-PEO$_{48}$$^{8.8}$ (PE$_{110}$ (15.4 nm^{-1}) and PEO$_{120}$ (13.8 nm^{-1})), and (**c**) PE$_{37}$ $^{9.5}$-b-PEO$_{34}$$^{8.8}$-$b$-PCL$_{29}$$^{7.6}$ (PE$_{200}$ (16.9 nm^{-1}), (PCL$_{1200}$ (16.7nm^{-1}) and (PEO$_{032}$ (16.4 nm^{-1})) with colored data points and lines indicating crystallization of the corresponding blocks. Empty data points correspond to the molten state.

In this case, Figure 7a shows that the crystallization of the homopolymer PE$^{9.5}$ starts at 112 °C (PE$_{110}$ at 15.4 nm^{-1}), and the second scattering peak appears at 100 °C, PE$_{200}$ (16.9 nm^{-1}) (see violet arrows). Figure 8a shows the broad temperature range at which PE crystallizes since a plateau is not reached until approximately 60 °C, determining this way that PE crystallizes in between 112 and 60 °C.

Continuing with Figure 7b, the first reflection at 103 °C ((110) reflection at 15.4 nm^{-1}) corresponds to the PE block, along with the (200) reflection (16.9 nm^{-1}) at 100 °C (see violet arrows). The second block to crystallize in this diblock copolymer at 39 °C is the PEO block (red arrows), identified due to the presence of the (120) reflection at 13.8 nm^{-1} and ((032)/(112)/(132)/(212)) reflections at 16.4 nm^{-1}. Once again, normalized intensities in Figure 8b confirm the temperature ranges at which both the PE and the PEO blocks start to crystallize due to the sharp increase in the intensity of the corresponding peaks.

To conclude, Figure 7c shows the WAXS patterns for the PE$_{37}$$^{9.5}$-b-PEO$_{34}$$^{8.8}$-b PCL$_{29}$$^{7.6}$ (T2) triblock terpolymer. The crystallization sequence remains the same as in the previous triblock terpolymer discussed above (Figure 5c): the PE block first (violet arrows) at 110 °C ((110) and (200) reflections at 15.4 and 16.9 nm^{-1}); then the PCL block (blue arrows) at 46 °C ((110) and (200) reflections at 15.5 and 16.7 nm^{-1}); and finally, the PEO block (red arrows) at 34 °C ((120) and (032)/(112)/(132)/(212) reflections at 13.8 and 16.4 nm^{-1}). In addition, the normalized intensities shown in Figure 8c demonstrate the crystallization of the three blocks by analyzing the joint reflection that the three blocks show at q values between 16.4 and 16.9 nm^{-1}. Note that as the PE content is higher in this triblock terpolymer (T2) (37 wt% vs. 22 wt%), the increase in intensity is clearer than in the previous triblock terpolymer (T1), in which it was very low (Figure 6c).

In addition, to confirm the crystallization of every single block in the cooling scans, results for the subsequent heating scans are shown in the Supporting Information. Figures S1–S4 report WAXS diffraction patterns and normalized intensity measurements of both triblock terpolymers here analyzed (T1 and T2).

3.4. Polarized Light Optical Microscopy (PLOM) Observations

PLOM was employed to follow crystallization of the blocks and to give evidence of the final morphology. Micrographs taken at room temperature (after cooling the samples at 20 °C/min) are shown in Figures 9–12.

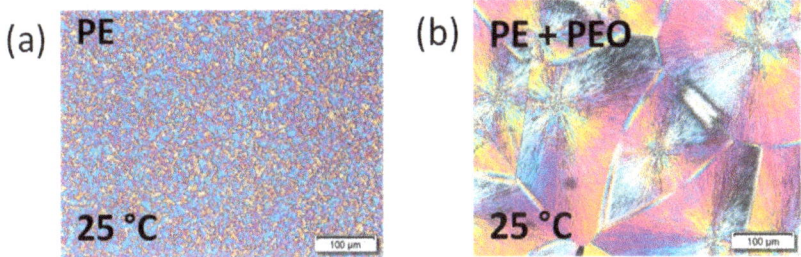

Figure 9. PLOM micrographs taken at room temperature after cooling from the melt at 20 °C/min for (a) PE$^{7.1}$ and (b) PE$_{32}$$^{7.1}$-b-PEO$_{68}$$^{15.1}$, indicating the crystalline phases at 25 °C.

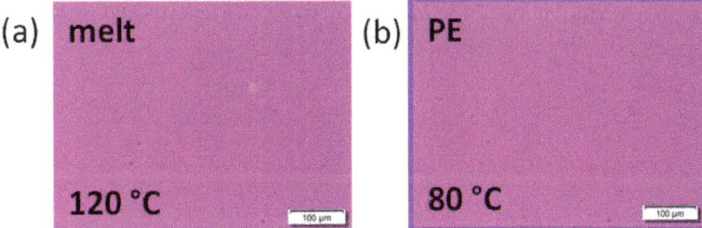

Figure 10. Cont.

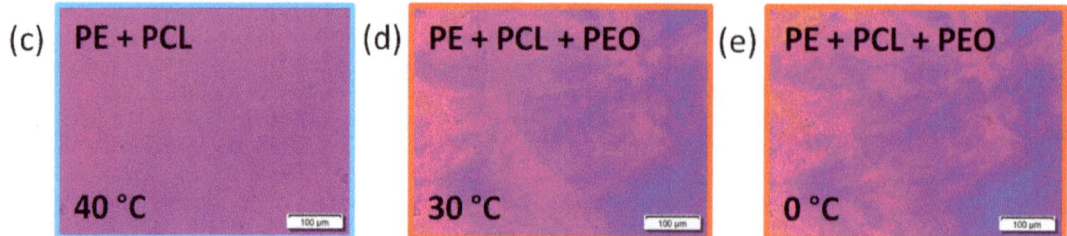

Figure 10. PLOM micrographs of the triblock terpolymer PE$_{22}^{7.1}$-b-PEO$_{46}^{15.1}$-b-PCL$_{32}^{10.4}$ (T1) cooling the sample from the melt at 20 °C/min. Colored squares (violet for PE, blue for PCL, and red for PEO) refer to the crystallized block at their corresponding temperature, indicated at the top of the micrographs for (**a**) molten state at 120 °C, (**b**) PE at 80 °C, (**c**) PE and PCL at 40 °C, (**d**) PE, PCL and PEO at 30 °C, and (**e**) PE, PCL, and PEO at 0 °C.

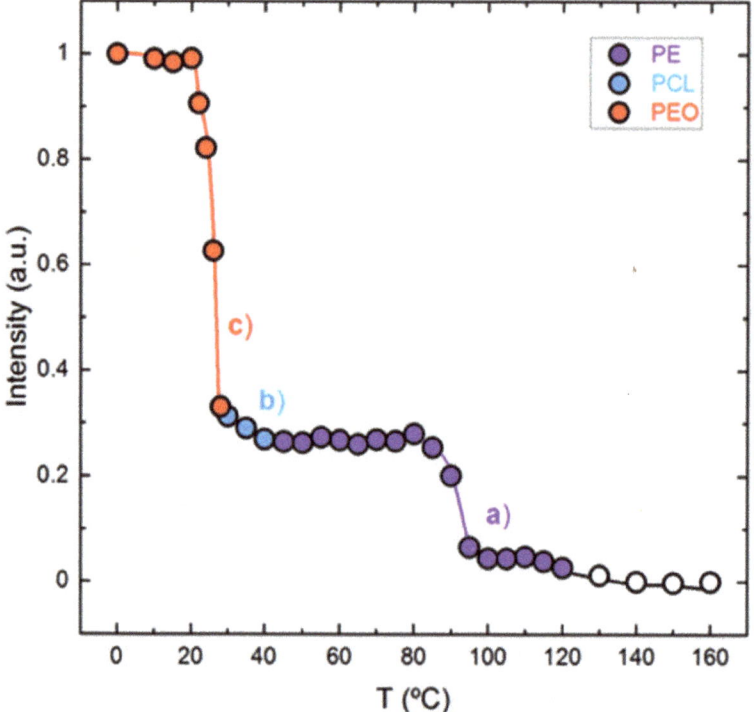

Figure 11. PLOM intensity measurement calculation from data in Figure 10 as a function of temperature during cooling from the melt at 20 °C/min, showing crystallization of (a) the PE block, (b) the PCL block, and (c) the PEO block for the triblock terpolymer PE$_{22}^{7.1}$-b-PEO$_{46}^{15.1}$-b-PCL$_{32}^{10.4}$ (T1). Colored data points and lines (violet for PE, blue for PCL, and red for PEO) are employed in order to follow the crystallization of the blocks. Empty data points correspond to the molten state of the sample.

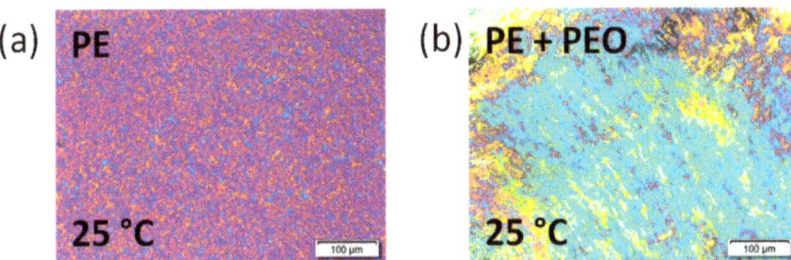

Figure 12. PLOM micrographs taken at room temperature after cooling the samples from the melt at 20 °C/min for (**a**) $PE^{9.5}$ and (**b**) $PE_{52}{}^{9.5}\text{-}b\text{-}PEO_{48}{}^{8.8}$, indicating the crystallized blocks at room temperature.

Figure 9a corresponds to the homopolymer $PE^{7.1}$, showing very small spherulites. In Figure 9b, the $PE_{32}{}^{7.1}\text{-}b\text{-}PEO_{68}{}^{15.1}$ diblock copolymer shows large spherulites characteristic of PEO. According to the evidence gathered in the previous sections, the PE block crystallizes first, probably forming microspherulites that are later engulfed by the much larger PEO block spherulites.

The triple crystalline morphology of the $PE_{22}{}^{7.1}\text{-}b\text{-}PEO_{46}{}^{15.1}\text{-}b\text{-}PCL_{32}{}^{10.4}$ (T1) triblock terpolymer is shown in Figure 10, in which the whole cooling process at 20 °C/min was followed. Figure 10a indicates that the sample at 120 °C is in the molten state. Cooling to 80 °C (Figure 10b), the first block to start to crystallize is the PE block, forming very small and barely observable microspherulites. Due to this difficulty, light intensity measurements as a function of temperature were measured since slight changes in the PLOM micrographs can be better detected.

Figure 11 shows all intensity changes that occur during the cooling scan of this sample. Curve a of Figure 11 shows the increase in intensity related to the crystallization of the PE block, which crystallizes until saturation at 80 °C. Going back to Figure 10c, the second block to crystallize is the PCL block at 40 °C. A slight change is appreciable in this micrograph, but the difference in intensity in curve b of Figure 11 confirms the PCL block crystallization. Finally, Figure 10d,e shows the crystallization of the PEO block, which corresponds to the sharp increase in intensity in curve c of Figure 11. Due to the crystallization of the three blocks, a triple crystalline block copolymer is obtained.

The micrographs taken during the subsequent heating of this $PE_{22}{}^{7.1}\text{-}b\text{-}PEO_{46}{}^{15.1}\text{-}b\text{-}PCL_{32}{}^{10.4}$ (T1) triblock terpolymer are provided in Figure S5 in the SI, along with the normalized intensity calculations as a function of temperature also in the SI (Figure S6). These graphs show the melting of all blocks, demonstrating the triple crystalline behavior of the sample. In addition, all PLOM observations match very well with DSC (Figure 3) and WAXS (Figures 5 and 6) results previously discussed.

Regarding the second system listed in Table 1, the same PLOM observations were performed in order to compare the crystalline behavior of both series of samples. Figure 12 shows the PLOM micrographs at 25 °C of the precursors of the $PE_{37}{}^{9.5}\text{-}b\text{-}PEO_{34}{}^{8.8}\text{-}b\text{-}PCL_{29}{}^{7.6}$ (T2) triblock terpolymer after cooling the samples at a constant rate of 20 °C/min. Figure 12a corresponds to the $PE^{9.5}$ homopolymer, in which very small PE spherulites can be observed. The micrograph in Figure 12b, on the contrary, refers to the $PE_{52}{}^{9.5}\text{-}b\text{-}PEO_{48}{}^{8.8}$ diblock copolymer. Although there are no clear PEO spherulites, it shows a double crystalline morphology at room temperature.

Figure 13 shows the cooling process employing as cooling rate 20 °C/min for the triblock terpolymer $PE_{37}{}^{9.5}\text{-}b\text{-}PEO_{34}{}^{8.8}\text{-}b\text{-}PCL_{29}{}^{7.6}$ (T2). As indicated in Figure 13a, at 118 °C, the sample is melted. Decreasing temperature to 110 °C (Figure 13b), a slight change in the micrograph indicates that the crystallization of the PE block occurred. In addition, Figure 13c shows that all PE has crystallized until saturation at 50 °C. Once again, it is challenging to notice meaningful changes in the micrographs, so the normalized intensity calculations as a function of temperature are provided in Figure 14. The first increase in intensity shows the crystallization of the PE block (curve a of Figure 14). The following slight increase in intensity corresponds to the crystallization of the PCL block (curve b of Figure 14), also shown in Figure 13d at 40 °C. Cooling down the sample, the last block to crystallize is the PEO block (Figure 13e,f), and its crystallization continues until saturation is obtained at approximately 0 °C (Figure 13g). Curve c in Figure 14 indicates that the crystallization of the PEO block starts at around 28 °C and continues with further decreases in temperature.

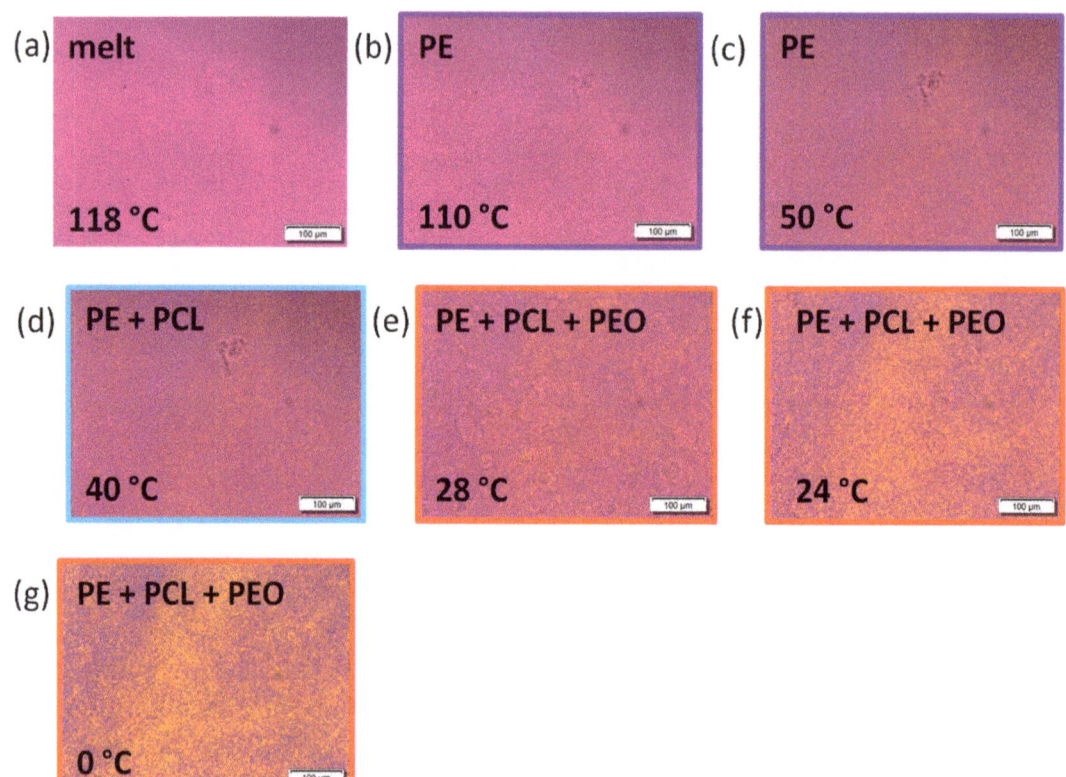

Figure 13. PLOM micrographs of the triblock terpolymer $PE_{37}^{9.5}$-b-$PEO_{34}^{8.8}$-b-$PCL_{29}^{7.6}$ (T2) cooling the sample from the melt at a constant rate of 20 °C/min. Colored squares (violet for PE, blue for PCL, and red for PEO) refer to the crystallized block at the corresponding temperature indicated on the top of the micrographs for (**a**) molten state at 118 °C, (**b**) PE at 110 °C, (**c**) PE at 50 °C, (**d**) PE and PCL at 40 °C, (**e**) PE, PCL, and PEO at 28 °C, (**f**) PE, PCL, and PEO at 24 °C, and (**g**) PE, PCL, and PEO at 0 °C.

Figures S7 and S8 in the SI provide the subsequent heating scan and the normalized intensity measurements of the $PE_{37}^{9.5}$-b-$PEO_{34}^{8.8}$-b-$PCL_{29}^{7.6}$ (T2) triblock terpolymer, respectively. The discussed results agree well with DSC (Figure 4) and WAXS (Figures 7 and 8) according to the evidences discussed above.

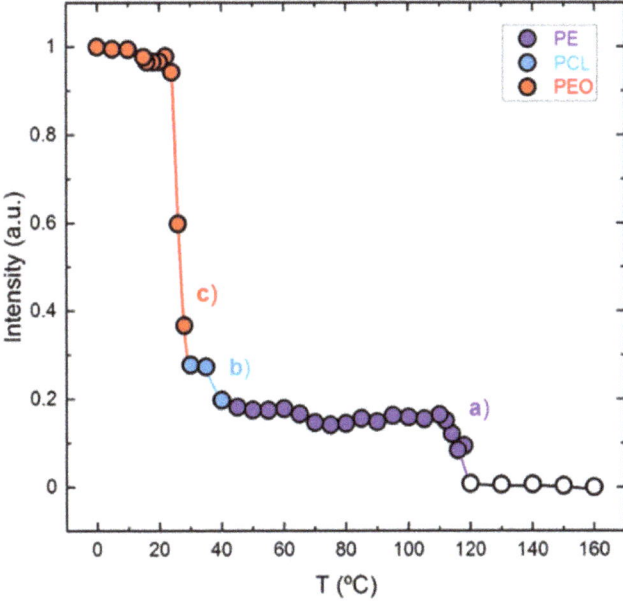

Figure 14. PLOM intensity measurement from data in Figure 13 as a function of temperature during cooling from the melt (20 °C/min), indicating crystallization of the following: (a) the PE block, (b) the PCL block, and (c) the PEO block for the triblock terpolymer PE$_{37}$$^{9.5}$-b-PEO$_{34}$$^{8.8}$-b-PCL$_{29}$$^{7.6}$ (T2). Colored data points and lines (violet for PE, blue for PCL, and red for PEO) are employed to follow the crystallization. Empty data points correspond to the molten state of the sample.

4. Conclusions

The main objective of this study is the analysis of the morphology and crystallization of triblock terpolymers with three potentially crystallizable blocks: the apolar PE and the polar PEO (biocompatible), and PCL (biodegradable) blocks, as well as their corresponding precursors. Although adding a third block to diblock copolymers makes the study more challenging, it was possible to ascertain the crystallization sequence of each of the blocks following the crystallization process by three complementary techniques: DSC, WAXS, and PLOM.

The aim of comparing two triblock terpolymers, PE$_{22}$$^{7.1}$-b-PEO$_{46}$$^{15.1}$-b-PCL$_{32}$$^{10.4}$ (T1) and PE$_{37}$$^{9.5}$-b-PEO$_{34}$$^{8.8}$-b-PCL$_{29}$$^{7.6}$ (T2), was to determine the effect of composition and molecular weight on the properties. Regarding melt miscibility, both triblock terpolymers (T1 and T2) show weak phase segregation, and the microstructure present in the melt is destroyed when crystallization of the first block starts (PE crystallization). Furthermore, the crystallization of the three blocks upon cooling from the melt employing 20 °C/min as cooling rate in both triblock terpolymers is identified. The crystallization sequence resulted as follows: the PE block crystallized first, followed by the PCL block and finally by the PEO block, as evidenced by DSC, in situ WAXS experiments, and PLOM observations with light intensity calculations.

The crystalline behavior of both triblock terpolymers (T1 and T2) is very similar regardless of the molecular weight and composition. However, for their corresponding diblock copolymer precursors, the effect of the PE block content and the molecular weight is significant. The PE$_{32}$$^{7.1}$-b-PEO$_{68}$$^{15.1}$ diblock copolymer is melt miscible, and the PE block crystallization is hindered due to its low content (32 wt%). Nevertheless, in the PE$_{52}$$^{9.5}$-b-PEO$_{48}$$^{8.8}$ diblock copolymer, the PE block crystallization is enhanced due to its higher content (52 wt%) and phase segregated nature in the melt.

The fact that three different blocks can crystallize in a triblock terpolymer forming a triple crystalline material opens a window for new applications, such as drug delivery devices. In this respect, a comprehensive understanding of these materials could be beneficial to tune their crystallizability and obtain new materials with enhanced properties.

Supplementary Materials: The following are available online at https://www.mdpi.com/article/10.3390/polym13183133/s1. Table S1: χ and χN values of diblock copolymers (precursors) and diblock copolymer pairs in the triblock terpolymers, calculated at 180 °C. Table S2: Thermal DSC cooling properties of the homopolymers PE, diblock copolymers PE-b-PEO, and triblock terpolymers PE-b-PEO-b-PCL (T1 and T2). Crystallization enthalpies are normalized according to block content. Table S3: Thermal DSC healing properties of the homopolymers PE, diblock copolymers PE-b-PEO, and triblock terpolymers PE-b-PEO-b-PCL (T1 and T2). Melting enthalpies are normalized according to block content in each of the samples. Table S4: Crystallinity values (%) of the samples calculated from DSC heating scans taking into account the mass fractions of each of the blocks and using $X_c = (\Delta H_m / \Delta H_{m,100\%}) \cdot 100$ and enthalpy of fusion of 100% crystalline polymers ($\Delta H_{m,100\%}$) is taken from literature: 293 J/g for PE [107], 139 J/g for PCL [108] and 214 J/g for PEO [109]. Table S5: WAXS indexation for all the samples [19,92]. Figure S1: WAXS patterns taken during subsequent heating at 20 °C/min for (a) $PE^{7.1}$, (b) $PE_{32}{}^{7.1}$-b-$PEO_{68}{}^{15.1}$, and (c) $PE_{22}{}^{7.1}$-b-$PEO_{46}{}^{15.1}$-b-$PCL_{32}{}^{10.4}$ (T1) at different temperatures with arrows indicating transitions for each block (violet for PE, blue for PCL, and red for PEO) and the corresponding (hkl) planes of the blocks. Figure S2: Normalized WAXS intensities as a function of temperature calculated from heating WAXS data in Figure S1 with close-ups for (a) $PE^{7.1}$, (b) $PE_{32}{}^{7.1}$-b-$PEO_{68}{}^{15.1}$, and (c) $PE_{22}{}^{7.1}$-b-$PEO_{46}{}^{15.1}$-b-$PCL_{32}{}^{10.4}$ (T1). Colored data points and lines (violet for PE, blue for PCL, and red for PEO) are employed to follow the crystallization of each block. Empty data points represent the molten state of the corresponding block in the samples. Figure S3: WAXS patterns taken during subsequent heating at 20 °C/min for (a) $PE^{9.5}$, (b) $PE_{52}{}^{9.5}$-b-$PEO_{48}{}^{8.8}$, and (c) $PE_{37}{}^{9.5}$-b-$PEO_{34}{}^{8.8}$-b-$PCL_{29}{}^{7.6}$ (T2) at different temperatures with arrows indicating transitions for each block (violet for PE, blue for PCL and red for PEO) and the corresponding (hkl) planes of the blocks. Figure S4: Normalized WAXS intensities as a function of temperature calculated from heating WAXS data in Figure S3 for (a) $PE^{9.5}$, (b) $PE_{52}{}^{9.5}$-b-$PEO_{48}{}^{8.8}$, and (c) $PE_{37}{}^{9.5}$-b-$PEO_{34}{}^{8.8}$-b-$PCL_{29}{}^{7.6}$ (T2). Colored data points and lines (violet for PE, blue for PCL, and red for PEO) are employed to follow the crystallization of each block. Empty data points represent the molten state of the corresponding block in the samples. Figure S5: PLOM subsequent heating micrographs from 0 °C to the melt at 20 °C/min for the triblock $PE_{22}{}^{7.1}$-b-$PEO_{46}{}^{15.1}$-b-$PCL_{32}{}^{10.4}$ (T1) with colored boxes indicating the crystallization of each of the blocks (violet for PE, blue for PCL and red for PEO) and the crystallized blocks in each of the micrographs for (a) PE, PCL, and PEO at 0 °C, (b) PE, PCL, and PEO at 25 °C, (c) PE, PCL, and PEO at 50 °C, (d) PE, PCL, and PEO at 70 °C, (e) PE at 72 °C, (f) PE at 125 °C, and (g) molten state at 130 °C. Figure S6: PLOM intensity measurements from micrographs of Figure S5 as a function of temperature indicating melting of the (a) PEO block, (b) PCL block, and (c) PE block for the triblock terpolymer $PE_{22}{}^{7.1}$-b-$PEO_{46}{}^{15.1}$-b-$PCL_{32}{}^{10.4}$ (T1) with colored data points and lines (red for PEO, blue for PCL and violet for PE) to follow the crystallization of each block. Empty data points represent the molten state of the sample. Figure S7: PLOM subsequent heating micrographs from 10 °C to the melt at 20 °C/min for the triblock $PE_{37}{}^{9.5}$-b-$PEO_{34}{}^{8.8}$-b-$PCL_{29}{}^{7.6}$ (T2) with colored boxes indicating the crystallization of each of the blocks (violet for PE, blue for PCL and red for PEO) and the crystallized blocks in each of the micrographs for (a) PE, PCL, and PEO at 10 °C, (b) PE, PCL, and PEO at 60 °C, (c) PE and PCL at 65 °C, (d) PE and PCL at 70 °C, (e) PE at 75 °C, (f) PE at 130 °C, (g) PE a 145 °C, and h) molten state at 150 °C. Figure S8: PLOM intensity measurements from micrographs of Figure S7 as a function of temperature indicating melting of the (a) PEO block, (b) PCL block, and (c) PE block for the triblock terpolymer $PE_{37}{}^{9.5}$-b-$PEO_{34}{}^{8.8}$-b-$PCL_{29}{}^{7.6}$ (T2) with colored data points and lines (red for PEO, blue for PCL and violet for PE) to follow the crystallization of each block. Empty data points represent the molten state of the sample.

Author Contributions: This work and its conceptualization was designed by A.J.M. and N.H. All materials (homopolymers, diblock copolymers, and triblock terpolymers) were synthesized by V.L. and G.Z. under the supervision of N.H. Experiments were performed at the UPV/EHU labs and the ALBA synchrotron facility by E.M. under the supervision of A.M. and A.J.M., D.C. and M.Z. helped with the interpretation of the data. The article was written by E.M. and A.J.M., and it was revised by all co-authors. All authors have read and agreed to the published version of the manuscript.

Funding: This research received funding from MINECO through projects MAT2017-83014-C2-1-P, from the Basque Government through grant IT1309-19, and from ALBA synchrotron facility through granted proposal u2020084441 (March 2020). We would like to thank the financial support provided by the BIODEST project; this project has received funding from the European Union's Horizon 2020 research and innovation program under the Marie Sklodowska-Curie grant agreement no. 778092. GZ, VL, and NH wish to acknowledge the support of KAUST.

Institutional Review Board Statement: Not applicable.

Informed Consent Statement: Not applicable.

Data Availability Statement: The present data in this research is available upon request from the corresponding author.

Conflicts of Interest: The authors declare no conflict of interest.

References

1. Hamley, I. *Crystallization in Block Copolymers*; Advances in Polymer Science; Springer: Cham, Switzerland, 1999; Volume 148.
2. Abetz, V.; Simon, P.F.W. *Phase Behavior and Morphologies of Block Copolymers*; Advances in Polymer Science; Springer: Cham, Switzerland, 2005; Volume 189, pp. 125–212.
3. Müller, A.J.; Balsamo, V.; Arnal, M.L. Nucleation and crystallization in diblock and triblock copolymers. In *Block Copolymers II*; Abetz, V., Ed.; Advances in Polymer Science; Springer: Berlin/Heidelberg, Germany, 2005; Volume 190, pp. 1–63.
4. Müller, A.J.; Arnal, M.L.; Balsamo, V. Crystallization in block copolymers with more than one crystallizable block. In *Progress in Understanding of Polymer Crystallization*; Reiter, G., Strobl, G.R., Eds.; Lecture Notes in Physics; Springer: Berlin/Heidelberg, Germany, 2007; Volume 714, pp. 229–259.
5. Michell, R.M.; Müller, A.J. Confined crystallization of polymeric materials. *Prog. Polym. Sci.* **2016**, *54–55*, 183–216. [CrossRef]
6. Nakagawa, S.; Marubayashi, H.; Nojima, S. Crystallization of polymer chains confined in nanodomains. *Eur. Polym. J.* **2015**, *70*, 262–275. [CrossRef]
7. Castillo, R.V.; Müller, A.J. Crystallization and morphology of biodegradable or biostable single and double crystalline block copolymers. *Prog. Polym. Sci.* **2009**, *34*, 516–560. [CrossRef]
8. Li, S.; Register, A. Crystallization in Copolymers. In *Handbook of Polymer Crystallization*; Piorkowska, E., Rutledge, G.C., Eds.; John Wiley and Sons: Hoboken, NJ, USA, 2013; p. 327.
9. Huang, S.; Jiang, S. Structures and morphologies of biocompatible and biodegradable block copolymers. *RSC Adv.* **2014**, *4*, 24566–24583. [CrossRef]
10. Hamley, I.W. *The Physics of Block Copolymers*; Oxford University Press: Oxford, UK, 1998.
11. Arif, M.; Kalarikkal, N.; Thomas, S. Introduction on crystallization in multiphase polymer systems. In *Crystallization in Multiphase Polymer Systems*; Thomas, S., Arif, M., Gowd, E.B., Eds.; Elsevier: Amsterdam, The Netherlands, 2018; pp. 1–13.
12. Hadjichristidis, N.; Pitsikalis, M.; Iatrou, H. *Synthesis of Block Copolymers*; Advances in Polymer Science; Springer: Cham, Switzerland, 2005; Volume 189, pp. 1–124.
13. Barthel, M.J.; Schacher, F.H.; Schubert, U.S. Poly (ethylene oxide)(PEO)-based ABC triblock terpolymers—Synthetic complexity vs. application benefits. *Polym. Chem.* **2014**, *5*, 2647. [CrossRef]
14. Guo, X.; Wang, L.; Wei, X.; Zhou, S. Polymer-based drug delivery systems for cancer treatment. *J. Polym. Sci. Part A Polym. Chem.* **2016**, *54*, 3525–3550. [CrossRef]
15. Van Horn, R.M.; Steffen, M.R.; O'connor, D. Recent progress in block copolymer crystallization. *Polym. Cryst.* **2018**, *1*, e10039. [CrossRef]
16. Palacios, J.P.; Mugica, A.; Zubitur, M.; Müller, A.J. Crystallization and morphology of block copolymers and terpolymers with more than one crystallizable block. In *Crystallization in Multiphase Polymer Systems*; Sabu, T., Mohammed, A.P., Bhoje, G.E., Nandajumar, K., Eds.; Elsevier: Amsterdam, The Netherlands, 2018; pp. 123–171.
17. Palacios, J.K.; Zhang, H.; Zhang, B.; Hadjichristidis, N.; Müller, A.J. Direct identification of three crystalline phases in PEO-*b*-PCL-*b*-PLLA triblock terpolymer by in situ hot-stage atomic force microscopy. *Polymer* **2020**, *205*, 122863. [CrossRef]
18. Müller, A.J.; Arnal, M.L.; Lorenzo, A.T. Crystallization in nano-confined polymeric systems. In *Handbook of Polymer Crystallization*; Piorkowska, E., Rutledge, G.C., Eds.; John Wiley and Sons: Hoboken, NJ, USA, 2013; pp. 347–372.
19. Castillo, R.V.; Müller, A.J.; Lin, M.C.; Chen, H.L.; Jeng, U.S.; Hillmyer, M.A. Confined crystallization and morphology of melt segregated PLLA-*b*-PE and PLDA-*b*-PE diblock copolymers. *Macromolecules* **2012**, *45*, 4254–4261. [CrossRef]
20. Müller, A.J.; Castillo, R.V.; Hillmyer, M. Nucleation and crystallization of PLDA-*b*-PE and PLLA-*b*-PE diblock copolymers. *Macromol. Symp.* **2006**, *242*, 174–181. [CrossRef]
21. Müller, A.J.; Lorenzo, A.T.; Castillo, R.V.; Arnal, M.L.; Boschetti-de-Fierro, A.; Abetz, V. Crystallization kinetics of homogeneous and melt segregated PE containing diblock copolymers. *Macromol. Symp.* **2006**, *245–246*, 154–160. [CrossRef]

22. Lin, M.C.; Wang, Y.C.; Chen, J.H.; Chen, H.L.; Müller, A.J.; Su, C.J.; Jeng, U.S. Orthogonal crystal orientation in double-crystalline block copolymer. *Macromolecules* **2011**, *44*, 6875–6884. [CrossRef]
23. Bao, J.; Dong, X.; Chen, S.; Lu, W.; Zhang, X.; Chen, W. Confined crystallization, melting behavior and morphology in PEG-*b*-PLA diblock copolymers: Amorphous versus crystalline PLA. *J. Polym. Sci.* **2020**, *58*, 455–456. [CrossRef]
24. Bao, J.; Dong, X.; Chen, S.; Lu, W.; Zhang, X.; Chen, W. Fractionated crystallization and fractionated melting behaviors of poly(ethylene glycol) induced by poly(lactide) stereocomplex in their block copolymers and blends. *Polymer* **2020**, *190*, 122189. [CrossRef]
25. Ring, J.O.; Thomann, R.; Mülhaupt, R.; Raquez, J.-M.; Degée, P.; Dubois, P. Controlled synthesis and characterization of Poly[ethylene-*block*-(L,L-lactide)]s by combining catalytic ethylene oligomerization with "Coordination-insertion" ring-opening polymerization. *Macromol. Chem. Phys.* **2007**, *208*, 896–902. [CrossRef]
26. Müller, A.J.; Albuerne, J.; Marquez, L.; Raquez, J.M.; Degée, P.; Dubois, P.; Hobbs, J.; Hamley, I.W. Self-nucleation and crystallization kinetics of double crystalline poly(p-dioxanone)-*b*-poly(ε-caprolactone) diblock copolymers. *Faraday Discuss* **2005**, *128*, 231–252. [CrossRef]
27. Müller, A.J.; Albuerne, J.; Esteves, L.M.; Marquez, L.; Raquez, J.-M.; Degée, P.; Dubois, P.; Collins, S.; Hamley, I.W. Confinement effects on the crystallization kinetics and self-nucleation of double crystalline poly(p-dioxanone)-*b*-poly(ε-caprolactone) diblock copolymers. *Macromol. Symp.* **2004**, *215*, 369–382. [CrossRef]
28. Albuerne, J.; Máquez, L.; Müller, A.J.; Raquez, J.M.; Degée, P.; Dubois, P.; Castelleto, V.; Hamley, I.W. Nucleation and crystallization in double crystalline poly(p-dioxanone)-*b*-poly(ε-caprolactone) diblock copolymers. *Macromolecules* **2003**, *36*, 1633–1644. [CrossRef]
29. Hamley, I.W.; Parras, P.; Castelletto, V.; Castillo, R.V.; Müller, A.J.; Pollet, E.; Dubois, P.; Martin, C.M. Melt structure and its transformation by sequential crystallization of the two blocks within poly(L-lactide)-*block*- poly(ε-caprolactone) double crystalline diblock copolymers. *Macromol. Chem. Phys.* **2006**, *207*, 941–953. [CrossRef]
30. Castillo, R.V.; Müller, A.J.; Raquez, J.M.; Dubois, P. Crystallization kinetics and morphology of biodegradable double crystalline PLLA-*b*-PCL diblock copolymers. *Macromolecules* **2010**, *43*, 4149–4160. [CrossRef]
31. Laredo, E.; Prutsky, N.; Bello, A.; Grimau, M.; Castillo, R.V.; Müller, A.J.; Dubois, P. Miscibility in poly(L-lactide)-*b*-poly(ε-caprolactone) double crystalline diblock copolymers. *Eur. Phys. J. E* **2007**, *23*, 295–303. [CrossRef]
32. Hamley, I.W.; Castelletto, V.; Castillo, R.W.; Müller, A.J.; Martin, C.M.; Pollet, E.; Dubois, P. Crystallization in poly(ε-lactide)-*b*-poly(ε-caprolactone) double crystalline diblock copolymers: A study using X-ray scattering, differential scanning calorimetry and polarized optical microscopy. *Macromolecules* **2005**, *38*, 463–472. [CrossRef]
33. Myers, S.B.; Register, R.A. Crystalline-crystalline diblock copolymers of linear polyethylene and hydrogenated polynorbornene. *Macromolecules* **2008**, *41*, 6773–6779. [CrossRef]
34. Ponjavic, M.; Nikolic, M.S.; Jevtic, S.; Rogan, J.; Stevanovic, S.; Djonlagic, J. Influence of a low content of PEO segment on the thermal, surface and morphological properties of triblock and diblock PCL copolymers. *Macromol. Res.* **2016**, *24*, 323–335. [CrossRef]
35. Li, L.; Meng, F.; Zhong, Z.; Byelov, D.; De Jeu, W.H.; Feijen, J. Morphology of a highly asymmetric double crystallizable poly(ε-caprolactone-*b*-ethylene oxide) block copolymer. *J. Chem. Phys.* **2007**, *126*, 024904. [CrossRef]
36. Van Horn, R.M.; Zheng, J.X.; Sun, H.J.; Hsiao, M.S.; Zhang, W.B.; Dong, X.H.; Xu, J.; Thomas, E.L.; Lotz, B.; Chen, S.Z.D. Solution crystallization behavior of crystalline-crystalline diblock copolymers of poly(ethylene oxide)-*block*-poly(ε-caprolactone). *Macromolecules* **2010**, *43*, 6113–6119. [CrossRef]
37. Vivas, M.; Contreras, J.; López-Carrasquero, F.; Lorenzo, A.T.; Arnal, M.L.; Balsamo, V.; Müller, A.J.; Laredo, E.; Schmalz, H.; Abetz, V. Synthesis and characterization of triblock terpolymers with three potentially crystallisable blocks: Polyethylene-*b*-poly(ethylene oxide)-*b*-poly(ε-caprolactone). *Macrolecular Symp.* **2006**, *239*, 58–67. [CrossRef]
38. Jiang, S.; He, C.; An, L.; Chen, X.; Jiang, B. Crystallization and ring-banded spherulite morphology of poly(ethylene oxide)-*block*-poly(ε-caprolactone) diblock copolymer. *Macromol. Chem. Phys.* **2004**, *205*, 2229–2234. [CrossRef]
39. Arnal, M.L.; López-Carrasquero, F.; Laredo, E.; Müller, A.J. Coincident or sequential crystallization of PCL and PEO blocks within polystyrene-*b*-poly(ethylene oxide)-*b*-poly(ε-caprolactone) linear triblock copolymers. *Eur. Polym. J.* **2004**, *40*, 1461–1476. [CrossRef]
40. Nojima, S.; Ono, M.; Ashida, T. Crystallization of block copolymers II. Morphological study of poly(ethylene glycol)-poly(ε-caprolactone) block copolymers. *Polym. J.* **1992**, *24*, 1271–1280. [CrossRef]
41. Wei, Z.; Liu, L.; Yu, F.; Wang, P.; Qi, M. Synthesis and characterization of poly(ε-caprolactone)-*b*-poly(ethylene glycol)-*b*-poly(ε-caprolactone) triblock copolymers with dibutylmagnesium as catalyst. *J. Appl. Polym. Sci.* **2009**, *111*, 429–436. [CrossRef]
42. Arnal, M.L.; Balsamo, V.; López-Carrasquero, F.; Contreras, J.; Carrillo, M.; Schmalz, H.; Abetz, V.; Laredo, E.; Müller, A.J. Synthesis and characterization of polystyrene-*b*-poly(ethylene oxide)-*b*-poly(ε-caprolactone) block copolymers. *Macromolecules* **2001**, *34*, 7973–7982. [CrossRef]
43. Li, Y.; Zhou, J.; Zhang, J.; Gou, Q.; Gu, Q.; Wang, Z. Morphology of poly(ethylene oxide)-*b*-poly(ε-caprolactone) spherulites formed under compressed CO_2. *J. Macromol. Sci. Part B Phys.* **2014**, *53*, 1137–1144. [CrossRef]
44. Li, Y.; Huang, H.; Wang, Z.; He, T. Tuning radial lamellar packing and orientation into diverse ring-banded spherulites: Effects of structural feature and crystallization condition. *Macromolecules* **2014**, *47*, 1783–1792. [CrossRef]

45. Xue, F.-F.; Chen, X.-S.; An, L.-J.; Funari, S.S.; Jiang, S.-C. Confined lamella formation in crystalline-crystalline poly(ethylene-oxide)-*b*-poly(ε-caprolactone) diblock copolymers. *Chin. J. Polym. Sci.* **2013**, *31*, 1260–1270. [CrossRef]
46. Xue, F.; Chen, X.; An, L.; Funari, S.S.; Jiang, S. Soft nanoconfinement effects on the crystallization behavior of asymmetric poly(ethylene oxide)-*block*-poly(ε-caprolactone) diblock copolymers. *Polym. Int.* **2012**, *61*, 909–917. [CrossRef]
47. Sun, J.; He, C.; Zhuang, X.; Jing, X.; Chen, X. The crystallization behavior of poly(ethylene glycol)-poly(ε-caprolactone) diblock copolymers with asymmetric block compositions. *J. Polym. Res.* **2011**, *18*, 2161–2168. [CrossRef]
48. Hua, C.; Dong, C.-M. Synthesis, characterization, effect of architecture on crystallization of biodegradable poly(ε-caprolactone)-*b*-poly(ethylene oxide) copolymers with different arms and nanoparticles thereof. *J. Biomed. Mater. Res. Part A* **2007**, *82*, 689–700. [CrossRef]
49. He, C.; Sun, J.; Zhao, T.; Hong, Z.; Zhuang, X.; Chen, X.; Jin, X. Formation of a unique crystal morphology for the poly(ethylene glycol)-poly(ε-caprolactone) diblock copolymer. *Biomacromolecules* **2006**, *7*, 252–258. [CrossRef]
50. Piao, L.; Dai, Z.; Deng, M.; Chen, X.; Jing, X. Synthesis and characterization of PCL/PEG/PCL triblock copolymers by using calcium catalyst. *Polymer* **2003**, *44*, 2025–2031. [CrossRef]
51. He, C.; Sun, J.; Deng, C.; Zhao, T.; Deng, M.; Chen, X.; Jing, X. Study of the synthesis, crystallization, and morphology of poly(ethylene glycol)-poly(ε-caprolactone) diblock copolymers. *Biomacromolecules* **2004**, *5*, 2040–2047. [CrossRef]
52. Arnal, M.L.; Boissé, S.; Müller, A.J.; Meyer, F.; Raquez, J.M.; Dubois, P.; Prud'homme, R.E. Interplay between poly(ethylene oxide) and poly(ε-lactide) blocks during diblock copolymer crystallization. *CrystEngComm* **2016**, *18*, 3635–3649. [CrossRef]
53. Zhou, D.; Sun, J.; Shao, J.; Bian, X.; Huang, S.; Li, G.; Chen, X. Unusual crystallization and melting behavior induced by microphase separation in MPEG-*b*-PLLA diblock copolymer. *Polymer* **2015**, *80*, 123–129. [CrossRef]
54. Yang, J.; Liang, Y.; Han, C.C. Effect of crystallization temperature on the interactive crystallization behavior of poly(L-lactide)-*block*-poly (ethylene glycol) copolymer. *Polymer* **2015**, *79*, 56–64. [CrossRef]
55. Huang, S.; Li, H.; Jiang, S.; Chen, X.; An, L. Morphologies and structures in poly(L-lactide-*b*-ethylene oxide) copolymers determined by crystallization, microphase separation and vitrification. *Polym. Bull.* **2011**, *67*, 885–902. [CrossRef]
56. Huang, L.; Kiyofuji, G.; Matsumoto, J.; Fukagawa, Y.; Gong, C.; Nojima, S. Isothermal crystallization of poly (b-propiolactone) blocks starting from lamellar microdomain structures of double crystalline poly (b-propiolactone)-*block*-polyethylene copolymers. *Polymer* **2012**, *53*, 5856–5863. [CrossRef]
57. Sun, J.; Hong, Z.; Yang, L.; Tang, Z.; Chen, X.; Jing, X. Study on crystalline morphology of poly(L-lactide)-poly(ethylene glycol) diblock copolymer. *Polymer* **2004**, *45*, 5969–5977. [CrossRef]
58. Shin, D.; Shin, K.; Aamer, K.A.; Tew, G.N.; Russell, T.P.; Lee, J.H.; Jho, J.Y. A morphological study of a semicrystalline poly(ε-lactic acid-*b*-ethylene oxide-*b*-L-lactic acid) triblock copolymer. *Macromolecules* **2005**, *38*, 104–109. [CrossRef]
59. Xue, F.; Chen, X.; An, L.; Funari, S.S.; Jiang, S. Crystallization induced layer-to-layer transitions in symmetric PEO-*b*-PLLA block copolymer with synchrotron simultaneous SAXS/WAXS investigations. *RSC Adv.* **2014**, *4*, 56346–56354. [CrossRef]
60. Huang, S.; Jiang, S.; An, L.; Chen, X. Crystallization and morphology of poly(ethylene oxide-*b*-lactide) crystalline-crystalline diblock copolymers. *J. Polym. Sci. Part B Polym. Phys.* **2008**, *46*, 1400–1411. [CrossRef]
61. Yang, J.; Zhao, T.; Zhou, Y.; Liu, L.; Li, G.; Zhou, E.; Chen, X. Single crystals of the poly(L-lactide) block and the poly(ethylene glycol) block in poly(L-lactide)-poly(ethylene glycol) diblock copolymer. *Macromolecules* **2007**, *40*, 2791–2797. [CrossRef]
62. Cai, C.; Wang, L.U.; Dong, C.M. Synthesis, characterization, effect of architecture on crystallization, and spherulitic growth of poly(L-lactide)-*b*-poly(ethylene oxide) copolymers with different branch arms. *J. Polym. Sci. Part A Polym. Chem.* **2006**, *44*, 2034–2044. [CrossRef]
63. Yang, J.; Zhao, T.; Cui, J.; Liu, L.; Zhou, Y.; Li, G.; Zhou, E.; Chen, X. Nonisothermal crystallization behavior of the poly(ethylene glycol) block in poly(L-lactide)-poly(ethylene glycol) diblock copolymers: Effect of the poly(ε-lactide) block length. *J. Polym. Sci. Part B Polym. Phys.* **2006**, *44*, 3215–3226. [CrossRef]
64. Huang, C.I.; Tsai, S.H.; Chen, C.M. Isothermal crystallization behavior of poly(L-lactide) in poly(L-lactide)-*block*-poly(ethylene glycol) diblock copolymers. *J. Polym. Sci. Part B Polym. Phys.* **2006**, *44*, 2438–2448. [CrossRef]
65. Kim, K.S.; Chung, S.; Chin, I.J.; Kim, M.N.; Yoon, J.S. Crystallization behavior of biodegradable amphiphilic poly(ethylene glycol)-poly(L-lactide) block copolymers. *J. Appl. Polym. Sci.* **1999**, *72*, 341–348. [CrossRef]
66. Wang, J.L.; Dong, C.M. Synthesis, sequential crystallization and morphological evolution of well-defined star-shaped poly(ε-caprolactone)-*b*-poly(L-lactide) block copolymer. *Macromol. Chem. Phys.* **2006**, *207*, 554–562. [CrossRef]
67. Liénard, R.; Zaldua, N.; Josse, T.; Winter, J.D.; Zubitur, M.; Mugica, A.; Iturrospe, A.; Arbe, A.; Coulembier, O.; Müller, A.J. Synthesis and characterization of double crystalline cyclic diblock copolymers of poly(ε-caprolactone) and poly(L(D)-lactide) (c(PCL-*b*-PL(D)LA)). *Macromol. Rapid Commun.* **2016**, *37*, 1676–1681. [CrossRef] [PubMed]
68. Navarro-Baena, I.; Marcos-Fernández, A.; Fernández-Torres, A.; Kenny, J.M.; Peponi, L. Synthesis of PLLA-*b*-PCL-*b*-PLLA linear tri-block copolymers and their corresponding poly(ester-urethane)s: Effect of the molecular weight on their crystallisation and mechanical properties. *RSC Adv.* **2014**, *4*, 8510–8524. [CrossRef]
69. Peponi, L.; Navarro-Baena, I.; Báez, J.E.; Kenny, J.M.; Marcos-Fernández, A. Effect of the molecular weight on the crystallinity of PCL-*b*-PLLA di-block copolymers. *Polymer* **2012**, *53*, 4561–4568. [CrossRef]
70. Ho, R.-M.; Hsieh, P.-Y.; Tseng, W.-H.; Lin, C.-C.; Huang, B.-H.; Lotz, B. Crystallization-induced orientation for microstructures of Poly(L-lactide)-*b*-poly(ε-caprolactone) diblock copolymers. *Macromolecules* **2003**, *36*, 9085–9092. [CrossRef]

71. Kim, J.K.; Park, D.-J.; Lee, M.-S.; Ihn, K.J. Synthesis and crystallization behavior of poly(L-lactide)-*block*-poly(ε-caprolactone) copolymer. *Polymer* **2001**, *42*, 7429–7441. [CrossRef]
72. Yan, D.; Huang, H.; He, T.; Zhang, F. Coupling of microphase separation and dewetting in weakly segregated di-block copolymer ultrathin films. *Langmuir* **2011**, *27*, 11973–11980. [CrossRef]
73. Casas, M.T.; Puiggalí, J.; Raquez, J.M.; Dubois, P.; Córdova, M.E.; Müller, A.J. Single crystals morphology of biodegradable double crystalline PLLA-*b*-PCL diblock copolymers. *Polymer* **2011**, *52*, 5166–5177. [CrossRef]
74. Jeon, O.; Lee, S.H.; Kim, S.H.; Lee, Y.M.; Kim, Y.H. Synthesis and characterization of poly(L-lactide)-poly(ε-caprolactone) multiblock copolymers. *Macromolecules* **2003**, *36*, 5585–5592. [CrossRef]
75. Danafar, H. Applications of copolymeric nanoparticles in drug delivery systems. *Drug Res.* **2016**, *66*, 506–519. [CrossRef] [PubMed]
76. Ostacolo, L.; Marra, M.; Ungaro, F.; Zappavigna, S.; Maglio, G.; Quaglia, F.; Abbruzzese, A.; Caraglia, M. In vitro anticancer activity of docetaxel-loaded micelles based on poly(ethylene oxide)-poly(ε-caprolactone) block copolymers: Do nanocarrier properties have a role? *J. Control. Release* **2010**, *148*, 255–263. [CrossRef] [PubMed]
77. Zhou, S.; Deng, X.; Yang, H. Biodegradable poly(ε-caprolactone)-poly(ethylene glycol) block copolymers: Characterization and their use as drug carriers for a controlled delivery system. *Biomaterials* **2003**, *24*, 3563–3570. [CrossRef]
78. Lim, D.W.; Park, T.G. Stereocomplex formation between enantiomeric PLA-PEG-PLA triblock copolymers: Characterization and use as protein delivery microparticulate carriers. *J. Appl. Polym. Sci.* **2000**, *75*, 1615–1623. [CrossRef]
79. Chen, J.; Huang, W.; Xu, Q.; Tu, Y.; Zhu, X.; Chen, E. PBT-*b*-PEO-*b*-PBT triblock copolymers: Synthesis, characterization and double crystalline properties. *Polymer* **2013**, *54*, 6725–6731. [CrossRef]
80. Schmalz, H.; Van Guldener, M.; Gabriëlse, W.; Lange, R.; Abetz, V. Morphology, surface structure, and elastic properties of PBT-based copolyesters with PEO-*b*-PEB-*b*-PEO triblock copolymer soft segments. *Macromolecules* **2002**, *35*, 5491–5499. [CrossRef]
81. Voet, V.S.D.; van Ekenstein, G.O.R.A.; Meereboer, N.L.; Hoffman, A.H.; Brinke, G.T.; Loos, K. Double crystalline PLLA-*b*-PVDF-*b*-PLLA triblock copolymers: Preparation and crystallization. *Polym. Chem.* **2014**, *5*, 2219–2230. [CrossRef]
82. Balsamo, V.; Müller, A.J.; Von Gyldenfeldt, F.; Stadler, R. Ternary ABC block copolymers based on one glassy and two crystallizable blocks: Polystyrene-*block*-polyethylene-*block*-poly(ε-caprolactone). *Macromol. Chem. Phys.* **1998**, *199*, 1063–1070.
83. Balsamo, V.; Paolini, Y.; Ronca, G.; Müller, A.J. Crystallization of the polyethylene block in polystyrene-*b*-polyethylene-*b*-polycaprolactone triblock copolymers, 1: Self-nucleation behavior. *Macromol. Chem. Phys.* **2000**, *201*, 2711–2720. [CrossRef]
84. Balsamo, V.; Müller, A.J.; Stadler, R. Antinucleation effect of the polyethylene block on the polycaprolactone block in ABC triblock copolymers. *Macromolecules* **1998**, *31*, 7756–7763. [CrossRef]
85. Müller, A.J.; Balsamo, V.; Arnal, M.L.; Jakob, T.; Schmalz, H.; Abetz, V. Homogeneous nucleation and fractionated crystallization in block copolymers. *Macromolecules* **2002**, *35*, 3048–3058. [CrossRef]
86. Chiang, Y.W.; Hu, Y.Y.; Li, J.N.; Huang, S.H.; Kuo, S.W. Trilayered single crystals with epitaxial growth in poly(ethylene oxide)-*block*-poly(ε-caprolactone)-*block*-poly(ε-lactide)thin films. *Macromolecules* **2015**, *48*, 8526–8533. [CrossRef]
87. Zhao, J.; Pahovnik, D.; Gnanou, Y.; Hadjichristidis, N. Sequential polymerization of ethylene oxide, ε-caprolactone and L-lactide: A one-pot metal-free route to tri- and pentablock terpolymers. *Polym. Chem.* **2014**, *5*, 3750–3753. [CrossRef]
88. Guillerm, B.; Lemaur, V.; Ernould, B.; Cornil, J.; Lazzaroni, R.; Gohy, J.F.; Dubois, P.; Coulembier, O. A one-pot two-step efficient metal-free process for the generation of PEO-PCL-PLA amphiphilic triblock copolymers. *RSC Adv.* **2014**, *4*, 10028–10038. [CrossRef]
89. Sun, L.; Shen, L.J.; Zhu, M.Q.; Dong, C.M.; Wei, Y. Synthesis, self-assembly, drug-release behavior, and cytotoxicity of triblock and pentablock copolymers composed of poly(ε-caprolactone), poly(L-lactide), and poly(ethylene glycol). *J. Polym. Sci. Part A Polym. Chem.* **2010**, *48*, 4583–4593. [CrossRef]
90. Tamboli, V.; Mishra, G.P.; Mitra, A.K. Novel pentablock copolymer (PLA-PCL-PEG-PCL-PLA)-based nanoparticles for controlled drug delivery: Effect of copolymer composition on the crystallinity of copolymers and in vitro drug release profile from nanoparticles. *Colloid Polym. Sci.* **2013**, *291*, 1235–1245. [CrossRef] [PubMed]
91. Hadjichristidis, N.; Iatrou, H.; Pitsikalis, M.; Pispas, S.; Avgeropoulos, A. Linear and non-linear triblock terpolymers. Synthesis, self-assembly in selective solvents and in bulk. *Prog. Polym. Sci.* **2005**, *30*, 725–782. [CrossRef]
92. Palacios, J.K.; Múgica, A.; Zubitur, M.; Iturrospe, A.; Arbe, A.; Liu, G.; Wang, D.; Zhao, J.; Hadjichristidis, N.; Müller, A.J. Sequential crystallization and morphology of triple crystalline biodegradable PEO-*b*-PCL-*b*-PLLA triblock terpolymers. *R. Soc. Chem. Adv.* **2016**, *6*, 4739–4750. [CrossRef]
93. Palacios, J.K.; Zhao, J.; Hadjicrhistidis, N.; Müller, A.J. How the complex interplay between different blocks determines the isothermal crystallization kinetics of triple-crystalline PEO-*b*-PCL-*b*-PLLA triblock terpolymers. *Macromolecules* **2017**, *50*, 9683–9695. [CrossRef]
94. Palacios, J.K.; Tercjak, A.; Liu, G.; Wang, D.; Zhao, J.; Hadjichristidis, N.; Müller, A.J. Trilayered morphology of an ABC triple crystalline triblock terpolymer. *Macromolecules* **2017**, *50*, 7268–7281. [CrossRef]
95. Palacios, J.K.; Liu, G.; Wang, D.; Hadjichristidis, N.; Müller, A.J. Generating triple crystalline superstructures in melt-miscible PEO-*b*-PCL-*b*-PLLA triblock terpolymers by controlling thermal history and sequential crystallization. *Macromol. Chem. Phys.* **2019**, *220*, 1900292. [CrossRef]
96. Matxinandiarena, E.; Múgica, A.; Zubitur, M.; Zhang, B.; Ladelta, V.; Zapsas, G.; Hadjichristidis, N.; Müller, A.J. The effect of the cooling rate on the morphology and crystallization of triple crystalline PE-*b*-PEO-*b*-PLLA and PE-*b*-PCL-*b*-PLLA triblock terpolymers. *ACS Appl. Polym. Mater.* **2020**, *2*, 4952–4963. [CrossRef]

97. Loo, Y.L.; Register, R.A.; Ryan, A.J. Modes of crystallization in block copolymer microdomains: Breakout, templated and confined. *Macromolecules* **2002**, *35*, 2365–2374. [CrossRef]
98. He, W.N.; Xu, J.T. Crystallization assisted self-assembly of semicrystalline block copolymers. *Prog. Polym. Sci.* **2012**, *37*, 1350–1400. [CrossRef]
99. Schmalz, H.; Knoll, A.; Müller, A.J.; Abetz, V. Synthesis and characterization of ABC triblock copolymers with two different crystalline end blocks: Influence of confinement on crystallization behavior and morphology. *Macromolecules* **2002**, *35*, 10004–10013. [CrossRef]
100. Cui, D.; Tang, T.; Bi, W.; Cheng, J.; Chen, W.; Huang, B. Ring-opening polymerization and block copolymerization of L-lactide with divalent samarocene complex. *J. Polym. Sci. Part A Polym. Chem.* **2003**, *41*, 2667–2675. [CrossRef]
101. Wang, D.; Zhang, Z.; Hadjichristidis, N. C1 polymerization: A unique tool towards polyethylene-based complex macromolecular architectures. *Polym. Chem.* **2017**, *8*, 4062–4073. [CrossRef]
102. Ladelta, V.; Zapsas, G.; Abou-Hamad, E.; Gnanou, Y.; Hadjichristidis, N. Tetracrystalline tetrablock quarterpolymers: Four different crystallites under the same roof. *Angew. Chem. Int. Ed.* **2019**, *58*, 16267–16274. [CrossRef]
103. Carmeli, E.; Wang, B.; Moretti, P.; Tranchida, D.; Cavallo, D. Estimating the nucleation ability of various surfaces towards isotactic polypropylene via light intensity induction time measurements. *Entropy* **2019**, *21*, 1068. [CrossRef]
104. Hiemenz, P.C.; Lodge, T.P. *Polymer Chemistry*, 2nd ed.; CRC Press: Boca Raton, FL, USA, 2007.
105. Maglio, G.; Migliozzi, A.; Palumbo, R. Thermal properties of di- and triblock copolymers of poly (L-lactide) with poly(oxyethylene) or poly (ε-caprolactone). *Polymer* **2003**, *44*, 369–375. [CrossRef]
106. Zhang, J.; Tashiro, K.; Tsuji, H.; Domb, J. Disorder-to-order phase transition and multiple melting behavior of poly(L-lactide) investigated by simultaneous measurements of WAXD and DSC. *Macromolecules* **2008**, *41*, 1352–1357. [CrossRef]
107. Ren, M.; Tang, Y.; Gao, D.; Ren, Y.; Yao, X.; Shi, H.; Zhang, T.; Wu, C. Recrystallization of biaxially oriented polyethylene film from partially melted state within crystallite networks. *Polymer* **2020**, *191*, 122291. [CrossRef]
108. Izquierdo, R.; Garcia-Giralt, N.; Rodríguez, M.T.; Cáceres, E.; García, S.J.; Gómez Ribelles, J.M.; Monleón, M.; Monllau, J.C.; Suay, J. Biodegradable PCL scaffolds with an interconnected spherical pore network for tissue engineering. *J. Biomed. Mater. Res.* **2007**, *85*, 25–35. [CrossRef] [PubMed]
109. Ebers, L.S.; Auvergne, R.; Boutevin, B.; Laborie, M.P. Impact of PEO structure and formulation on the properties of a Lignin/PEO blend. *Ind. Crop. Prod.* **2020**, *143*, 111883. [CrossRef]

Article

Phase Transitions in Poly(vinylidene fluoride)/Polymethylene-Based Diblock Copolymers and Blends

Nicolás María [1], Jon Maiz [1,2,3,*], Daniel E. Martínez-Tong [2,4], Angel Alegria [2,4], Fatimah Algarni [5], George Zapzas [5], Nikos Hadjichristidis [5,*] and Alejandro J. Müller [1,3,4,*]

1. POLYMAT, University of the Basque Country UPV/EHU, Avenida de Tolosa 72, 20018 Donostia-San Sebastián, Spain; nicolas.maria@polymat.eu
2. Centro de Física de Materiales (CFM) (CSIC-UPV/EHU)-Matrials Physics Center (MPC), Paseo Manuel de Lardizabal 5, 20018 Donostia-San Sebastián, Spain; danielenrique.martinezt@ehu.eus (D.E.M.-T.); angel.alegria@ehu.eus (A.A.)
3. IKERBASQUE—Basque Foundation for Science, Plaza Euskadi 5, 48009 Bilbao, Spain
4. Department of Polymers and Advanced Materials: Physics, Chemistry and Technology, University of the Basque Country UPV/EHU, Paseo Manuel de Lardizabal 3, 20018 Donostia-San Sebastián, Spain
5. KAUST Catalysis Center, Polymer Synthesis Laboratory, Physical Sciences and Engineering Division, King Abdullah University of Science and Technology (KAUST), Thuwal 23955-6900, Saudi Arabia; fatimah.algarni@kaust.edu.sa (F.A.); georgios.zapzas@kaust.edu.sa (G.Z.)
* Correspondence: jon.maizs@ehu.eus (J.M.); nikolaos.hadjichristidis@kaust.edu.sa (N.H.); alejandrojesus.muller@ehu.es (A.J.M.)

Citation: María, N.; Maiz, J.; Martínez-Tong, D.E.; Alegria, A.; Algarni, F.; Zapzas, G.; Hadjichristidis, N.; Müller, A.J. Phase Transitions in Poly(vinylidene fluoride)/Polymethylene-Based Diblock Copolymers and Blends. *Polymers* **2021**, *13*, 2442. https://doi.org/10.3390/polym13152442

Academic Editors: Holger Schmalz and Volker Abetz

Received: 6 July 2021
Accepted: 22 July 2021
Published: 24 July 2021

Publisher's Note: MDPI stays neutral with regard to jurisdictional claims in published maps and institutional affiliations.

Copyright: © 2021 by the authors. Licensee MDPI, Basel, Switzerland. This article is an open access article distributed under the terms and conditions of the Creative Commons Attribution (CC BY) license (https://creativecommons.org/licenses/by/4.0/).

Abstract: The crystallization and morphology of two linear diblock copolymers based on polymethylene (PM) and poly(vinylidene fluoride) (PVDF) with compositions PM_{23}-b-$PVDF_{77}$ and PM_{38}-b-$PVDF_{62}$ (where the subscripts indicate the relative compositions in wt%) were compared with blends of neat components with identical compositions. The samples were studied by SAXS (Small Angle X-ray Scattering), WAXS (Wide Angle X-ray Scattering), PLOM (Polarized Light Optical Microscopy), TEM (Transmission Electron Microscopy), DSC (Differential Scanning Calorimetry), BDS (broadband dielectric spectroscopy), and FTIR (Fourier Transform Infrared Spectroscopy). The results showed that the blends are immiscible, while the diblock copolymers are miscible in the melt state (or very weakly segregated). The PVDF component crystallization was studied in detail. It was found that the polymorphic structure of PVDF was a strong function of its environment. The number of polymorphs and their amount depended on whether it was on its own as a homopolymer, as a block component in the diblock copolymers or as an immiscible phase in the blends. The cooling rate in non-isothermal crystallization or the crystallization temperature in isothermal tests also induced different polymorphic compositions in the PVDF crystals. As a result, we were able to produce samples with exclusive ferroelectric phases at specific preparation conditions, while others with mixtures of paraelectric and ferroelectric phases.

Keywords: poly(vinylidene fluoride)/polymethylene; blends; diblock copolymers; ferroelectric phase

1. Introduction

Nowadays, polymers are important materials that may be used to enhance the safety and the quality of the environment and reduce the human impact. For example, they may take relevance in the field of renewable energies or self-powered applications where new polymeric materials can substitute inorganic devices having the same or better yield at a lower cost and with less environmental impact [1–3]. Therefore, the current development in new technologies requires the research of new materials to achieve a balance between evolution and pollution.

During the last years, poly(vinylidene fluoride) (PVDF) [4,5] and its copolymers [6–8] have been the most used polymers in electronic devices or renewable energies. PVDF has good mechanical properties, such as flexibility and low cost. Its biocompatibility with other polymers and/or an extremely high chemical resistance make this polymer a great option for this kind of applications [9]. The most important characteristics of PVDF, apart from the properties commented above, are its ferroelectricity, piezoelectricity, and pyroelectricity, resulting from the polarization of its C-F bonds [10]. Therefore, the most used applications for this kind of fluoropolymers are data storage devices [11,12], sensors [13] and/or energy harvesting devices [14].

Another relevant characteristic of PVDF is its polymorphism: PVDF can crystallize in at least four different phases (α, β, γ, and δ), and not all of these phases have the same polar or non-polar properties [15,16]. When PVDF crystallizes from the melt, the most common and stable phase is the α-phase. This phase has a trans-gauche conformation, $Tg^+Tg^-Tg^+Tg^-$, and it is paraelectric. The drawback of this non-polar phase is that the PVDF crystallizing in this crystalline form is not very useful for the applications mentioned above [17]. 0 In contrast, the β-phase, with a conformation in all the carbons are in trans configuration, TTTT, has the highest dipole moment and is a piezoelectric and ferroelectric material [18,19]. 1 Unfortunately, this phase is not the most stable one, and it is difficult to obtain. A lot of methods and efforts have been developed during these last years to crystallize this ferroelectric phase, from mechanical stress to PVDF-based mixtures or blends (either with other polymers and/or fillers) and the synthesis of different copolymers [20–22]. The γ-phase has a higher melting temperature than the two phases mentioned above, and it is also ferroelectric, but it has less polarity, and its chain conformation is three trans and one gauche conformation $TTTg^+TTTg^-$ [23]. Finally, the δ-phase has the same chain conformation than the α-phase. The only difference is that the δ-phase has each second chain rotated 180° around the chain axis, and this small change provides the ferroelectric property to this phase compared to the paraelectric α-phase [24,25].

There are several papers in the literature in which ferroelectric and piezoelectric properties were obtained in PVDF and in PVDF-based materials. One of the most employed methods to obtain the β-phase in PVDF films is stretching [20,26], where mechanical stress is applied to transform polymer crystals from an α-phase to a β-phase. In this process, the stretching temperature is one of the important parameters to be considered [27], but the conversion from α- to β-phase obtained by this method is not complete, and both phases coexist simultaneously in the PVDF films [28]. The preparation of PVDF-based blends is another method to achieve the polar β-phase in PVDF [21,29] directly. PVDF blended with poly(methyl methacrylate) (PMMA), for example, crystallizes directly in the β-phase when the crystallization process occurs from the melt [30,31]. The addition of different fillers PVDF is another alternative, for example, samples of PVDF-TrFE (polyvinylidene-trifluoroethylene) with modified ZnO particles can promote the crystallization of the β-phase in the copolymer [32], and when PVDF is mixed with less than 0.2% of multi-walled carbon nanotubes, it can crystallize in an almost pure β-phase [33].

Other alternatives are to produce PVDF-based graft or block copolymers [34]. Graft copolymers based on PVDF were studied in order to improve the crystallization of the β-phase. Synthesis of PVDF grafted with poly (butylene succinate-co-adipate) (PVDF-g-PBSA) or poly (methyl methacrylate-co-acrylic acid) [PVDF-g-(PMMA-co-AA)] with previous ozonation of the PVDF induces the crystallization of the β-phase in almost 100%, thanks to the covalent links formed in the PVDF-OH groups [35]. Moreover, also block copolymers with PVDF were investigated to induce the β-phase. Beuermann et al. demonstrated by Fourier Transform Infrared Spectroscopy (FTIR) and Wide-Angle X-Ray Scattering (WAXS) that the PVDF crystallizes in the ferroelectric phase when PVDF-b-PMMA and PVDF-b-PS (Polystyrene) are synthesized [36,37]. In addition, in a previous work published by us, we have demonstrated that PVDF-b-PEO (Polyethylene oxide) block copolymers can crystallize only in the β-phase when the crystallization happens from the melt at low cooling rates, for instance, 1 °C/min [38].

In general, the properties of the blends and/or copolymers are different depending on the synthesis and the form in which they are present in the sample [39–41]. If the polymers are not compatible, the segregation observed in the material is different for blends and for copolymers. Segregation in blends happens on a larger scale due to the macro-phase segregation behavior [42]. Immiscible block copolymers cannot segregate into macro-phases due to their covalent bonds, but micro-phase segregation into regular domain patterns can occur [43]. Daoulas et al. have demonstrated by mesoscopic simulations that the differences between the block copolymers and blends in poly (*p*-phenylene vinylene) (PPV) and polyacrylate systems are due to this segregation phenomenon that makes the materials different for light-emitting diodes, so the final applications of both materials are not the same [44].

In this work, we study the crystallization of a polymethylene (PM) and PVDF system, polymers that are not miscible. We compare the PVDF homopolymer with two PM/PVDF blends and two PM-*b*-PVDF block copolymers in the same proportion in order to see the relevance of the segregation in the final properties of both materials. Using Differential Scanning Calorimetry (DSC), we study the behavior of these samples during the non-isothermal crystallization and during an isothermal process. Microscopy techniques and Small-Angle X-Ray Scattering (SAXS) are employed to study the miscibility between both polymers. Finally, the samples are fully characterized by Broadband Dielectric Spectroscopy (BDS), Fourier Transform Infrared Spectroscopy (FTIR), and Wide-Angle X-Ray Scattering (WAXS).

2. Materials and Methods

2.1. Materials

The diblock copolymers of polymethylene (PM) and poly(vinylidene fluoride) (PVDF) have been synthetized by Hadjichristidis et al. and published in a previous work [45]. In brief, the synthesis involves the following steps: (a) polyhomologation of dimethylsulfoxonium methylide using triethylborane as initiator followed by oxidation/hydrolysis to afford PM-OH, (b) esterification of the OH group with 2,2-bromoisobutyrylbromide to introduce bromide at the chain end, (c) halide exchange (Br→I) using sodium iodine to produce the macro-chain transfer agent (macro-CTA), and (d) Iodine transfer polymerization (ITP) of VDF with the macro-CTA and 1,1-bis(tert-butylperoxy)cyclohexa as the initiator (Scheme S1). The synthesis of polyvinylidene fluoride (PVDF) homopolymer has been accomplished via reversible addition−fragmentation chain-transfer polymerization (RAFT) polymerization by using (S-benzyl O-ethylxathate) as CTA and 1,1-bis(tert-butylperoxy)cyclohexane (Luperox 331P80, Sigma-Aldrich, Munich, Germany) as initiator. The synthesis and the characterization of the linear PVDF used in this study are given in the SI.

Blends were prepared by mixing the block copolymers with linear homopolymers, PM-OH and PVDF. The blends were prepared in the same compositions used for the block copolymers so that they could be compared. First, the PVDF and the PM mixtures were stirred until the total dissolution in cyclohexane during 24 h at 50 °C. Then, each mixture was drop-casted onto Teflon holders. Afterward, a fume hood was used to slowly evaporate the solvent, and finally, under vacuum conditions, the samples were well-dried in an oven at 40 °C for 72 h. All the polymers used in this work and their molecular characteristics are listed in Table 1.

Table 1. Principal characteristics of all samples employed during this work. The subscripts indicate the wt% of each block.

Sample	Topology	M_n (g/mol) [a]	M_n PM (g/mol) [a]	M_n PVDF (g/mol) [a]	Đ [b]
PM$_{23}$-b-PVDF$_{77}$	Linear diblock copolymer	28.6 K	6.6 K	22.0 K	PM: 1.12 PVDF: 1.29
PM$_{38}$-b-PVDF$_{62}$	Linear diblock copolymer	17.6 K	6.6 K	11.0 K	PM: 1.12 PVDF: 1.25
PM$_{23}$PVDF$_{77}$	Blend	-	5.6 K	7.6 K	
PM$_{38}$PVDF$_{62}$	Blend	-	5.6 K	7.6 K	
PVDF	Linear homopolymer	7.6 K	-	7.6 K	1.50 [c]
PM-OH	Linear homopolymer	5.6 K	5.6 K	-	1.12 [d]

[a] All M_n were determined by ^{1}H NMR, toluene-d$_8$, and DMF-d$_7$ mixture; [b] Direct GPC characterization of PM-b-PVDF copolymers was impossible due to the difficulty in finding a common solvent for both blocks. The results given in the Table correspond to each block after hydrolysis of the junction point; [c] HT-GPC (trichlorobenzene as eluent, 145 °C, PS standards) for PM-OH and [d] GPC (dimethylformamide as eluent, 35 °C, PS standards).

2.2. Methods

2.2.1. Differential Scanning Calorimetry (DSC)

A Perkin Elmer DSC 8000 equipment was used to carry out the DSC experiments. This equipment uses an Intracooler II as a cooling system. Before the measurements were performed, the equipment was calibrated using indium and tin standards.

For the non-isothermal procedure, first, the samples were heated up to 20 °C above the highest melting temperature and held there for 3 min to ensure that the thermal history of the materials was completely erased. Then, samples were cooled at different cooling rates (60, 20, 5, and 1 °C/min) from the melt to 25 °C and then heated again to the molten state at a constant rate of 20 °C/min.

The protocol used to carry out the isothermal crystallization procedure was the same followed by Lorenzo et al. [46]. First, the minimum crystallization temperature ($T_{c,min}$) was searched. To find it, samples were heated up to 20 °C above the melting temperature and held there for 3 min. Then, samples were cooled fast (at 60 °C/min) to a previously selected T_c. When this T_c was reached, samples were heated at 20 °C/min to the same melting temperature chosen in the previous step. When no peaks were observed in the subsequent heating scan, the T_c mentioned in the second step was considered to be the minimum isothermal crystallization temperature [46].

The isothermal crystallization procedure consisted in a series of different steps. First, samples were melted at 20 °C above the melting temperature and held there for 3 min to erase the thermal history of the material. Then, samples were cooled down at 60 °C/min to the selected isothermal crystallization temperature and held at this T_c for 40 min to achieve crystallization saturation. Once this crystallization process was finished, samples were heated at 20 °C/min to the previously selected melting temperature, and the process was reinitiated to the next programmed T_c [46].

2.2.2. X-ray Diffraction

Block copolymer samples were analyzed using Wide-Angle X-Ray Scattering (WAXS) and Small-Angle X-Ray Scattering (SAXS). These experiments were carried out in the ALBA Synchrotron facility using synchrotron radiation at the BL11-NCD beamline. Samples were measured in capillaries using a Linkam hot-stage system equipped with liquid nitrogen to control the temperature. The samples were melted at 200 °C for 3 min, then cooled down at the chosen cooling rate. The energy of the X-ray source was 12.4 keV (λ = 1.0 Å). The WAXS system configuration employed was a Rayonix LX255-HS sample detector with an active area of 230.4 mm × 76.8 mm. A sample to detector distance of 15.5 mm with a tilt angle of 27.3° was employed. The resulting pixel size was 44 µm^2. For the SAXS experiments, the configuration was a Pilatus 1M sample detector, which had the following characteristics:

active image area = 168.7 mm × 179.4 mm, the total number of pixels = 981 × 1043, pixel size = 172 µm × 172 µm, rate = 25 frames/sec and the distance used was 6463 mm.

2.2.3. Polarized Light Optical Microscopy (PLOM)

All samples were analyzed by an Olympus BX51 polarized optical microscope coupled to a Linkam hot-stage that uses nitrogen to control the temperature and manages the cooling rate. An Olympus SC50 camera linked to the microscope was employed to observe the samples and take micrographs. Samples were dissolved in acetone or cyclohexane, and drops of the solutions were placed on a glass substrate and dried at room temperature.

2.2.4. Fourier Transform Infrared Spectroscopy (FTIR)

A Nicolet 6700 Fourier Transform Infrared Spectrometer equipped with an Attenuated Total Reflectance (ATR) Golden Gate MK II with a diamond crystal was employed to analyze the samples. Samples were melted directly from the bulk at 200 °C in a Linkam hot-stage and then cooled down at 1 °C/min employing N_2 in the cooling process. FTIR measurements were carried out after the cooling process at room temperature.

2.2.5. Transmission Electron Microscopy (TEM)

All samples were stained with RuO_4 before the measurements by immersing thin strips of material in this solution for 16 h. Then, the samples were cut in ultra-thin sections at room temperature with a diamond knife on a Leica EMFC6 ultra-microtome device. These 90 nm thick ultra-thin sections were mounted on a 200 mesh copper grid and then observed by a TECNAI G2 20 TWIN TEM equipped with a LaB_6 filament operating at an accelerating voltage of 120 kV.

2.2.6. Broadband Dielectric Spectroscopy (BDS)

The complex dielectric permittivity, $\varepsilon^*(\omega) = \varepsilon'(\omega) - i\varepsilon''(\omega)$, where ε' is the real part and ε'' is the imaginary part, was obtained as a function of the frequency (ω) and temperature (T) by using a Novocontrol high-resolution dielectric analyzer (Alpha analyzer) (Novocontrol, Montabaur, Germany). The sample cell was set in a cryostat, whose temperature was controlled via a nitrogen gas jet stream coupled with a Novocontrol Quatro controller. Samples were placed between two flat gold-plated electrodes (10 and 20 mm in diameter) forming a parallel plate capacitor with a 0.1 mm thick Teflon spacer. Frequency sweeps were performed at a constant temperature with a stability of ±0.1 °C. BDS measurements were carried out as follows. Samples were heated up to 200 °C inside the cryostat. This temperature was held for 5 min to ensure a homogeneous filling of the capacitor and to obtain a *fully* amorphous initial state. Then, measurements started at 200 °C, cooling the samples in isothermal steps of 10 °C down to −100 °C, and subsequently heating them up to 200 °C, again in 10 °C steps. Samples were tested at different temperatures over a frequency range of 10^{-1} to 10^7 Hz.

3. Results and Discussion

3.1. Miscibility between PM and PVDF

The final properties of materials that are made up of more than one component can be affected by their miscibility. The Flory interaction parameter χ_{12} can be estimated by the following semi-empirical equation (Equation (1)) [47],

$$\chi_{12} = 0.34 + \frac{V_1}{RT}(\delta_1 - \delta_2)^2 \tag{1}$$

where χ_{12} is the interaction parameter, V_1 is the molar volume of the matrix component (PVDF in our case) calculated through the molar mass of the repeating unit (M = 64.03 g/mol) and the amorphous density (ρ = 1.68 g/cm³), in this case, V_1 = 38.1 cm³/mol, R is a constant the value of which is 1.987 cal/mol K, T is the temperature chosen to calculate the miscibility (473 K in order to know the miscibility in the molten state), and δ_1 (8.57 (cal/cm³)$^{1/2}$) and

δ_2 (7.9 (cal/cm^3)$^{1/2}$) are the solubility parameters. In our case, the calculated χ_{12} is 0.36 at 200 °C.

To calculate the segregation strength in the case of block copolymers, the χ_{12} value is multiplied by N, the degree of polymerization. When the value obtained is below 10, the polymers are miscible with each other; if the estimated value is between 10 and 30, there is a weak segregation; and if it is between 30 and 50, there is a medium segregation. Only when the calculated value is above of 50, it is possible to predict that there will be a strong segregation. For our samples, we have calculated that the segregation strength is 117 for the PM$_{23}$-b-PVDF$_{77}$ and 72 in the case of PM$_{38}$-b-PVDF$_{62}$. Therefore, we can expect a strong segregation in the melt for both samples.

Nevertheless, SAXS results do not show any evidence of phase segregation in the melt. Figure 1 shows the SAXS curves for both block copolymers at different temperatures during a heating sweep at 20 °C/min. When the copolymers are in the molten state (above 165 °C), there is not any segregation peak observed, indicating that either the electron density contrast in the melt is not enough to produce a signal or that the copolymers are either very weakly segregated or melt-mixed. The prominent SAXS peaks observed at temperatures below the melting point of PVDF are due to the average long period values of the constituent crystalline lamellae. As expected, they shift to lower q values (i.e., larger long periods) as temperature increases.

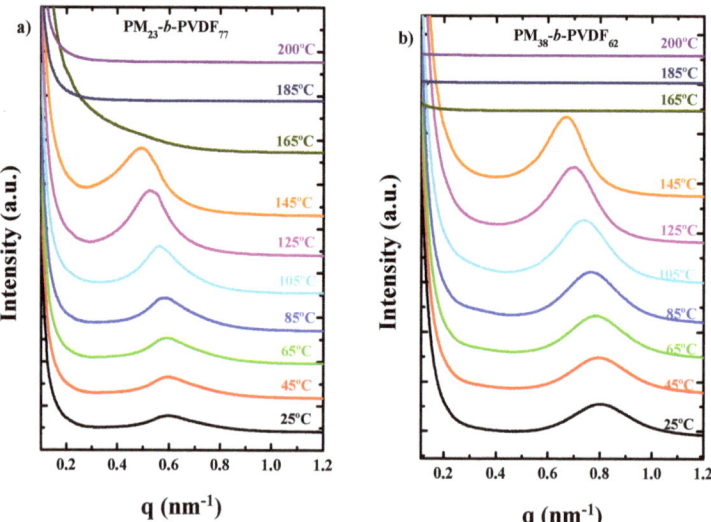

Figure 1. SAXS analysis at different temperatures during heating scans at 20 °C/min after a cooling process also at 20 °C/min of (a) PM$_{23}$-b-PVDF$_{77}$ sample and (b) PM$_{38}$-b-PVDF$_{68}$ sample.

PLOM was used to observe the crystallization process in the different samples and to check if the segregation behavior is different between block copolymers and blends. Figure 2a shows the crystallization of PM$_{38}$-b-PVDF$_{62}$ during a cooling sweep from the melt at 20 °C/min. In a strongly segregated diblock copolymer with this composition, the expected microphase separated morphology in the melt would be that of a lamellar assembly. Additionally, if the segregation is strong, each block has to crystallize within the confined microdomain morphology produced during the phase segregation in the melt. As a result, it would be impossible to observe spherulites.

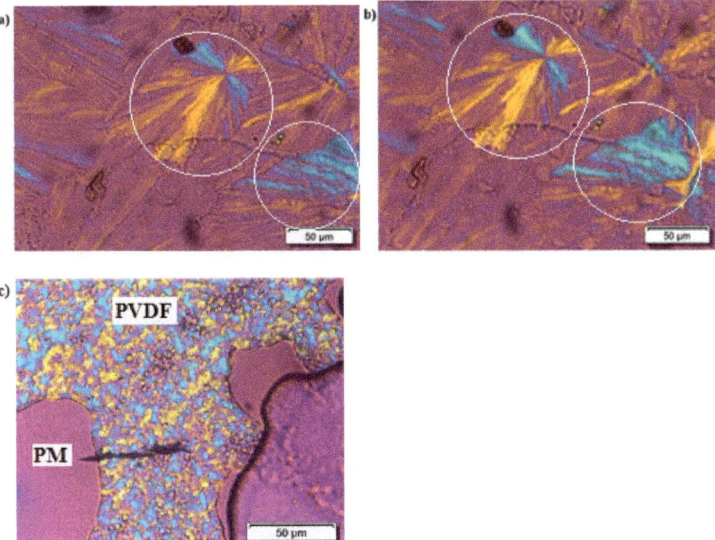

Figure 2. PLOM images of (**a**) PVDF block spherulites in the PM_{38}-b-$PVDF_{62}$ diblock copolymer sample after having been cooled at 20 °C/min to a T = 130 °C and (**b**) crystallization of the PM block in the PM_{38}-b-$PVDF_{62}$ sample after having been cooled at 20 °C/min to T = 25 °C. (**c**) Evident phase segregation of the PVDF and PM phases in a $PM_{23}PVDF_{77}$ blend sample after a cooling process at 20 °C/min down to T = 25 °C.

The micrograph shown in Figure 2a was taken at a temperature higher than the melting point of the PM block in the copolymer (i.e., T = 130 °C). The PVDF block crystallizes as spherulites in this case. This observation indicates that the diblock copolymer crystallizes either from a weakly segregated melt, from which break out leads to spherulites formation or from a melt mixed state, which can also explain the observation of spherulites. As shown in Figure 2b, when the temperature is lower than the PM block crystallization temperature (micrograph taken at 25 °C), a subtle change in the birefringence is observed. This change in birefringence has been highlighted by surrounding the most noticeable areas with a white circle. In order to quantify this, change in the transmitted light intensity during the cooling process was measured using the ImageJ software [48]. The results obtained are plotted in Figure S5 in the Supporting Information, and they conclusively show the sequential crystallization of the PVDF and PM blocks upon cooling from the melt. This change happens as the PM block crystallizes within the already formed PVDF spherulites, just within the intraspherulitic amorphous regions, as has been observed before for other block copolymer systems, such as PCL-b-PLLA or PEO-b-PCL [49,50]. The PLOM results obtained in Figure 2a,b indicate that these copolymers are either miscible or weakly segregated. These results are consistent with the lack of phase segregation observed by SAXS.

On the other hand, Figure 2c shows the complete crystallization of both phases (PM and PVDF) in the blends after a cooling scan at 20 °C/min at T = 25 °C from the molten state. The phase segregation between the phases is evident. PVDF crystallizes as spherulites, and PM crystallizes in microaxialites (difficult to see in the micrograph due to their small size). This result suggests that there is evident macrophase segregation in the blends.

TEM was used to see the differences in the miscibility and in the lamellar structure between the block copolymers and the blends. Figure 3 shows the TEM images for the PM_{23}-b-$PVDF_{77}$ diblock copolymer sample (Figure 3a) and the $PM_{23}PVDF_{77}$ blend sample (Figure 3b), respectively. Figure 3a shows a close-up region of a spherulite whose center is located to the right of the micrograph. A large number of lamellae that have grown from the right to the left of the micrograph can be observed. We were not able to distinguish the lamellae belonging to the PVDF block or to the PM block, as they seem to have similar sizes. Their co-existence without any discontinuity suggests that both blocks crystallize from a miscible melt. No signs of phase separation were observed for the block copolymer samples.

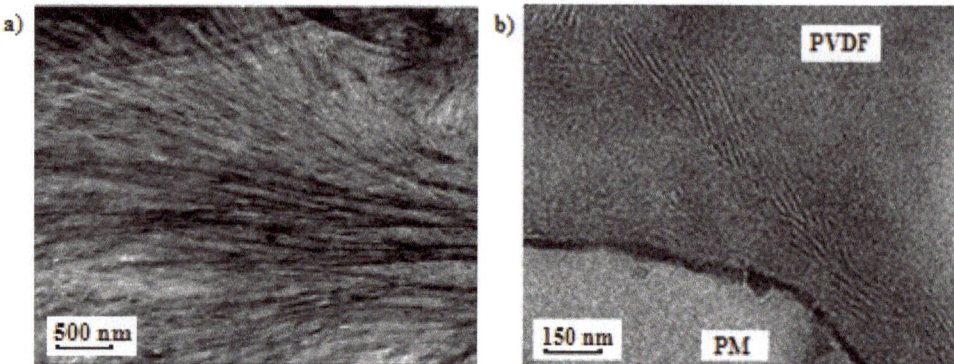

Figure 3. TEM images for (**a**) PM$_{23}$-*b*-PVDF$_{77}$ linear diblock copolymer and (**b**) PM$_{23}$PVDF$_{77}$ blend after cooling the samples at 20 °C/min to 25 °C.

On the other hand, in Figure 3b, it is possible to observe the evident phase segregation between PVDF and PM phases in the PM$_{23}$PVDF$_{77}$ blend. In summary, taking into account the collected evidence by PLOM and TEM, we can conclude that the PM and PVDF samples employed here are miscible when they form diblock copolymers, but they are immiscible when they are physically blended. This aspect is important to take into account in the next sections.

3.2. How the Cooling Rate Affects the Crystallization of the PVDF Phase in Block Copolymers and Blends

Blends and block copolymers were studied at different cooling rates in order to observe how this parameter affects the crystallization of PVDF in both systems. The cooling rates employed were 1, 5, 20, and 60 °C/min, and the heating rate used after the cooling process was always 20 °C/min. A PVDF homopolymer was also studied for comparative purposes.

Figure 4a shows the DSC cooling scans at 20 °C/min of the PVDF homopolymer, the PM homopolymer (PM-OH), the two different diblock copolymers, and their respective blends at the same composition. The crystallization (Figure 4a) peaks located at higher temperatures correspond to the PVDF component. In the blends, the PVDF component crystallizes at higher temperatures than the PVDF homopolymer (which is one of the components used to formulate the blend). This corresponds to a nucleating effect of the molten PM-OH phase, which can be explained by a transference of impurities from the PM phase to the PVDF phase during blending, as already described for other systems [51–53]. On the other hand, the PVDF blocks in the diblock copolymers have lower T_c values than the PVDF homopolymer sample, a possible sign of miscibility between the blocks. The other crystallization peak, at lower temperatures, corresponds to the PM blocks. In this case, the crystallization of the PM in the diblock copolymers is bimodal and occurs at higher temperatures than those observed for the blends and for the PM homopolymer. This higher crystallization temperature could be related to a nucleating effect of the PVDF block crystals.

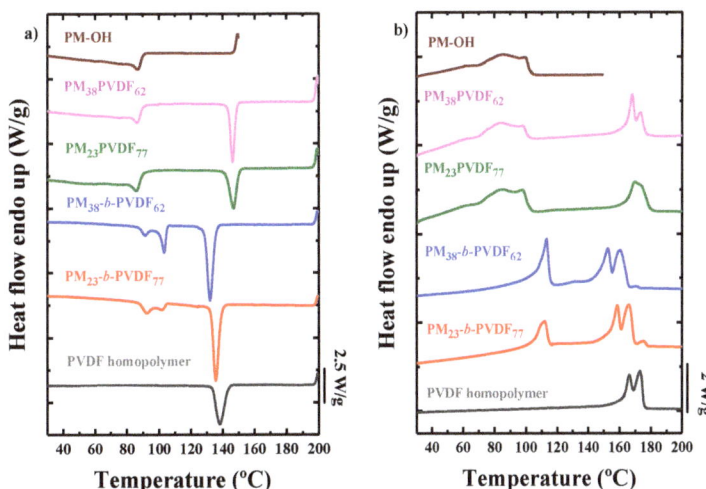

Figure 4. DSC scans of the blends, the diblock copolymers, and homopolymers samples. (**a**) Cooling curves at 20 °C/min and (**b**) heating curves at 20 °C/min after the previous cooling process.

The DSC subsequent DSC heating curves taken at 20 °C/min are plotted in Figure 4b and show that the melting peak that corresponds to the PM crystalline phase shows up at lower temperatures than that one observed for the PVDF. It is clear that the blends are totally immiscible, and the melting points of the PM phase (which shows a bimodal character) in the blends are very similar and located at the same temperatures as in the PM homopolymer. On the other hand, in the block copolymers, the PM block melting peak is a monomodal sharp endotherm that peaks at significantly higher values than that of the PM homopolymer or the PM phase in the blends. Regarding the melting peaks associated to the PVDF phases in the blends, these are located in the same temperature range as those of the PVDF homopolymer, once again suggesting that PM and PVDF are immiscible. In summary, due to the phase segregation encountered in the blends, the melting peaks of the blends correspond to those observed for their homopolymers in the same temperature range.

For the PVDF phase, melting is characterized by two main peaks. Due to the polymorphism observed in PVDF, different phases can form in the same sample [54]. In the case of the diblock copolymers, even a third minor peak appears at higher temperatures. This peak could be either a third crystalline phase or the result of a crystal reorganization that occurred during the heating process. The first melting peak in PVDF usually corresponds to the less stable, ferroelectric β-phase, and the second melting peak, to the paraelectric α-phase [30].

Figure 5 shows the comparison of the DSC heating scans of the samples (all performed at 20 °C/min) in the PVDF melting range obtained after using different cooling rates. The PM_{23}-b-$PVDF_{77}$ diblock copolymer (Figure 5a) shows three melting peaks at all the cooling rates studied, except at 1 °C/min, where only one main peak with a low temperature shoulder is observed. The third peak that can be observed at around 175 °C seems to be related to a crystal reorganization process, and Figure 5a shows that it does not depend on the cooling rate used (except for the experiment performed at 1 °C/min). The height and the area of the other two peaks seem to remain constant at all the cooling rates except at 1 °C/min, where the behavior of the subsequent melting curve is completely different. First, there is not a third peak, and second, the first peak, probably the β-phase peak, has almost disappeared, so at 1 °C/min, the α-phase peak is promoted. This is a common behavior reported in the literature for the PVDF: at low cooling rates, the formation of the most stable phase is promoted [55,56].

The second diblock copolymer (Figure 5b), PM$_{38}$-b-PVDF$_{62}$, shows different behavior. At high cooling rates, the α-phase peak is larger than the β-phase peak, but when the cooling rate is decreased, the α-phase peak also decreases, and the β-phase peak is the majority phase in the copolymer. For instance, at 1 °C/min, the promotion of the β-phase is evident. The crystallization behavior of the PVDF at 1 °C/min is completely different than the behavior shown by the PM$_{23}$-b-PVDF$_{77}$ copolymer: the formation of the less stable phase is promoted in this case.

On the other hand, both PM/PVDF blends exhibit similar behavior (Figure 5c,d). In this case, it seems that the amount of PM in the blend has no effect on the crystallization of the PVDF phase. The formation of the β-phase is always promoted in the blends, even at high cooling rates, where it coexists with the α-phase. When the cooling rate is decreased (5 °C/min), the α-phase almost disappears, and a new high temperature peak appears, which is associated to a different crystalline phase that is more stable than the last two ones explained. It has been reported in the literature that at these high temperatures (higher than 175 °C) the γ-phase, which is also polar, crystallizes [57,58]. When samples are cooled at 1 °C/min, the α-phase peak completely disappears, and the β-phase and the γ-phase coexist. For comparative purposes, a PVDF homopolymer was also studied at different cooling rates (Figure 5e). As can be seen at high cooling rates, the α-phase and the β-phase coexist; however, when the cooling rate is decreased, the PVDF tends to crystallize preferentially in the β-phase. At 1 °C/min, the three crystalline phases mentioned above coexist, and the β-phase is the main crystalline phase. A small shoulder at high temperatures corresponds to the α-phase, and finally, the new stable melting peak appears, which probably corresponds to the previously mentioned γ-phase. All the calorimetric parameters obtained by DSC are listed in Table 2.

DSC heating scans performed after cooling the samples at 1 °C/min show that the crystalline phase obtained depends on the sample and the origin of the sample. Samples cooled at 1 °C/min were analyzed by FTIR to verify which phases the PVDF block crystallizes in. Figure 6 shows the FTIR results for the PM homopolymer, the PVDF homopolymer, both diblock copolymers, and both blends, at room temperature after the samples were cooled from the melt at 1 °C/min. The wavenumber range studied was 1400–600 cm^{-1}, which is where the most useful information for PVDF can be observed. There is a large band located at 720 cm^{-1} and a smaller one at 1377 cm^{-1}, where the main characteristic bands for the PM polymer are observed [59]. There is also a weak band located at 801 cm^{-1}. We can observe that the main peaks perceived for PM do not overlap with the main bands associated with PVDF.

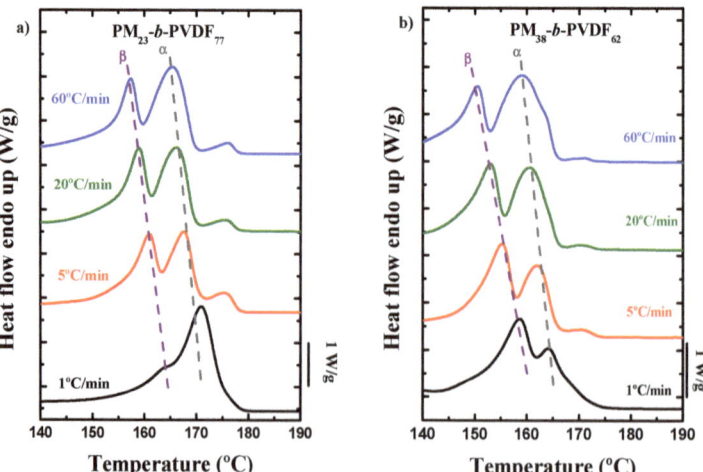

Figure 5. Cont.

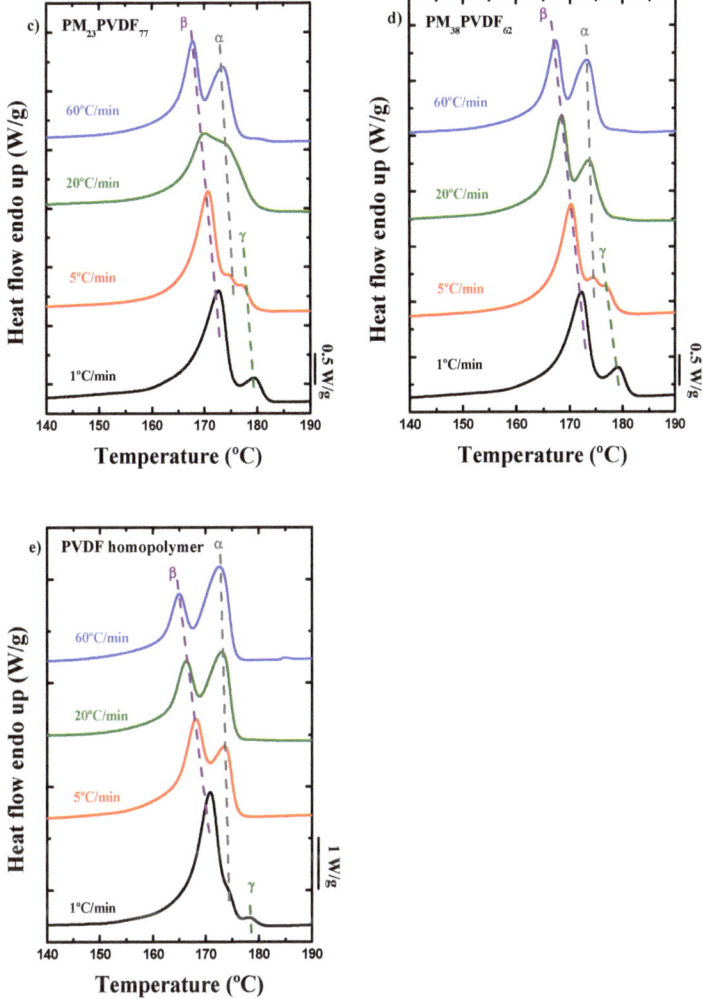

Figure 5. DSC heating scans for PVDF after different cooling rates were used: (**a**) PM$_{23}$-b-PVDF$_{77}$ and (**b**) PM$_{38}$-b-PVDF$_{62}$ block copolymers, (**c**) PM$_{23}$PVDF$_{77}$, (**d**) PM$_{30}$PVDF$_{62}$, and (**e**) PVDF homopolymer samples.

Table 2. Melting and crystallization temperatures and enthalpies for each block copolymer, blend, and homopolymer sample studied.

Sample	Polymer	Rate (°C/min)	$T_{m,PM}$ (°C)	$T_{m,\alpha}$ (°C)	$T_{m,\beta}$ (°C)	$T_{m,\gamma}$ (°C)	T_c (°C)	ΔH_m (J/g)	ΔH_c (J/g)
Homopolymer	PVDF	1	-	-	170.9	178.1	150.6	52.6	69.8
		5	-	173.5	168.2	-	144.0	53.8	60.4
		20	-	173.0	166.3	-	138.2	54.3	57.0
		60	-	172.5	165.0	-	129.3	53.8	58.5
PM$_{23}$-b-PVDF$_{77}$	PM	1	113.0	-	-	-	107.9	19.9	4.6
		5	112.2	-	-	-	105.6	25.1	3.4
		20	112.1	-	-	-	102.3	23.7	3.0
		60	111.9	-	-	-	98.3	24.3	1.6
	PVDF	1	-	170.9	-	-	147.8	67.1	67.0
		5	-	167.6	161.1	-	141.7	66.6	69.5
		20	-	166.1	158.9	-	135.9	70.6	71.6
		60	-	165.4	157.3	-	128.9	71.0	60.8

Table 2. Cont.

Sample	Polymer	Rate (°C/min)	$T_{m,PM}$ (°C)	$T_{m,\alpha}$ (°C)	$T_{m,\beta}$ (°C)	$T_{m,\gamma}$ (°C)	T_c (°C)	ΔH_m (J/g)	ΔH_c (J/g)
PM$_{38}$-b-PVDF$_{62}$	PM	1	114.4	-	-	-	108.4	38.4	25.6
		5	113.7	-	-	-	106.3	40.6	19.6
		20	113.4	-	-	-	103.4	43.2	18.8
		60	112.7	-	-	-	98.9	43.6	12.6
	PVDF	1	-	164.3	158.7	-	141.9	60.7	66.9
		5	-	162.1	155.2	-	137.6	57.4	72.3
		20	-	160.4	153.0	-	132.3	64.8	76.1
		60	-	159.1	150.6	-	124.2	70.4	65.7
PM$_{23}$PVDF$_{77}$	PM	1	100.5	-	-	-	92.7	24.6	37.5
		5	98.5	-	-	-	90.1	23.9	12.2
		20	97.8	-	-	-	86.1	13.1	10.9
		60	97.1	-	-	-	80.8	13.8	12.9
	PVDF	1	-	-	172.7	179.3	157.4	30.5	33.6
		5	-	174.4	170.7	176.9	152.5	37.3	38.9
		20	-	174.2	170.1	-	147.0	33.5	37.4
		60	-	173.4	167.8	-	141.0	35.0	37.7
PM$_{38}$PVDF$_{62}$	PM	1	100.9	-	-	-	94.3	17.2	17.9
		5	99.6	-	-	-	91.5	12.5	13.2
		20	98.1	-	-	-	86.8	20.5	14.1
		60	97.4	-	-	-	80.8	21.9	14.4
	PVDF	1	-	-	172.2	179.1	157.5	25.8	26.8
		5	-	174.5	170.2	176.7	151.5	25.8	29.5
		20	-	173.6	168.4	-	146.4	26.8	28.9
		60	-	173.1	167.3	-	139.8	27.1	29.7

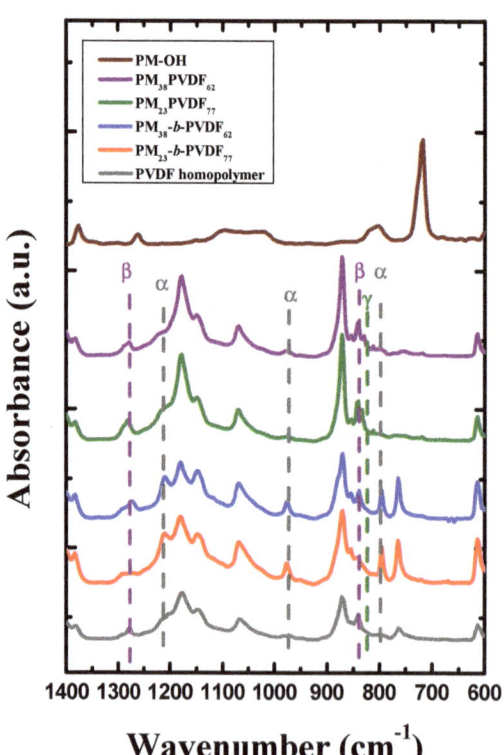

Figure 6. Sections of FTIR spectra of PM-OH, PVDF homopolymer, PM$_{23}$-b-PVDF$_{77}$, PM$_{38}$-b-PVDF$_{62}$, PM$_{23}$PVDF$_{77}$, and PM$_{38}$PVDF$_{62}$ samples after a cooling sweep at 1 °C/min. The grey dashed line shows the bands for the α-phase; the purple dashed line is for the β-phase, and the green dashed line corresponds to the γ-phase.

When the crystallization of the PVDF homopolymer happens at a low cooling rate, three very weak bands can be seen at 1214, 976, and 796 cm^{-1}, which correspond to the α-phase. This means that the formation of the α-phase is not really promoted in the homopolymer. Moreover, there are two additional, more intense, main bands, at 1275 and 840 cm^{-1}, which correspond to the crystalline β-phase. This means that, surprisingly, the PVDF homopolymer is able to crystalize in the ferroelectric β-phase when the polymer is crystallized slowly from the melt.

The spectra for both diblock copolymers show bands for the crystalline α-phase and β-phase. The PM$_{23}$-b-PVDF$_{77}$ shows only one small band located at 1278 cm^{-1}, corresponding to the β-phase, but there is not any band at 840 cm^{-1}. This indicates the presence of a small amount of β-phase in the copolymer. In addition, the FTIR spectrum of this sample clearly shows the bands corresponding to the α-phase, which indicates that the crystallization observed at 1 °C/min corresponds mainly to the paraelectric α-phase, which confirms the DSC results.

On the other hand, the spectrum of the PM$_{38}$-b-PVDF$_{62}$ sample shows the α-crystals bands mentioned before and the band located at 1278 cm^{-1} that corresponds to the β-phase. The FTIR analysis of this diblock copolymer demonstrates that the α-phase and the β-phase coexist simultaneously after samples have been cooled at 1 °C/min. Again, this behavior confirms the DSC results: at low cooling rates, the formation of the β-phase is promoted, but the α-phase remains present.

The FTIR spectra for the two blends (Figure 6) show the two main bands corresponding to the β-phase plus a new band located at 811 cm^{-1}, which corresponds to the γ-phase crystals [60]. All the characteristic bands for PM and PVDF are shown in Table 3.

Table 3. Values and description of the main FTIR bands for α, β, γ-phases for PVDF and PM.

Wavenumber (cm^{-1})	Phase	Description [61,62]
720	PM	C-C rocking deformation
796	α-PVDF	CH$_2$ rocking
811	γ-PVDF	-
840	β-PVDF	CH$_2$,CF$_2$ asymmetric stretching vibration
976	α-PVDF	CH out of plane deformation
1214	α-PVDF	CF stretching
1232	γ-PVDF	CF out of plane deformation
1275	β-PVDF	CF out of plane deformation
1377	PM	CH$_3$ symmetric deformation

WAXS experiments were performed to investigate what phases crystallized during the cooling process at 1 °C/min from the molten state (Figure 7). The main reflections for the PM are located at 15.2 and 16.7 nm^{-1} as can be seen in the pattern of the PM-OH sample. PM crystallizes in an orthorhombic unit cell with parameters a = 0.742 nm, b = 0.495 nm, c = 0.255 nm, and β = 90°, with a P-D$_{2h}$ space group [63,64]. The crystallographic planes for these peaks are (110) and (200), respectively [65,66].

PVDF has different crystalline phases, which appear as WAXS reflections at different q-values (see Figure 7). The peaks that are located at q-values of 12.6, 13.1, 14.2, and 18.9 nm^{-1} correspond to the crystalline α-phase, and the reflections of this paraelectric phase have the following crystallographic planes: (100), (020), (110), and (120/021) [67–69]. The α-phase of PVDF is characterized by a pseudo-orthorhombic unit cell with a = 0.496 nm, b = 0.964 nm, c = 0.462 nm, and β = 90° and has a P2/C space group [70,71]. In our case, these reflections appear for the diblock copolymers, the blends, and the homopolymer. These reflexions are more intense in the homopolymer and in the PM$_{23}$-b-PVDF$_{77}$ than in the other samples. Based on this result and the FTIR spectra, we can conclude that during the crystallization of the PM$_{23}$-b-PVDF$_{77}$ the formation of the α-phase is always promoted at low cooling rates.

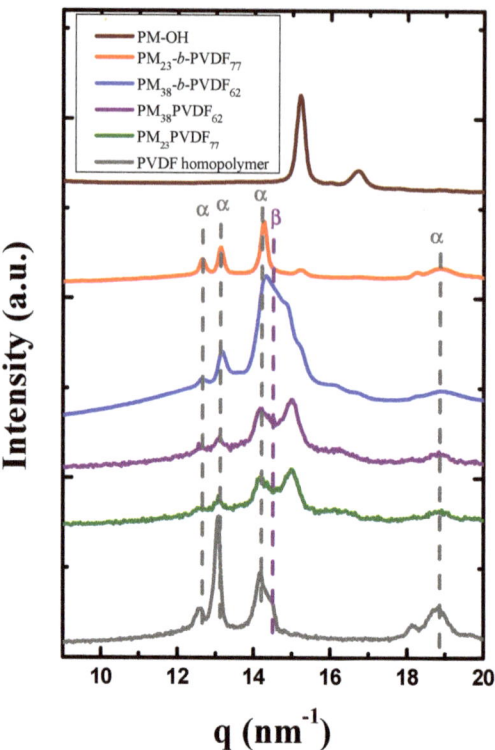

Figure 7. WAXS diffraction patterns of PM-OH, PVDF homopolymer, both blends, and both block copolymers at room temperature after a crystallization process at 1 °C/min. The grey dashed lines indicate the peaks associated to the α-phase, and the purple dashed line indicates the peak of the β-phase.

However, apart from the characteristic peaks of the α-phase, the other samples containing PVDF display one extra peak or shoulder in their patterns at 13.5 nm^{-1} (Figure 7). This new reflection corresponds to the crystallization of the β-phase, which has the (200/110) crystal plane [18]. The β-phase of PVDF is characterized by an orthorhombic unit cell, which has a $Cm2m$ space group and the following dimensions: a = 0.847 nm, b = 0.490 nm, and c = 0.256 nm [72]. The presence of this peak is in agreement with the results obtained before in the DSC analysis, which suggests that the formation of the β-phase is promoted in samples that were previously cooled at 1 °C/min and coexists with a small amount of crystalline α-phase. It seems that the amount of PM in the diblock copolymer can affect the PVDF crystallization in order to promote the desired β-phase.

3.3. Dielectric Spectroscopy Studies in PVDF and Its Copolymers

Figure 8 shows the BDS results for PVDF and its copolymers with PM. In particular, Figure 8a–c displays dielectric spectra: the imaginary part of the complex dielectric permittivity as a function of the frequency. The data presented correspond to the one collected by isotherms from −100 to 0 °C in steps of 10 °C (measured on heating). The corresponding experiments on cooling were nearly indistinguishable. In general, the relaxation processes were characterized by a single maximum, which shifted towards higher frequencies and increased in intensity as the temperature was increased.

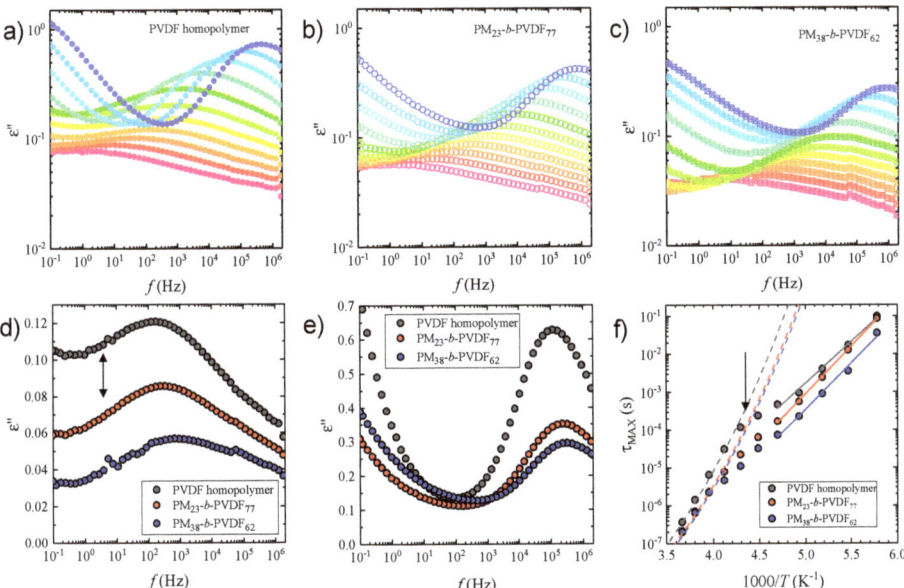

Figure 8. Dielectric spectra (imaginary part of the complex dielectric permittivity as a function of the frequency) for (**a**) PVDF homopolymer, (**b**) PM$_{23}$-b-PVDF$_{77}$, (**c**) PM$_{38}$-b-PVDF$_{62}$, as well as dielectric relaxations of the studied samples at (**d**) $-70\,°C$ and (**e**) $-10\,°C$ and (**f**) relaxation map of the studied samples.

At low temperatures ($-100\,°C$ to $-60\,°C$), a weak and broad peak was observed for all samples, although with different characteristics. PVDF displayed the highest intensity peaks, reaching ε'' values of around 0.1. In the case of the diblock copolymers, the intensity of the relaxations decreased with PM content. We also observed that, as PM content increased, the relaxation peaks maxima shifted towards higher frequencies. As an example, Figure 8d shows the dielectric relaxations of the samples at $-70\,°C$. In addition to the differences already discussed, PVDF displayed a pronounced asymmetry towards low frequencies (black arrow in Figure 8d). However, the relative intensity of this low-frequency signal decreased for the samples containing PM blocks.

Comparing with previous literature reports, and taking into consideration the intensity and position of the peaks, we were able to assign the low-temperature process to the local β-relaxation of PVDF related to local motions of polar groups in the polymer [73–76]. As the temperature was further increased ($T > -60\,°C$), the relaxation peaks suffered important changes. In all cases, as the maxima moved towards higher frequencies, the peaks were narrower and showed a dramatic intensity increase. These changes in the dielectric relaxation occurred at temperatures close to the glass transition of PVDF (-43 to $-23\,°C$) [77]. Thus, we could relate the changes to the α-relaxation of the PVDF. This relaxation process is related to the segmental motion of the PVDF polymer chain taking place at temperatures above the glass transition (T_g), as widely reported [73–75,78,79]. Please notice that our experimental results showed a continuous change in the dielectric spectra, going from the β- to α-relaxation, instead of separated peaks observed in previous works [73–75,78,79]. Nonetheless, although in our current work the α-relaxation peak could not be well resolved at low frequencies, the data showed an increased broadness at $T = -50$ to $-40\,°C$. The peak was better resolved in the PVDF sample than in PM-b-PVDF copolymers, which indicates that the PVDF segmental relaxation was affected by the presence of PM units. In fact, in the -50–$0\,°C$ temperature range, PM-b-PVDF copolymers showed lower segmental relaxation intensities and slightly faster dynamics compared to the PVDF. Figure 8e presents a comparison of the datasets at −10 °C where this evidence can be observed.

Figure 8f shows the relaxation map of the samples. The relaxation time (τ_{MAX}) was calculated from the maxima of the dielectric relaxation peaks. In all cases, we observed two trends in the temperature dependence of relaxation times. At low temperatures ($-100 \leq T$ (°C) ≤ -60), the relaxation times followed an Arrhenius behavior, as described by [80]:

$$\tau_{MAX} = \tau_0 \exp\left[\frac{E_A}{kT}\right] \quad (2)$$

where E_A is the activation energy, k is Boltzmann's constant, and τ_0 a pre-exponential factor. The obtained results are shown in Figure 8f as continuous lines and are summarized in Table 4. For PVDF, we found E_A = 42 kJ/mol, which increased slightly for the PM-b-PVDF systems (~48 kJ/mol). These values are quite similar to the one reported before by Sy and Mijovic (~43 kJ/mol) [73] for the local relaxation of PVDF, while slightly lower than that observed by Linares and collaborators (~60 kJ/mol) [74].

Table 4. Arrhenius fit results for PVDF and its copolymers with PM.

Sample	τ_0 (s)	E_A (kJ/mol)	τ_0 (s)	D	T_{VFT} (°C)	T_{g-BDS} (°C)
PVDF	$2 \times 10^{-14\pm1}$	42 ± 1			-151 ± 1	-80 ± 1
PM$_{23}$-b-PVDF$_{77}$	$3 \times 10^{-16\pm1}$	48 ± 1	10^{-14}	21 ± 1	-154 ± 1	-85 ± 1
PM$_{38}$-b-PVDF$_{62}$	$2 \times 10^{-16\pm1}$	47 ± 1			-155 ± 1	-86 ± 1

At temperatures above −60 °C, the relaxation times of the samples showed a deviation from the low-temperature Arrhenius trend. In all the studied samples, a sort of "kink" appeared at temperatures around −50 to −60 °C (see arrow in Figure 8f). We related these changes to the effect of the segmental relaxation of PVDF on the relaxation times. We also observed that the kink's intensity was reduced in the block copolymer as the PM content increased. These sorts of trends, or anomalies, have been reported before for PVDF-based systems. For example, Sy and Mijovic observed a similar behavior in local motions of semicrystalline PVDF/PMMA blends [73]. In that work, the temperature dependence of the relaxation times of PVDF/PMMA blends was described as a gradual crossover from local to segmental motions, which was clearly different from an α-β merging. The 90/10 PVDF/PMMA showed the most pronounced kink, which decreased as the PMMA content increased. However, the neat PVDF did not show this signature. Martínez-Tong et al. [81] also observed a continuous transition in the dielectric relaxation map of a PVDF copolymer with trifluoroethylene (PVDF-TrFE), with a VDF mol content of 76%. In that work, the authors observed a crossover from the segmental relaxation to the ferroelectric-paraelectric relaxation of the polymer. Just at the transition temperatures (~47–57 °C), a small kink can be detected in the relaxation plot. Finally, very recently, Napolitano and collaborators observed an anomalous behavior in the local relaxation of PVDF copolymers with hexafluoropropylene (HFP) [76]. In their work, the dielectric relaxation experiments showed that, in the vicinity of T_g, the PVDF-HFP copolymers displayed a so-called "anomalous minimum" in the local relaxation. The authors related their findings to the bonds formed by fluorine entities, similar to those observed in propylene glycol systems. Moreover, the authors also observed that the anomalous process weakened when the PVDF-HFP samples were prepared as ultrathin polymer films. This nanoconfinement-induced reduction in the anomaly was explained by means of the minimal model and related to an asymmetry in the well potential describing the molecular motion. In this work, we observed that PM-b-PVDF samples showed a reduction of the observed kink, whose intensity decreased as PM content was increased. This could indicate that the PM block is inducing local confinement effects on the samples.

Finally, we attempted to model the data points in the −50–0 °C temperature range using the Vogel–Fulcher–Tamman (VFT) equation, described by [80]:

$$\tau_{MAX} = \tau_\infty \exp\left[\frac{DT_{VFT}}{T - T_{VFT}}\right] \quad (3)$$

where τ_∞ is a pre-exponential factor, D is a dimensionless parameter related to the dynamic fragility [82], and T_{VFT} the Vogel temperature. The results obtained are summarized in Table 4 and the fits are shown in Figure 8f by dashed lines. We highlight that the value of τ_∞ was set at 10^{-14} s, based on the discussion of Angell [82,83]. For all samples, we obtained a $D = 21$, indicating a small deviation from an Arrhenius process. This value was slightly larger than the ones reported before ($D = 12$–15) for PVDF [73,79]. However, it was fairly comparable to the one obtained by Martínez-Tong and collaborators for the PVDF-TrFE copolymer ($D = 21.6$). Finally, we were able to predict the dynamic glass transition temperature (T_{g-BDS}) of the samples in our study from the VFT fit. This parameter was defined as the temperature where the segmental relaxation time reached 100 s. The results obtained, shown in Table 4, allowed to determine a $T_{g-BDS} = -80$ °C for PVDF. This value decreased for the PM-b-PVDF samples with increasing PM content, which was in line with the faster dynamics observed. The T_{g-BDS} obtained were lower than the usual ones reported for PVDF by different methods ($T_g = -63$ to 23 °C) [84,85]. However, we emphasize that both the PVDF and PM-b-PVDF copolymers have low molecular weights (6–8 kDa), which would explain the obtained results. In addition, we should take into account that, in semicrystalline polymers, the dynamics in the more amorphous environments dominate the dielectric relaxation peak frequency position [86].

3.4. How the Isothermal Crystallization Affects PVDF Blends and Block Copolymer Samples

Figure 9 shows the spherulitic growth rate of PVDF, its copolymers and the prepared blends as a function of the isothermal crystallization temperature. The high nucleation density observed in the blends only allow us to measure spherulites at relatively high crystallization temperatures. Experiments were performed by cooling the samples from the melt to a chosen crystallization temperature in the range from 131 to 164 °C. Spherulitic growth rates for each sample, G (µm/min), were determined at different crystallization temperatures from the slope of radius versus time plots (which were always linear).

Figure 9a shows the spherulitic growth rate G (µm/min) as a function of T_c. As can be seen, the growth rate is faster in the copolymers than in the blends and the homopolymer sample in the low temperature range. However, the comparison is difficult, as the crystallization ranges of the sample do not overlap. G dramatically decreases when the PVDF is blended with PM. The supercooling required for crystallization increases when the PVDF is blended with PM, as a result of the change in the equilibrium melting temperature. When G is plotted as a function of supercooling ($\Delta T = T_m^0 - T_c$), using the equilibrium melting temperatures (T_m^0) determined by the Hoffman–Weeks method, in Figure 9b, the curves are now shifted along the x-axis reducing the differences between the overall crystallization curves versus T_c. In this representation as a function of supercooling, it is easier to observe the above mentioned trends.

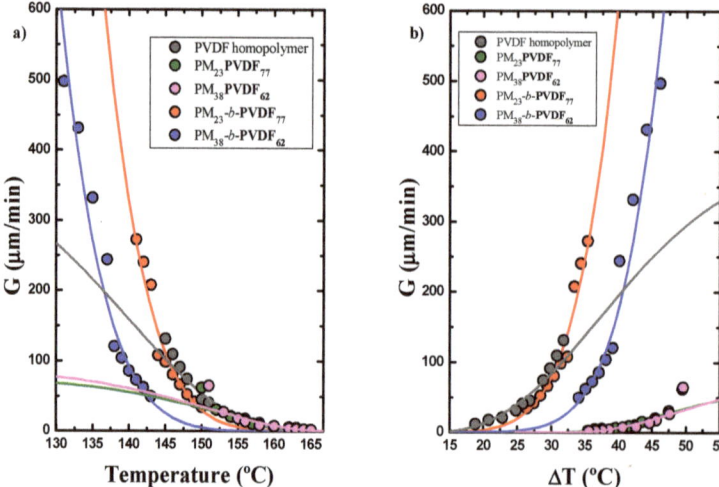

Figure 9. (a) Spherulitic growth rates determined by PLOM for homopolymer PVDF, the PVDF block of the diblock copolymers, and the PVDF phase within the blends studied and (b) spherulitic growth rates as a function of supercooling. The solid lines are the fits to the Lauritzen–Hoffman (LH) theory.

It is unexpected that the growth rate (Figure 9b) of the PVDF component decreases in the blends as compared to the neat PVDF. One possible explanation could be that even though the blends are immiscible (as indicated by the DSC results), the molten PM-OH is capable of interacting with the PVDF (though the OH group) reducing the PVDF diffusion to the growth front.

In the diblock copolymers case, the growth rate of the PVDF block decreases as the PM content in the copolymer decreases. It can also be noted that the temperature dependence of the growth rate between the neat PVDF homopolymer and the PVDF blocks in the diblock copolymers is very different. This is easily captured by the Lauritzen and Hoffman fits, which are represented as solid lines in Figure 9.

Isothermal crystallization experiments were performed by DSC to determine the overall crystallization rate of the samples (which include both nucleation and growth contributions). Differences in the PVDF polymorphism and its crystallization kinetics were observed depending on the structural forms of the respective samples. The Avrami theory and the Lauritzen and Hoffman theory were employed [87,88] to describe the primary crystallization process in polymers and to plot several kinetic crystallization parameters as a function of the crystallization temperature.

Figure 10a shows the inverse of the induction time (t_0) versus the isothermal crystallization temperature (T_c) for the different PVDF samples. The induction time is equivalent to the primary nucleation time before any crystallization is detected by the DSC. The inverse of the induction time is proportional to the primary nucleation rate of the PVDF components in the different samples. The nucleation rate depends on the composition and the nature of the samples. The nucleation rate of the PVDF block within the PM_{38}-b-$PVDF_{62}$ sample is faster than in the homopolymer sample, while in the blend, the PVDF phase has a slower nucleation rate.

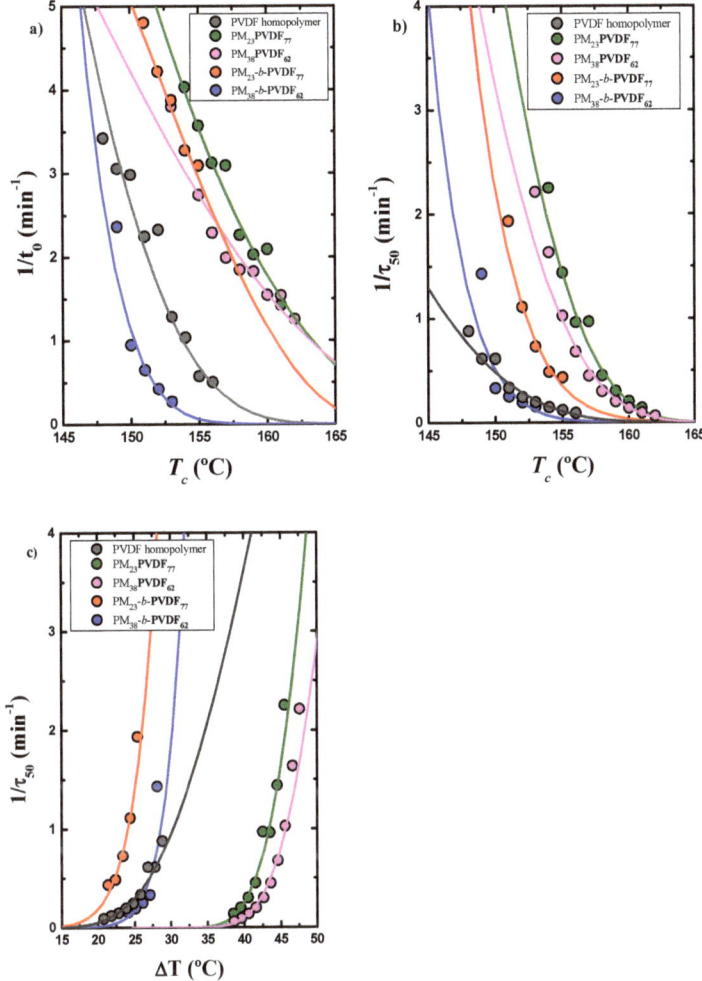

Figure 10. (a) $1/t_0$ as a function of crystallization temperature and inverse of half-crystallization time for the PVDF component of all samples shown as a function of (b) T_c and (c) ΔT for all the PVDF samples measured by DSC. The solid lines are the fits to the Lauritzen–Hoffman (LH) theory.

Figure 10b shows the inverse of the half crystallization rate ($\tau_{50\%}$) versus the isothermal crystallization temperature (T_c). The $1/\tau_{50\%}$ value is the inverse of the time needed to achieve the 50% of the total transformation to the semi-crystalline state during the isothermal crystallization process and represents an experimental measure of the overall crystallization rate, which includes both growth and nucleation contributions.

Figure 10b reflects a combined trend of the observed nucleation behavior (Figure 9a) and the spherulitic growth behavior (Figure 9a). Both the PVDF homopolymer and the PM$_{23}$PVDF$_{77}$ blend exbibit the lowest overall crystallization rates. However, as in the overall crystallization, both nucleation and spherulitic growth rate contribute; in this case, $1/\tau_{50\%}$ does not decrease as dramatically as G for the rest of the materials. Therefore, the changes in nucleation density strongly affect the overall crystallization rates determined by DSC in these PVDF-based blend samples. Figure 10c shows these results when they are plotted against the supercooling (ΔT) and the curves are shifted in the x-axis standardizing the differences in crystallization temperature exhibited by the different samples.

The Avrami theory is a useful tool to fit the overall crystallization kinetics of polymers during the primary crystallization regime [89–91]. The Avrami theory is given by the following equation:

$$1 - V_c(t - t_0) = exp(-k(t - t_0)^n) \qquad (4)$$

where V_c is the relative volumetric transformed fraction, t is the time of the experiment, t_0 is the induction time before the crystals start to grow, k is the overall crystallization rate constant, and n is the Avrami index, which is related to the time dependence of the nucleation and the crystal growth geometry.

By applying the Avrami equation to the isothermal crystallization curves at each chosen crystallization temperature, it is possible to calculate the Avrami index (n), but it is only possible when the crystallization starts at the isothermal temperature selected and not during the cooling, as happened in the case of the PM. Figure 11a shows all the n values for the crystallization of the PVDF component in all the samples studied during this work. Usually, for polymers, n is between 1.5 and 4. When this value is higher than 2.4, the crystals of the polymer grow as spherulites. In our case, all the samples have an n value higher than 2.5 with the exception of the PM_{38}-b-$PVDF_{62}$ sample. For the samples with an n value below 2.5, crystals grow in 2D, forming axialites. Figure 11b shows the evolution of the $k^{1/n}$ value at different crystallization temperatures, and these values are proportional to the overall crystallization rate. The comparison between Figures 9b and 11b demonstrates that the theoretical results obtained through the Avrami theory are really close to the experimental results obtained using the Lauritzen and Hoffman method as the trends in the data are similar($1/\tau_{50\%}$).

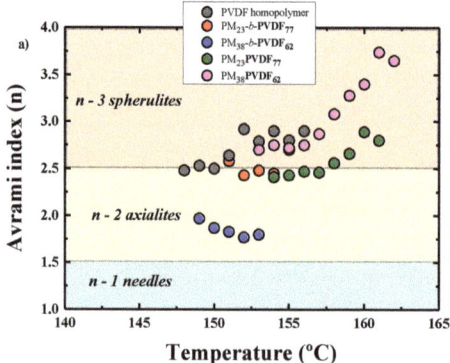

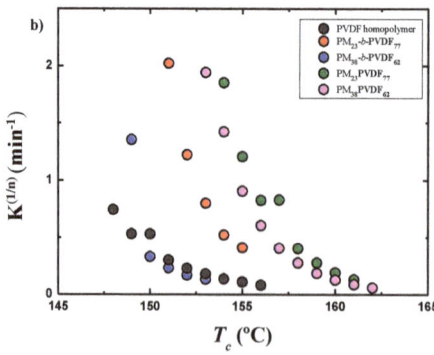

Figure 11. (a) PVDF Avrami index values for all the temperatures used in the isothermal crystallization and (b) isothermal crystallization rate obtained by the Avrami model.

The value of the equilibrium melting temperature of each sample was calculated using the Hoffman–Weeks method; see the Supporting Information. The values obtained are listed in Table S2 in the Supporting Information.

The analysis of the heating curves after the isothermal crystallization processes may allow us to know how the PVDF crystallizes and which crystalline phase is obtained after these procedures. Figure 12 shows all the melting curves for the PVDF component in each sample at all the isothermal crystallization temperatures studied. The T_c selected through the $T_{c,min}$ method are similar for the block copolymers and the homopolymer sample, while the blends have higher T_c values.

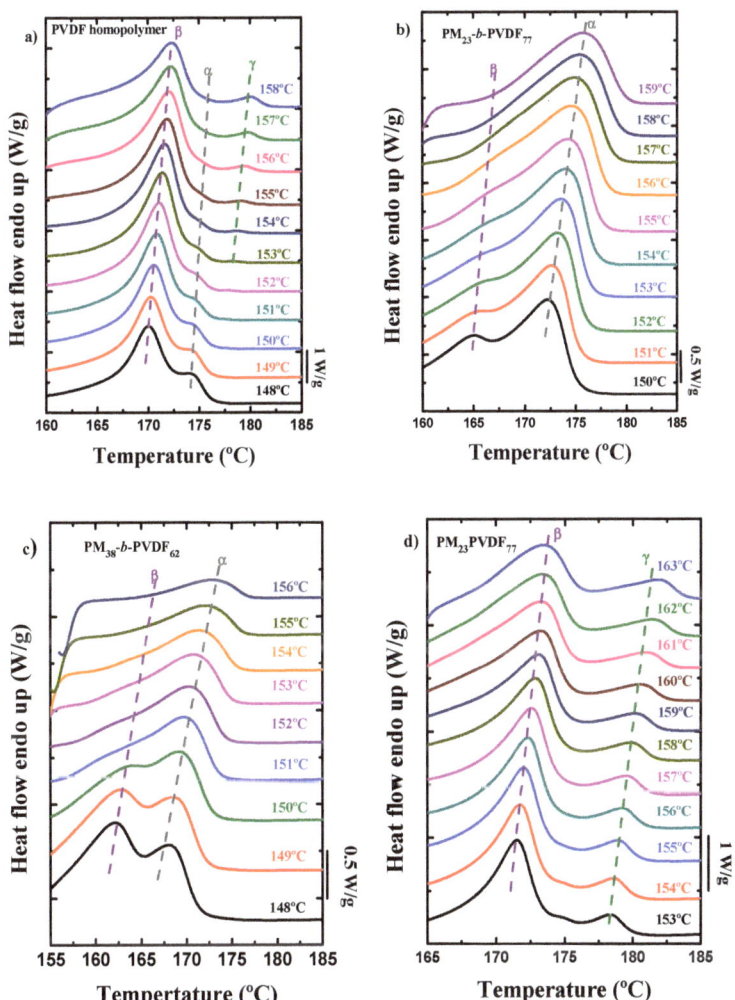

Figure 12. Cont.

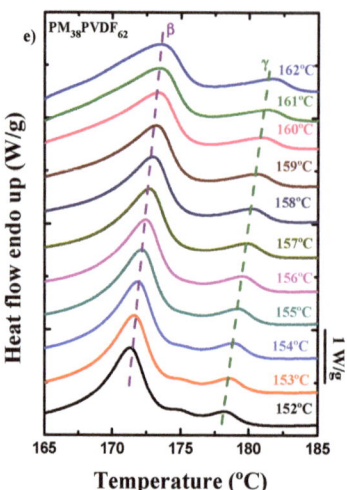

Figure 12. DSC PVDF melting curves after the isothermal crystallization at different temperatures of (**a**) PVDF homopolymer, (**b**) PM_{23}-b-$PVDF_{77}$, (**c**) PM_{38}-b-$PVDF_{62}$, (**d**) $PM_{23}PVDF_{77}$, and (**e**) $PM_{38}PVDF_{62}$ samples.

The PVDF homopolymer (Figure 12a) has two melting peaks when the isothermal crystallization temperature used was low: one main peak at low temperatures and another small peak at higher temperatures. The main peak corresponds to the β-phase, and the second peak to the α-phase. When the crystallization temperature increases, the peak from the α-phase starts decreasing until it disappears and a new peak appears at even higher temperatures. This new peak corresponds to the crystalline γ-phase. This means that the PVDF low molecular weight homopolymer sample can crystallize in all ferroelectric phases when submitted to low cooling rates and also during an isothermal process at high crystallization temperatures.

The behavior of the PVDF block in the diblock copolymers (Figure 12b,c) is completely different from the homopolymer sample. In this case, only two melting peaks are observed when the isothermal crystallization temperature used was low. In the case of the PM_{23}-b-$PVDF_{77}$ sample, the main peak is observed at higher temperatures. When the crystallization temperature increases, the first peak tends to disappear and only the main peak, which belongs to the α-phase, remains.

For the PM_{38}-b-$PVDF_{62}$ sample, at low crystallization temperatures, the first melting peak is promoted (β phase), but as the isothermal crystallization temperature is increased, the size of this peak starts to decrease, and at high crystallization temperatures, only one peak is observed, which also corresponds to the α-phase.

Both PM/PVDF blends (Figure 12d,e) have similar melting curves regardless of the PM content. Both blends show three peaks at low isothermal crystallization temperatures: The largest one is located at low temperatures and corresponds to the β-phase; then, there is a shoulder at about 175 °C, which is the melting peak of the α-phase, and finally, the last one at higher temperatures is the melting peak of the γ-phase. When the crystallization temperature is increased, only the shoulder of the α-phase disappears, while both ferroelectric phases remain.

As during isothermal crystallization, the PVDF component develops a complex polymorphic structure that changes with crystallization temperature; this helps to explain the complex trends observed in the growth kinetics (Figure 9), nucleation rate (Figure 10a), and overall crystallization rate (Figure 10c).

4. Conclusions

The complex crystallization of PVDF was found to depend on the nature of its chemical environment. We found significant differences in crystallization and polymorphic structure depending on whether the PVDF was a homopolymer (the homopolymer of the diblock copolymers), present as a block in the studied diblock copolymers, and present as a phase in the blends. The crystallization conditions were also found to dramatically affect the number and amount of the polymorphic crystalline phases produced.

DSC, PLOM, and TEM results clearly indicated that the blends prepared here are immiscible and phase segregate. On the other hand, the linear diblock copolymers crystallize from a mixed melt or very weakly segregated melt according to SAXS, TEM, and PLOM.

We were able to clearly identify the different crystalline phases form by the PVDF component in the different samples examined (i.e., α, β, and γ phases) by DSC, FTIR, and WAXS. Their number and content varied depending on sample composition, cooling rate employed, or isothermal crystallization temperature used during isothermal crystallization tests.

The BDS results indicated that the PVDF block in the copolymers has lower T_g values than the homopolymer, which was in line with the faster chain dynamics observed in them. The spherulitic growth rates, nucleation rates, and overall crystallization rates were determined, and different values were obtained depending on the sample. This is not surprising considering that the melting after isothermal crystallization revealed that the polymorphic structure of each sample varied during isothermal crystallization.

Supplementary Materials: The following are available online at https://www.mdpi.com/article/10.3390/polym13152442/s1, Figure S1: ^{1}H NMR (500 MHz) spectrum of *CTA* in $(CD_3)_2CO$ at 40 °C. Figure S2: ^{1}H NMR (500 MHz) spectrum of Linear PVDF in DMF-d_7 at 25 °C. Figure S3: ^{19}F NMR (500 Hz) spectrum of Linear PVDF in DMF-d_7 at 25 °C. Figure S4: GPC trace (DMF, 40 °C, PS standard) of linear PVDF (negative refractive index increment). Figure S5: Values of the intensity of the colours against the temperature during the cooling process at 20 °C/min using the Image J software of (a) PM$_{23}$-b-PVDF$_{77}$ and (b) PM$_{38}$-b-PVDF$_{62}$. Figure S6: Melting temperatures against crystallization temperatures with their respective linear fit to calculate the equilibrium melting temperature using the Hoffman–Weeks method. Table S1: Polymerization conditions and molecular characteristics of the linear PVDF synthesised by RAFT polymerization. Table S2: Equilibrium melting temperature (T_m^0) for the PVDF homopolymer, PVDF blends and PVDF block copolymers. Scheme S1: Synthesis of PM-b-PVDF Diblock Copolymer by Polyhomologation and ITP.

Author Contributions: A.J.M. and J.M. designed the work and its conceptualization. F.A. and G.Z. synthesized the block copolymers and their homopolymers under the supervision of N.H. The experiments were performed by N.M. at the UPV/EHU labs under the supervision of A.J.M. and J.M. while the BDS experiments were carried out in the CFM labs by D.E.M.-T. under the supervision of A.A. The article was written by N.M., A.J.M. and J.M. and revised by all the co-authors. All authors have read and agreed to the published version of the manuscript.

Funding: Technical and human support by SGIker (UPV/EHU) is gratefully acknowledged for the TEM images acquired. We would also like to acknowledge funding from MINECO through grant MAT2017-83014-C2-1-P and from the Basque Government through grant IT1309-19. This research was funded by ALBA synchrotron facility through granted proposal 2020084441.

Institutional Review Board Statement: Not applicable.

Informed Consent Statement: Not applicable.

Data Availability Statement: The data presented in this study are available upon request from the corresponding author.

Conflicts of Interest: The authors declare no conflict of interest.

References

1. Düerkop, D.; Widdecke, H.; Schilde, C.; Kunz, U.; Schmiemann, A. Polymer Membranes for All-Vanadium Redox Flow Batteries: A Review. *Membranes* **2021**, *11*, 214. [CrossRef]
2. Liao, C.-Y.; Hsiao, Y.-T.; Tsai, K.-W.; Teng, N.-W.; Li, W.-L.; Wu, J.-L.; Kao, J.-C.; Lee, C.-C.; Yang, C.-M.; Tan, H.-S.; et al. Photoactive Material for Highly Efficient and All Solution-Processed Organic Photovoltaic Modules: Study on the Efficiency, Stability, and Synthetic Complexity. *Solar RRL* **2021**, *5*, 2000749. [CrossRef]
3. Patnam, H.; Dudem, B.; Graham, S.A.; Yu, J.S. High-performance and robust triboelectric nanogenerators based on optimal microstructured poly(vinyl alcohol) and poly(vinylidene fluoride) polymers for self-powered electronic applications. *Energy* **2021**, *223*, 120031. [CrossRef]
4. Trevino, J.E.; Mohan, S.; Salinas, A.E.; Cueva, E.; Lozano, K. Piezoelectric properties of PVDF-conjugated polymer nanofibers. *J. Appl. Polym. Sci.* **2021**, *138*, 50665. [CrossRef]
5. Bregar, T.; Starc, B.; Čepon, G.; Boltežar, M. On the Use of PVDF Sensors for Experimental Modal Analysis. In *Topics in Modal Analysis & Testing*; Springer: Cham, Switzerland, 2021; Volume 8, pp. 279–281.
6. Tonazzini, I.; Bystrenova, E.; Chelli, B.; Greco, P.; De Leeuw, D.; Biscarini, F. Human Neuronal SHSY5Y Cells on PVDF:PTrFE Copolymer Thin Films. *Adv. Eng. Mater.* **2015**, *17*, 1051–1056. [CrossRef]
7. Cardoso, V.F.; Correia, D.M.; Ribeiro, C.; Fernandes, M.M.; Lanceros-Méndez, S. Fluorinated Polymers as Smart Materials for Advanced Biomedical Applications. *Polymers* **2018**, *10*, 161. [CrossRef]
8. Voet, V.S.D.; ten Brinke, G.; Loos, K. Well-defined copolymers based on poly(vinylidene fluoride): From preparation and phase separation to application. *J. Polym. Sci. Part A Polym. Chem.* **2014**, *52*, 2861–2877. [CrossRef]
9. Liu, Z.H.; Pan, C.T.; Lin, L.W.; Lai, H.W. Piezoelectric properties of PVDF/MWCNT nanofiber using near-field electrospinning. *Sens. Actuators A Phys.* **2013**, *193*, 13–24. [CrossRef]
10. Kepler, R.G.; Anderson, R.A. Ferroelectric polymers. *Adv. Phys.* **1992**, *41*, 1–57. [CrossRef]
11. Scott, J.F.; Paz de Araujo, C.A. Ferroelectric Memories. *Science* **1989**, *246*, 1400. [CrossRef] [PubMed]
12. Li, H.; Wang, R.; Han, S.-T.; Zhou, Y. Ferroelectric polymers for non-volatile memory devices: A review. *Polym. Int.* **2020**, *69*, 533–544. [CrossRef]
13. Lee, Y.; Park, J.; Cho, S.; Shin, Y.-E.; Lee, H.; Kim, J.; Myoung, J.; Cho, S.; Kang, S.; Baig, C.; et al. Flexible Ferroelectric Sensors with Ultrahigh Pressure Sensitivity and Linear Response over Exceptionally Broad Pressure Range. *ACS Nano* **2018**, *12*, 4045–4054. [CrossRef] [PubMed]
14. Guyomar, D.; Pruvost, S.; Sebald, G. Energy harvesting based on FE-FE transition in ferroelectric single crystals. *IEEE Trans. Ultrason. Ferroelectr. Freq. Control* **2008**, *55*, 279–285. [CrossRef]
15. Lando, J.B.; Doll, W.W. The polymorphism of poly(vinylidene fluoride). I. The effect of head-to-head structure. *J. Macromol. Sci. Part B* **1968**, *2*, 205–218. [CrossRef]
16. Lovinger, A.J. Ferroelectric Polymers. *Science* **1983**, *220*, 1115. [CrossRef]
17. Cortili, G.; Zerbi, G. Further infra-red data on polyvinylidene fluoride. *Spectrochim. Acta Part A Mol. Spectrosc.* **1967**, *23*, 2216–2218. [CrossRef]
18. Lando, J.B.; Olf, H.G.; Peterlin, A. Nuclear magnetic resonance and x-ray determination of the structure of poly(vinylidene fluoride). *J. Polym. Sci. Part A-1 Polym. Chem.* **1966**, *4*, 941–951. [CrossRef]
19. Tashiro, K.; Kobayashi, M.; Tadokoro, H.; Fukada, E. Calculation of Elastic and Piezoelectric Constants of Polymer Crystals by a Point Charge Model: Application to Poly(vinylidene fluoride) Form I. *Macromolecules* **1980**, *13*, 691–698. [CrossRef]
20. Murayama, N.; Nakamura, K.; Obara, H.; Segawa, M. The strong piezoelectricity in polyvinylidene fluoride (PVDF). *Ultrasonics* **1976**, *14*, 15–24. [CrossRef]
21. Kaempf, G. Special Polymers for Data Memories. *Polym. J.* **1987**, *19*, 257–268. [CrossRef]
22. Voet, V.S.D.; van Ekenstein, G.O.R.A.; Meereboer, N.L.; Hofman, A.H.; ten Brinke, G.; Loos, K. Double-crystalline PLLA-b-PVDF-b-PLLA triblock copolymers: Preparation and crystallization. *Polym. Chem.* **2014**, *5*, 2219–2230. [CrossRef]
23. Bachmann, M.A.; Gordon, W.L.; Koenig, J.L.; Lando, J.B. An infrared study of phase-III poly(vinylidene fluoride). *J. Appl. Phys.* **1979**, *50*, 6106–6112. [CrossRef]
24. Bachmann, M.; Gordon, W.L.; Weinhold, S.; Lando, J.B. The crystal structure of phase IV of poly(vinylidene fluoride). *J. Appl. Phys.* **1980**, *51*, 5095–5099. [CrossRef]
25. Lovinger, A.J. Poly(Vinylidene Fluoride). In *Developments in Crystalline Polymers—1*; Bassett, D.C., Ed.; Springer: Dodrecht, The Netherlands, 1982; pp. 195–273. [CrossRef]
26. Gal'perin, Y.L.; Strogalin, Y.V.; Mlenik, M.P. Crystal structure of polyvinylidene fluoride. *Polym. Sci. USSR* **1965**, *7*, 1031–1039. [CrossRef]
27. Li, L.; Zhang, M.; Rong, M.; Ruan, W. Studies on the transformation process of PVDF from α to β phase by stretching. *RSC Adv.* **2014**, *4*, 3938–3943. [CrossRef]
28. Du, C.-h.; Zhu, B.-K.; Xu, Y.-Y. Effects of stretching on crystalline phase structure and morphology of hard elastic PVDF fibers. *J. Appl. Polym. Sci.* **2007**, *104*, 2254–2259. [CrossRef]
29. Kaempf, G.; Siebourg, W.; Loewer, H.; Lazear, N. Polymeric Data Memories and Polymeric Substrate Materials for Information Storage Devices. In *Polymers in Information Storage Technology*; Mittal, K.L., Ed.; Springer: Boston, MA, USA, 1989; pp. 77–104. [CrossRef]

30. Li, M.; Stingelin, N.; Michels, J.J.; Spijkman, M.-J.; Asadi, K.; Feldman, K.; Blom, P.W.M.; de Leeuw, D.M. Ferroelectric Phase Diagram of PVDF:PMMA. *Macromolecules* **2012**, *45*, 7477–7485. [CrossRef]
31. Domenici, C.; De Rossi, D.; Nannini, A.; Verni, R. Piezoelectric properties and dielectric losses in PVDF–PMMA blends. *Ferroelectrics* **1984**, *60*, 61–70. [CrossRef]
32. Li, J.; Zhao, C.; Xia, K.; Liu, X.; Li, D.; Han, J. Enhanced piezoelectric output of the PVDF-TrFE/ZnO flexible piezoelectric nanogenerator by surface modification. *Appl. Surf. Sci.* **2019**, *463*, 626–634. [CrossRef]
33. Kim, G.H.; Hong, S.M.; Seo, Y. Piezoelectric properties of poly(vinylidene fluoride) and carbon nanotube blends: β-phase development. *Phys. Chem. Chem. Phys.* **2009**, *11*, 10506–10512. [CrossRef] [PubMed]
34. Zapsas, G.; Patil, Y.; Gnanou, Y.; Ameduri, B.; Hadjichristidis, N. Poly(vinylidene fluoride)-based complex macromolecular architectures: From synthesis to properties and applications. *Prog. Polym. Sci.* **2020**, *104*, 101231. [CrossRef]
35. Gebrekrstos, A.; Prasanna Kar, G.; Madras, G.; Misra, A.; Bose, S. Does the nature of chemically grafted polymer onto PVDF decide the extent of electroactive β-polymorph? *Polymer* **2019**, *181*, 121764. [CrossRef]
36. Golzari, N.; Adams, J.; Beuermann, S. Inducing β Phase Crystallinity in Block Copolymers of Vinylidene Fluoride with Methyl Methacrylate or Styrene. *Polymers* **2017**, *9*, 306. [CrossRef]
37. Lederle, F.; Härter, C.; Beuermann, S. Inducing β phase crystallinity of PVDF homopolymer, blends and block copolymers by anti-solvent crystallization. *J. Fluor. Chem.* **2020**, *234*, 109522. [CrossRef]
38. María, N.; Maiz, J.; Rodionov, V.; Hadjichristidis, N.; Müller, A.J. 4-Miktoarm star architecture induces PVDF β-phase formation in (PVDF)2-b-(PEO)2 miktoarm star copolymers. *J. Mater. Chem. C* **2020**, *8*, 13786–13797. [CrossRef]
39. Imai, S.; Hirai, Y.; Nagao, C.; Sawamoto, M.; Terashima, T. Programmed Self-Assembly Systems of Amphiphilic Random Copolymers into Size-Controlled and Thermoresponsive Micelles in Water. *Macromolecules* **2018**, *51*, 398–409. [CrossRef]
40. Imai, S.; Takenaka, M.; Sawamoto, M.; Terashima, T. Self-Sorting of Amphiphilic Copolymers for Self-Assembled Materials in Water: Polymers Can Recognize Themselves. *J. Am. Chem. Soc.* **2019**, *141*, 511–519. [CrossRef]
41. Jiang, Z.; Liu, H.; He, H.; Ribbe, A.E.; Thayumanavan, S. Blended Assemblies of Amphiphilic Random and Block Copolymers for Tunable Encapsulation and Release of Hydrophobic Guest Molecules. *Macromolecules* **2020**, *53*, 2713–2723. [CrossRef]
42. Cho, J. Analysis of Phase Separation in Compressible Polymer Blends and Block Copolymers. *Macromolecules* **2000**, *33*, 2228–2241. [CrossRef]
43. Leibler, L. Theory of microphase separation in block copolymers. *Macromolecules* **1980**, *13*, 1602–1617. [CrossRef]
44. Zhang, J.; Kremer, K.; Michels, J.J.; Daoulas, K.C. Exploring Disordered Morphologies of Blends and Block Copolymers for Light-Emitting Diodes with Mesoscopic Simulations. *Macromolecules* **2020**, *53*, 523–538. [CrossRef]
45. Zapsas, G.; Patil, Y.; Bilalis, P.; Gnanou, Y.; Hadjichristidis, N. Poly(vinylidene fluoride)/Polymethylene-Based Block Copolymers and Terpolymers. *Macromolecules* **2019**, *52*, 1976–1984. [CrossRef]
46. Lorenzo, A.T.; Arnal, M.L.; Albuerne, J.; Müller, A.J. DSC isothermal polymer crystallization kinetics measurements and the use of the Avrami equation to fit the data: Guidelines to avoid common problems. *Polym. Test.* **2007**, *26*, 222–231. [CrossRef]
47. Hiemenz, P.C.; Lodge, T.P. *Polymer Chemistry*, 2nd ed.; CRC Press: Boca Raton, FL, USA, 2007.
48. Schneider, C.A.; Rasband, W.S.; Eliceiri, K.W. NIH Image to ImageJ: 25 years of image analysis. *Nat. Methods* **2012**, *9*, 671–675. [CrossRef]
49. Hamley, I.W.; Castelletto, V.; Castillo, R.V.; Müller, A.J.; Martin, C.M.; Pollet, E.; Dubois, P. Crystallization in Poly(l-lactide)-b-poly(ε-caprolactone) Double Crystalline Diblock Copolymers: A Study Using X-ray Scattering, Differential Scanning Calorimetry, and Polarized Optical Microscopy. *Macromolecules* **2005**, *38*, 463–472. [CrossRef]
50. Castillo, R.V.; Müller, A.J. Crystallization and morphology of biodegradable or biostable single and double crystalline block copolymers. *Prog. Polym. Sci.* **2009**, *34*, 516–560. [CrossRef]
51. Bartczak, Z.; Galeski, A.; Krasnikova, N.P. Primary nucleation and spherulite growth rate in isotactic polypropylene-polystyrene blends. *Polymer* **1987**, *28*, 1627–1634. [CrossRef]
52. Su, Z.; Dong, M.; Guo, Z.; Yu, J. Study of Polystyrene and Acrylonitrile−Styrene Copolymer as Special β-Nucleating Agents To Induce the Crystallization of Isotactic Polypropylene. *Macromolecules* **2007**, *40*, 4217–4224. [CrossRef]
53. Yang, B.; Ni, H.; Huang, J.; Luo, Y. Effects of Poly(vinyl butyral) as a Macromolecular Nucleating Agent on the Nonisothermal Crystallization and Mechanical Properties of Biodegradable Poly(butylene succinate). *Macromolecules* **2014**, *47*, 284–296. [CrossRef]
54. Roerdink, E.; Challa, G. Influence of tacticity of poly(methyl methacrylate) on the compatibility with poly(vinylidene fluoride). *Polymer* **1978**, *19*, 173–178. [CrossRef]
55. Bormashenko, Y.; Pogreb, R.; Stanevsky, O.; Bormashenko, E. Vibrational spectrum of PVDF and its interpretation. *Polym. Test.* **2004**, *23*, 791–796. [CrossRef]
56. Gradys, A.; Sajkiewicz, P.; Adamovsky, S.; Minakov, A.; Schick, C. Crystallization of poly(vinylidene fluoride) during ultra-fast cooling. *Thermochim. Acta* **2007**, *461*, 153–157. [CrossRef]
57. Soin, N.; Boyer, D.; Prashanthi, K.; Sharma, S.; Narasimulu, A.A.; Luo, J.; Shah, T.H.; Siores, E.; Thundat, T. Exclusive self-aligned β-phase PVDF films with abnormal piezoelectric coefficient prepared via phase inversion. *Chem. Commun.* **2015**, *51*, 8257–8260. [CrossRef]
58. Gregorio, R.; Capitão, R.C. Morphology and phase transition of high melt temperature crystallized poly(vinylidene fluoride). *J. Mater. Sci.* **2000**, *35*, 299–306. [CrossRef]

59. Gulmine, J.V.; Janissek, P.R.; Heise, H.M.; Akcelrud, L. Polyethylene characterization by FTIR. *Polym. Test.* **2002**, *21*, 557–563. [CrossRef]
60. Lanceros-Méndez, S.; Mano, J.F.; Costa, A.M.; Schmidt, V.H. FTIR and DSC studies of mechanically deformed β-PVDF films. *J. Macromol. Sci. Part. B* **2001**, *40*, 517–527. [CrossRef]
61. Boccaccio, T.; Bottino, A.; Capannelli, G.; Piaggio, P. Characterization of PVDF membranes by vibrational spectroscopy. *J. Membr. Sci.* **2002**, *210*, 315–329. [CrossRef]
62. Ince-Gunduz, B.S.; Alpern, R.; Amare, D.; Crawford, J.; Dolan, B.; Jones, S.; Kobylarz, R.; Reveley, M.; Cebe, P. Impact of nanosilicates on poly(vinylidene fluoride) crystal polymorphism: Part 1. Melt-crystallization at high supercooling. *Polymer* **2010**, *51*, 1485–1493. [CrossRef]
63. Tasumi, M.; Shimanouchi, T. Crystal Vibrations and Intermolecular Forces of Polymethylene Crystals. *J. Chem. Phys.* **1965**, *43*, 1245–1258. [CrossRef]
64. Hughes, D.J.; Mahendrasingam, A.; Oatway, W.B.; Heeley, E.L.; Martin, C.; Fuller, W. A simultaneous SAXS/WAXS and stress-strain study of polyethylene deformation at high strain rates. *Polymer* **1997**, *38*, 6427–6430. [CrossRef]
65. Lv, F.; Wan, C.; Chen, X.; Meng, L.; Chen, X.; Wang, D.; Li, L. Morphology diagram of PE gel films in wide range temperature-strain space: An in situ SAXS and WAXS study. *J. Polym. Sci. Part B Polym. Phys.* **2019**, *57*, 748–757. [CrossRef]
66. Bartczak, Z.; Argon, A.S.; Cohen, R.E.; Kowalewski, T. The morphology and orientation of polyethylene in films of sub-micron thickness crystallized in contact with calcite and rubber substrates. *Polymer* **1999**, *40*, 2367–2380. [CrossRef]
67. Bachmann, M.A.; Lando, J.B. A reexamination of the crystal structure of phase II of poly(vinylidene fluoride). *Macromolecules* **1981**, *14*, 40–46. [CrossRef]
68. Newman, B.A.; Yoon, C.H.; Pae, K.D.; Scheinbeim, J.I. Piezoelectric activity and field-induced crystal structure transitions in poled poly(vinylidene fluoride) films. *J. Appl. Phys.* **1979**, *50*, 6095–6100. [CrossRef]
69. Doll, W.W.; Lando, J.B. The polymorphism of poly(vinylidene fluoride) V. The effect of hydrostatic pressure on the melting behavior of copolymers of vinylidene fluoride. *J. Macromol. Sci. Part B* **1970**, *4*, 897–913. [CrossRef]
70. Geiss, D.; Hofmann, D. Investigation of structural changes in PVDF by modified X-ray texture methods. *IEEE Trans. Electr. Insul.* **1989**, *24*, 1177–1182. [CrossRef]
71. Hasegawa, R.; Takahashi, Y.; Chatani, Y.; Tadokoro, H. Crystal Structures of Three Crystalline Forms of Poly(vinylidene fluoride). *Polym. J.* **1972**, *3*, 600–610. [CrossRef]
72. Davis, G.T.; McKinney, J.E.; Broadhurst, M.G.; Roth, S.C. Electric-field-induced phase changes in poly(vinylidene fluoride). *J. Appl. Phys.* **1978**, *49*, 4998–5002. [CrossRef]
73. Sy, J.W.; Mijovic, J. Reorientational dynamics of poly (vinylidene fluoride)/poly (methyl methacrylate) blends by broad-band dielectric relaxation spectroscopy. *Macromolecules* **2000**, *33*, 933–946. [CrossRef]
74. Linares, A.; Nogales, A.; Rueda, D.R.; Ezquerra, T.A. Molecular dynamics in PVDF/PVA blends as revealed by dielectric loss spectroscopy. *J. Polym. Sci. Part B Polym. Phys.* **2007**, *45*, 1653–1661. [CrossRef]
75. Zhao, X.; Jiang, X.; Peng, G.; Liu, W.; Liu, K.; Zhan, Z. Investigation of the dielectric relaxation, conductivity and energy storage properties for biaxially oriented poly(vinylidene fluoride-hexafluoropropylene)/poly(methyl methacrylate) composite films by dielectric relaxation spectroscopy. *J. Mater. Sci. Mater. Electron.* **2016**, *27*, 10993–11002. [CrossRef]
76. Nieto Simavilla, D.; Abate, A.A.; Liu, J.; Geerts, Y.H.; Losada-Peréz, P.; Napolitano, S. 1D-Confinement Inhibits the Anomaly in Secondary Relaxation of a Fluorinated Polymer. *ACS Macro Lett.* **2021**, *10*, 649–653. [CrossRef]
77. Hilczer, B.; Kułek, J.; Markiewicz, E.; Kosec, M.; Malič, B. Dielectric relaxation in ferroelectric PZT–PVDF nanocomposites. *J. Non Cryst. Solids* **2002**, *305*, 167–173. [CrossRef]
78. Martín, J.; Iturrospe, A.; Cavallaro, A.; Arbe, A.; Stingelin, N.; Ezquerra, T.A.; Mijangos, C.; Nogales, A. Relaxations and relaxor-ferroelectric-like response of nanotubularly confined poly (vinylidene fluoride). *Chem. Mater.* **2017**, *29*, 3515–3525. [CrossRef]
79. Linares, A.; Nogales, A.; Sanz, A.; Ezquerra, T.A.; Pieruccini, M. Restricted dynamics in oriented semicrystalline polymers: Poly (vinilydene fluoride). *Phys. Rev. E* **2010**, *82*, 031802. [CrossRef] [PubMed]
80. Kremer, F.; Schönhals, A. *Broadband Dielectric Spectroscopy*; Springer: Berlin, Germany, 2003.
81. Martínez-Tong, D.E.; Soccio, M.; Sanz, A.; García, C.; Ezquerra, T.A.; Nogales, A. Ferroelectricity and molecular dynamics of poly (vinylidenefluoride-trifluoroethylene) nanoparticles. *Polymer* **2015**, *56*, 428–434. [CrossRef]
82. Angell, C.A. Formation of glasses from liquids and biopolymers. *Science* **1995**, *267*, 1924–1935. [CrossRef]
83. Angell, C. Why C1 = 16–17 in the WLF equation is physical—And the fragility of polymers. *Polymer* **1997**, *38*, 6261–6266. [CrossRef]
84. Nakagawa, K.; Ishida, Y. Annealing effects in poly(vinylidene fluoride) as revealed by specific volume measurements, differential scanning calorimetry, and electron microscopy. *J. Polym. Sci. Polym. Phys. Ed.* **1973**, *11*, 2153–2171. [CrossRef]
85. Grieveson, B.M. The glass transition temperature in homologous series of linear polymers. *Polymer* **1960**, *1*, 499–512. [CrossRef]
86. Arandia, I.; Mugica, A.; Zubitur, M.; Mincheva, R.; Dubois, P.; Müller, A.J.; Alegría, A. The Complex Amorphous Phase in Poly(butylene succinate-ran-butylene azelate) Isodimorphic Copolyesters. *Macromolecules* **2017**, *50*, 1569–1578. [CrossRef]
87. Lorenzo, A.T.; Müller, A.J. Estimation of the nucleation and crystal growth contributions to the overall crystallization energy barrier. *J. Polym. Sci. Part B Polym. Phys.* **2008**, *46*, 1478–1487. [CrossRef]

88. Hoffman, J.D.; Lauritzen, J.I., Jr. Crystallization of Bulk Polymers With Chain Folding: Theory of Growth of Lamellar Spherulites. *J. Res. Natl. Bur. Stand. A Phys. Chem.* **1961**, *65A*, 297–336. [CrossRef]
89. Reiter, G.; Strobl, G.R. *Progress in Understanding of Polymer Crystallization*; Springer: Berlin, Germany, 2007; Volume 714.
90. Avrami, M. Granulation, phase change, and microstructure kinetics of phase change. III. *J. Chem. Phys.* **1941**, *9*, 177–184. [CrossRef]
91. Avrami, M. Kinetics of phase change. II transformation-time relations for random distribution of nuclei. *J. Chem. Phys* **1940**, *8*, 212–224. [CrossRef]

Article

Double Crystallization and Phase Separation in Polyethylene—Syndiotactic Polypropylene Di-Block Copolymers

Claudio De Rosa *, Rocco Di Girolamo, Alessandra Cicolella, Giovanni Talarico and Miriam Scoti

Dipartimento di Scienze Chimiche, Università di Napoli Federico II, Complesso Monte S. Angelo, Via Cintia, I-80126 Naples, Italy; rocco.digirolamo@unina.it (R.D.G.); alessandra.cicolella@unina.it (A.C.); talarico@unina.it (G.T.); miriam.scoti@unina.it (M.S.)
* Correspondence: claudio.derosa@unina.it; Tel.: +39-081-674346

Citation: De Rosa, C.; Di Girolamo, R.; Cicolella, A.; Talarico, G.; Scoti, M. Double Crystallization and Phase Separation in Polyethylene— Syndiotactic Polypropylene Di-Block Copolymers. *Polymers* 2021, *13*, 2589. https://doi.org/10.3390/polym13162589

Academic Editor: Holger Schmalz

Received: 25 July 2021
Accepted: 1 August 2021
Published: 4 August 2021

Publisher's Note: MDPI stays neutral with regard to jurisdictional claims in published maps and institutional affiliations.

Copyright: © 2021 by the authors. Licensee MDPI, Basel, Switzerland. This article is an open access article distributed under the terms and conditions of the Creative Commons Attribution (CC BY) license (https://creativecommons.org/licenses/by/4.0/).

Abstract: Crystallization and phase separation in the melt in semicrystalline block copolymers (BCPs) compete in defining the final solid state structure and morphology. In crystalline–crystalline di-block copolymers the sequence of crystallization of the two blocks plays a definitive role. In this work we show that the use of epitaxial crystallization on selected crystalline substrates allows achieving of a control over the crystallization of the blocks by inducing crystal orientations of the different crystalline phases and a final control over the global morphology. A sample of polyethylene-*block*-syndiotactic polypropylene (PE-*b*-sPP) block copolymers has been synthesized with a stereoselective living organometallic catalyst and epitaxially crystallized onto crystals of two different crystalline substrates, p-terphenyl (3Ph) and benzoic acid (BA). The epitaxial crystallization on both substrates produces formation of highly ordered morphologies with crystalline lamellae of sPP and PE highly oriented along one direction. However, the epitaxial crystallization onto 3Ph should generate a single orientation of sPP crystalline lamellae highly aligned along one direction and a double orientation of PE lamellae, whereas BA crystals should induce high orientation of only PE crystalline lamellae. Thanks to the use of the two selective substrates, the final morphology reveals the sequence of crystallization events during cooling from the melt and what is the dominant event that drives the final morphology. The observed single orientation of both crystalline PE and sPP phases on both substrates, indeed, indicates that sPP crystallizes first onto 3Ph defining the overall morphology and PE crystallizes after sPP in the confined interlamellar sPP regions. Instead, PE crystallizes first onto BA defining the overall morphology and sPP crystallizes after PE in the confined interlamellar PE regions. This allows for discriminating between the different crystalline phases and defining the final morphology, which depends on which polymer block crystallizes first on the substrate. This work also shows that the use of epitaxial crystallization and the choice of suitable substrate offer a means to produce oriented nanostructures and morphologies of block copolymers depending on the composition and the substrates.

Keywords: semicrystalline block copolymers; phase separation and crystallization; epitaxial crystallization; nanostructures

1. Introduction

In semicrystalline block copolymers (BCPs) microphase separation arises from incompatibility of the blocks as in amorphous BCPs, or by crystallization of one or more blocks [1]. Microphase separation in the melt of dissimilar blocks and crystallization may compete and generate a wide range of morphologies [1–7]. The final morphology is path dependent and is the result of this competition and of the interplay between phase separation of the incompatible blocks and the crystallization of blocks [1–7]. Different morphologies are possible depending on the composition of the BCP, the crystallization and glass transition temperatures of blocks and the order–disorder transition temperature. Various structures are obtained depending on which process between crystallization and phase separation occurs first [8]. When crystallization occurs first, from a homogeneous melt, it drives the microphase separation and the final structure is defined by the crystal morphology. If microphase separation occurs first, crystallization occurs from a microphase separated heterogeneous melt, resulting in a crystallization confined within preformed microdomains, or breaking out of the microphase separated structure formed in the melt [7–19]. In crystalline–crystalline block copolymers the crystallization of the first block may define the final morphology or be modified by the subsequent crystallization of the other block [20–24].

BCPs containing blocks based on crystallizable stereoregular polyolefins have been synthesized only recently thanks to the development of metal-based insertion polymerization methods able to ensure a high stereochemical control in living olefin polymerization [25], and studies on the crystallization and phase separation of BCPs containing linear polyethylene and isotactic and syndiotactic polypropylene have been published [18,19,23,24,26,27].

Crystallizable block copolymers have been mainly studied in the past for their possible application as thermoplastic elastomers due to their improved mechanical properties and better thermal stability. Moreover, the presence of a crystallizable component can be exploited for controlling the final morphology through the control of crystallization and orientation of the crystals [7]. In particular, a method for controlling the crystallization and crystal orientation of semicrystalline polymers in thin films is the epitaxial crystallization on suitable crystalline substrates [28]. This method allows the inducing of preferred orientation of crystals of polymers on the substrate and/or crystallization of unstable crystal modifications [28]. Driving crystallization of specific polymorphic forms of polymers is of interest to tailor materials' properties [29]. Recently this method has been applied to crystalline BCPs [7], resulting in the formation of highly ordered nanostructures with highly aligned microdomains as a consequence of the orientation of the crystalline phase [7,15–19,23,24].

In this paper we report a study of the structure and morphology of a crystalline–crystalline BCP composed of blocks of crystallizable polyethylene (PE) and syndiotactic polypropylene (sPP) (PE-b-sPP). The two crystallizable PE and sPP components have been epitaxially crystallized on two different crystalline substrates, that is, crystals of p-terphenyl (3Ph) and benzoic acid (BA). The two different substrates induce selective and different orientations of the two PE and sPP crystalline phases with a final morphology composed of highly aligned lamellar domains with long crystalline sPP and PE lamellae aligned along one direction. Thanks to the use of the two selective substrates, the final morphology reveals the sequence of crystallization events during cooling from the melt and what is the dominant event that drives the final morphology. We also show that use of epitaxial crystallization and the choice of suitable substrate offer a means to produce different oriented nanostructures and morphologies of BCPs depending on the BCP composition and the substrates.

2. Materials and Methods

The sample of PE-b-sPP was prepared with a living organometallic catalyst, bis[N-(3-tert-butylsalicylidene)-2,3,4,5,6-pentafluoroanilinato]-titanium(IV) dichloride (from MCAT, Donaueschingen, Germany), activated with methylalumoxane (MAO) (from Lanxess, Cologne, Germany) [30,31]. The molecular mass and the polydispersity of the sample was determined by gel permeation chromatography (GPC), using a Polymer Laboratories GPC220 apparatus equipped with a Viscotek 220R viscometer (Agilent Company, Santa Clara CA, USA), on polymer solutions in 1,2,4-trichlorobenzene at 135 °C. The molecular structure was analyzed by ^{13}C NMR spectroscopy using a Varian VXR 200 spectometer (Varian Company, Palo Alto, CA, USA).

The sample PE-b-sPP has a total molecular mass M_n = 22,000 g/mol with M_w/M_n = 1.2 and a sPP block longer than the PE block ($M_{n(sPP)}$ = 18,900 and $M_{n(PE)}$ = 3100) with

20 mol% of ethylene, evaluated from ^{13}C NMR spectrum (corresponding to 14 wt% of PE). The molecular mass of the blocks was estimated from total M_n and wt% of PE or sPP, such that $M_{n(PE)} = M_n \times 0.14 \approx 3{,}100$ g/mol and $M_{n(sPP)} = M_n - M_{n(PE)} \approx 18{,}900$ g/mol. The volume fraction of the PE block is $f_{PE} = 13\%$ and was calculated from the molecular masses $M_{n(PE)}$ and $M_{n(sPP)}$ and the densities of PE (0.997 g/cm^3) and sPP (0.9 g/cm^3) [32] such that $f_{PE} = (M_{n(PE)} / 0.997) / (M_{n(sPP)} / 0.9 + M_{n(PE)} / 0.997)$. The ^{13}C NMR spectrum and the GPC trace of the sample PE-*b*-sPP are reported in the supporting information.

It is worth noting that the sample PE-*b*-sPP analyzed in this paper is different in terms of molecular mass and relative lengths of PE and sPP blocks from the samples reported in our previous paper [23]. The sample PE-*b*-sPP has, indeed, a PE block much shorter than the sPP block with 13% volume fraction of PE, whereas in [23] a nearly symmetric sample with $f_{PE} = 47\%$ and a sample with higher molecular mass and $f_{PE} = 25\%$ were analyzed.

Calorimetric measurements (DSC-822, Mettler Toledo, Columbus, OH, USA) were performed under flowing N$_2$ at heating and cooling rates of 10 °C/min. X-ray powder diffraction profiles were obtained with Ni-filtered CuKα radiation with X-Pert diffractometer (Panalytical, Malvern, UK). Diffraction profiles were also recorded in situ at different temperatures during heating and cooling from the melt at about 10 °C/min using an attached TTK Anton-Paar non-ambient stage. The sample was heated from 25 °C up to the melt at 150 °C at nearly 10 °C/min and the diffraction profiles were recorded every 5 degrees starting from 105 °C up to 150 °C. Then, the sample was cooled from the melt at 150 °C down to 25 °C still at 10 °C/min and the diffraction profiles were recorded every 5 degrees during cooling. The temperature was kept constant during recording of each diffraction profile during both heating and cooling.

Epitaxial crystallizations of the block copolymer on the crystals of *p*-terphenyl (3Ph) or benzoic acid (BA) were performed following the procedure used for the PE [33–35] and sPP [36] homopolymers. Thin films (thickness lower than 50 nm) of the BCP were cast at room temperature on microscope glass slides from a *p*-xylene solution (0.2 wt%–0.5 wt%). Slightly different procedures were used for producing crystals of 3Ph and BA substrates. Single crystals of 3Ph were produced independently by slow cooling of a boiling acetone solution; a drop of the suspension was deposited onto the polymer film at room temperature. After evaporation of the solvent, large ($\approx$ 10–100 µm), flat crystals of 3Ph delimited by large top and bottom (001) surfaces remain on the copolymer film (Figure 1A). This composite material was heated to $\approx$ 180 °C to melt the sPP and PE for a short time to limit sublimation of the 3Ph substrate, and then recrystallized by cooling at a controlled rate (10–15 °C/min) to room temperature. During cooling sPP and PE crystallize epitaxially at the interface with the 3Ph crystals. The 3Ph crystals were subsequently dissolved with hot acetone. In the case of BA, powder of BA was spread on the BCP films; then, the polymer film was melted along with BA (melting temperature of BA equal to 123 °C) at $\approx$180 °C to melt both the BCP and BA and then the mixtures were crystallized by moving the slide slowly down the temperature gradient of a hot bar (cooling rate 10–15 °C/min). On cooling, the BA substrate crystals grow first through directional crystallization forming large, flat, and elongated crystals aligned with the *b* axis parallel to the growth front direction (Figure 1B) [35]. Then, the polymer crystallizes at lower temperatures epitaxially onto the (001) exposed face of BA crystals. These crystals of BA were subsequently dissolved with hot ethanol and the polymer film left on the glass.

The so obtained thin films crystallized onto 3Ph and BA were carbon-coated under vacuum in an EMITECH K950X evaporator (Quorum Technologies, Lewes, UK). To improve contrast, the thin films were decorated with gold nanoparticles by vacuum evaporation and condensation. After evaporation, gold condensates and deposits mainly at amorphous–crystalline interface of the semicrystalline lamellae, allowing better visualization of crystalline phases. The films were then floated off on water with the help of a poly(acrylic acid) backing and mounted on copper grids. Transmission electron microscope (TEM) images in bright-field mode were taken in a FEI TECNAI G^2 200kV S-TWIN microscope (electron source with LaB$_6$ emitter) (FEI Company, Dawson Creek Drive, Hillsboro, OR, USA). Bright-field (BF) TEM images were acquired at 120 or 200 kV.

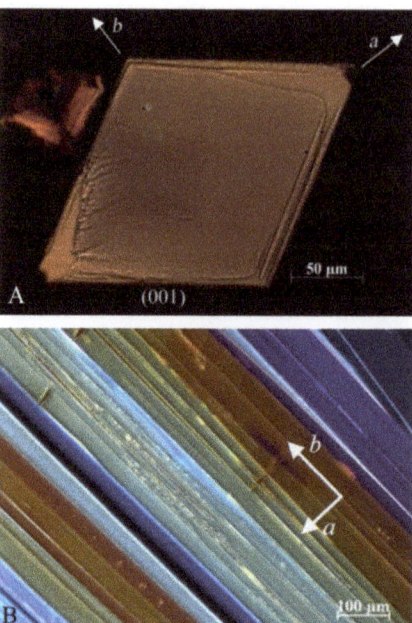

Figure 1. Polarized optical microscope images of flat crystal of 3Ph with exposed (001) face (**A**) and of directionally crystallized flat BA crystals (**B**). BA crystals are elongated and aligned with the b axis parallel to growth front direction. BA single crystals with various thicknesses lead to different colors under polarized light [37].

3. Results and Discussion

The X-ray powder diffraction profile of the as-polymerized sample PE-*b*-sPP is reported in Figure 2. The diffraction profile shows the 200, 020 and 121 reflections of form I of sPP at $2\theta = 12.2°$, $16°$ and $20.7°$ [38,39] and the 110 and 200 reflections at $2\theta = 21.4°$ and $23.9°$ of the orthorhombic form of PE [40] (profile a of Figure 2). This indicates that PE and sPP blocks crystallize in their most stable polymorphic forms with a total degree of crystallinity of nearly 40%.

The DSC thermograms of the sample PE-*b*-sPP recorded during first heating, successive cooling from the melt and second heating of the melt-crystallized samples, all recorded at 10 °C/min, are reported in Figure 3. The DSC heating curve of the as-prepared sample shows two melting peaks at 124 and 137 °C, which can probably be attributed to the melting of PE at low temperature and of sPP at high temperature. This agrees with the melting temperature of 144 °C (data not shown) of the sPP homopolymer synthesized with the same catalyst and in the same reaction conditions, consistent with a concentration of the syndiotactic pentad *rrrr* of 91%. Since a similar stereoregularity is expected for the PE-*b*-sPP copolymer, the slightly lower melting temperature (137 °C) is probably due to confinement phenomena due to phase separation, or confined crystallization inside crystalline lamellae of the other component [23,31].

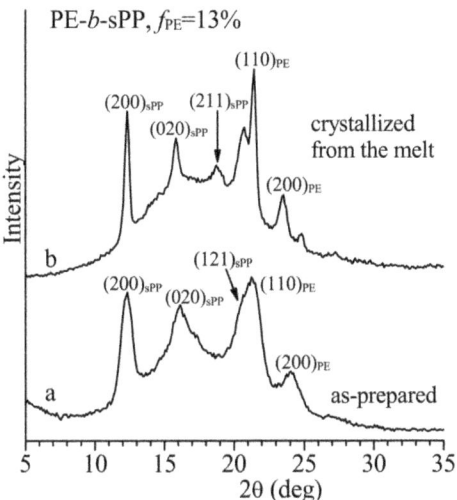

Figure 2. X-ray powder diffraction profiles of as-prepared specimen (a) and of sample crystallized from the melt by cooling the melt at 10 °C/min (b) of the BCP sample PE-b-sPP with f_{PE} = 13%. The (200)$_{sPP}$, (020)$_{sPP}$, (211)$_{sPP}$ and (121)$_{sPP}$ reflections of form I of sPP at 2θ = 12.2°, 16°, 18.8° and 20.7° and the (110)$_{PE}$ and (200)$_{PE}$ reflections at 2θ = 21.4° and 23.9° of the orthorhombic form of PE are indicated.

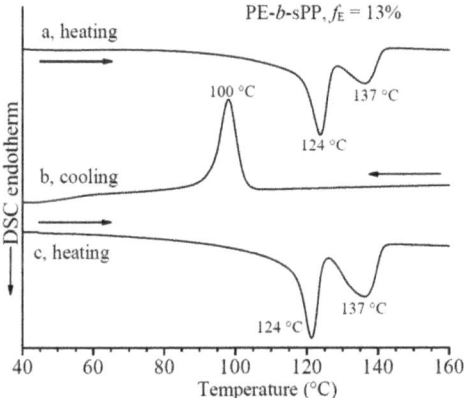

Figure 3. DSC thermograms of the sample PE-b-sPP with f_{PE} = 13% recorded at scanning rate of 10 °C/min during heating of the as-prepared sample (a), cooling from the melt to room temperature (b) and successive heating of the melt-crystallized sample (c).

It is worth noting that in our previous paper [23] different samples of PE-b-sPP BCPs with different relative lengths of PE and sPP blocks have shown only one broad melting peak due to the overlapping of PE and sPP melting. The shorter PE block of the sample here analyzed has been suitably designed to separate the melting endotherms of PE and sPP crystals, as actually occurs in the DSC heating curve of Figure 3a. However, also for this sample the DSC cooling curve from the melt shows only one crystallization peak (curve b of Figure 3), indicating overlapping of crystallization of PE and sPP blocks.

The X-ray diffraction profile of the sample crystallized from the melt in DSC at a cooling rate of 10 °C/min is shown in Figure 2 (profile b). The diffraction profile of Figure 2b shows the 200, 020 and 121 reflections of form I of sPP at 2θ = 12.2°, 16° and 20.7° and the 110 and 200 reflections at 2θ = 21.4° and 23.9° of the orthorhombic form of PE, which are sharper than those in the diffraction profile of the as-prepared sample of Figure 2a. The degree of crystallinity of the melt-crystallized sample is only slightly lower than that of the as-prepared sample (about 40%). Moreover, the diffraction profile of the melt-crystallized sample of Figure 2b shows, in addition, the presence of the 211 reflection at $2\theta = 18.8°$, typical of the ordered form I of sPP [38,39,41]. This indicates that the crystallization from the melt induces the crystallization of a more ordered modification of form I of sPP, characterized by a more ordered alternation of right-handed and left-handed 2/1 helical chains of sPP along the a and b axes of the orthorhombic unit cell of form I [39,41]. The absence of the 211 reflection in the diffraction profile of the as-prepared sample of Figure 2a indicates that this sample is instead crystallized in a disordered modification of form I characterized by disorder in the perfect alternation of enantiomophous helices along both axes of the unit cell [38,39,41].

The DSC melting curve of the melt-crystallized sample of Figure 3c still shows two separate melting endotherms at 124 and 137 °C of PE and sPP, respectively.

The X-ray diffraction profiles of the sample PE-b-sPP recorded at different temperatures during heating and cooling from the melt down to room temperature, are reported in Figure 4. The diffraction profiles of Figure 4A, recorded during first heating of the as-prepared sample, and of Figure 4C, recorded after cooling from the melt during heating of the melt-crystallized sample, show a decrease of the intensity of the diffraction peaks at $2\theta = 21°$ and $24°$, corresponding to the 110 and 200 reflections of PE, at temperatures higher than 120–125 °C (profiles e–g of Figure 4A and e–h of Figure 4C), while the intensities of the 200 and 020 reflections of sPP at $2\theta = 12$ and $16°$, respectively, do not change up to 140 °C. This clearly indicates that crystals of PE melt at low temperatures and confirms that the peak at 124 °C in the DSC heating curves of Figure 3a,c corresponds to the melting of PE and the peak at 137 °C corresponds to the melting of sPP.

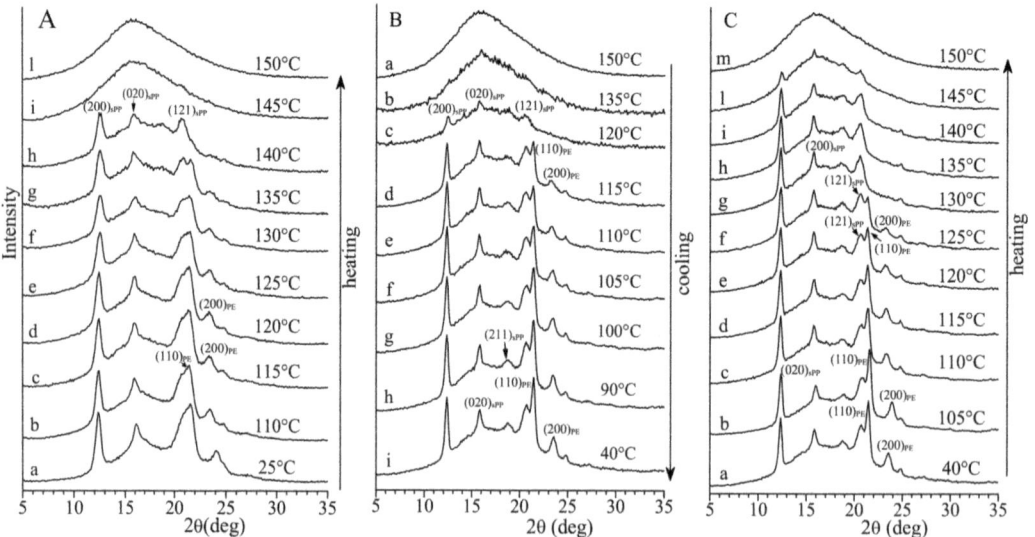

Figure 4. X-ray powder diffraction profiles of the sample PE-b-sPP with $f_{PE} = 13\%$ recorded at different temperatures during first heating of the as-prepared sample up to the melt (**A**), during cooling from the melt to room temperature (**B**) and during successive heating of the melt-crystallized sample up to the melt (**C**). The $(200)_{sPP}$, $(020)_{sPP}$, $(211)_{sPP}$ and $(121)_{sPP}$ reflections of form I of sPP at $2\theta = 12.2°$, $16°$, $18.8°$ and $20.7°$ and the $(110)_{PE}$ and $(200)_{PE}$ reflections at $2\theta = 21.4°$ and $23.9°$ of the orthorhombic form of PE are indicated.

The diffraction profiles recorded during cooling form the melt at 150 °C to room temperature of Figure 4B indicate that sPP and PE crystallize almost simultaneously, according to the single crystallization peak observed in the DSC cooling curve of Figure 3b, although the 200, 020 and 121 reflections of sPP at 2θ = 12, 16 and 20.7° appear first, already at 120 °C (profile c of Figure 4B), before the 110 and 200 reflections of PE that are well visible only at 115 °C, along with all reflections of sPP (profile d of Figure 4B). Therefore, during the slow cooling and the isothermal necessary to record the diffraction profile, sPP crystallizes first at high temperatures (nearly 120 °C). The intensities of reflections of both sPP and PE increase and become sharper upon further cooling and, as discussed above (Figure 2b), the 211 reflection at 2θ = 18.8° of the ordered form I of sPP develops (profiles e–i of Figure 4B).

The possible phase separation in the melt and the possible formation of phase separated structures for PE-b-sPP BCPs has been discussed in the ref [31]. According to mean-field theory, the order–disorder transition for symmetric BCPs occurs at a fixed interaction strength for calculated values of χN = 10.5, where χ is the Flory–Huggins interaction parameter and N is the total number of equivalent segments that constitute the macromolecules of the blocks of the BCP [31]. For non-symmetric BCPs the phase separation transition occurs for higher values of χN. For polyolefin-based BCPs, the equivalent segments are assumed as a portion of chains having the density of four CH_2 units (four carbon atoms segment). The Flory–Huggins interaction parameter χ between sPP and PE has been determined in [31] as: $\chi = 6.2/T - 0.0053$, with T the absolute temperature. For the sample PE-b-sPP with total M_n = 22,000 and f_{PE} = 13%, the total number of equivalent segments N that constitute the macromolecules of the blocks is $N = M_n/56 = 393$ (where 56 is the molecular mass of the four CH_2 carbon atoms segment). Therefore, for this sample the order–disorder transition temperature T_{ODT} may be calculated from $\chi N \geq 10.5 = (6.2/T - 0.0053)393$ and is expected to be lower than 0 °C. This indicates that crystallization of the sample PE-b-sPP most likely takes place from a homogeneous melt.

Thin films (thickness lower than 50 nm) of the sample PE-b-sPP have been epitaxially crystallized onto the (001) surfaces of crystals of p-terphenyl (3Ph) and benzoic acid (BA). Epitaxial crystallization of PE and sPP homopolymers onto crystals of various organic substances has been well-described and used as a tool for growing in thin films crystals of various polymorphic forms with single-crystal or fiber-like orientations [28,33–36,42]. Polymer–polymer epitaxy, involving heteroepitaxy of sPP with PE and homoepitaxy has been also described [43]. Epitaxial crystallizations of sPP and PE blocks when they are parts of crystalline/amorphous or crystalline–crystalline block copolymers have also been studied [7,15–19,23,24].

The TEM bright-field images of thin films of the sample PE-b-sPP crystallized by simple casting from the polymer solution (without epitaxy) and of films epitaxially crystallized onto 3Ph and BA are reported in Figure 5. The films have been coated with gold particles to improve the contrast in the TEM observation and reveal details of the morphology. The technique of gold decoration is used to visualize edge-on crystalline lamellae of polymers in TEM bright-field images, especially in the case of low TEM amplitude contrast between amorphous and crystalline phases, and to obtain a reliable value of the lamellar periodicity [44,45]. The vaporized gold gathers, indeed, in the ditches made by the interlamellar amorphous material and produces a regular pattern of gold particles, which is observed in bright-field imaging [44–46]. In the case of homopolymers this generally produces thin layers of gold particles at the interface between amorphous and crystalline lamellae, containing rows of essentially one gold particle thickness [44,45].

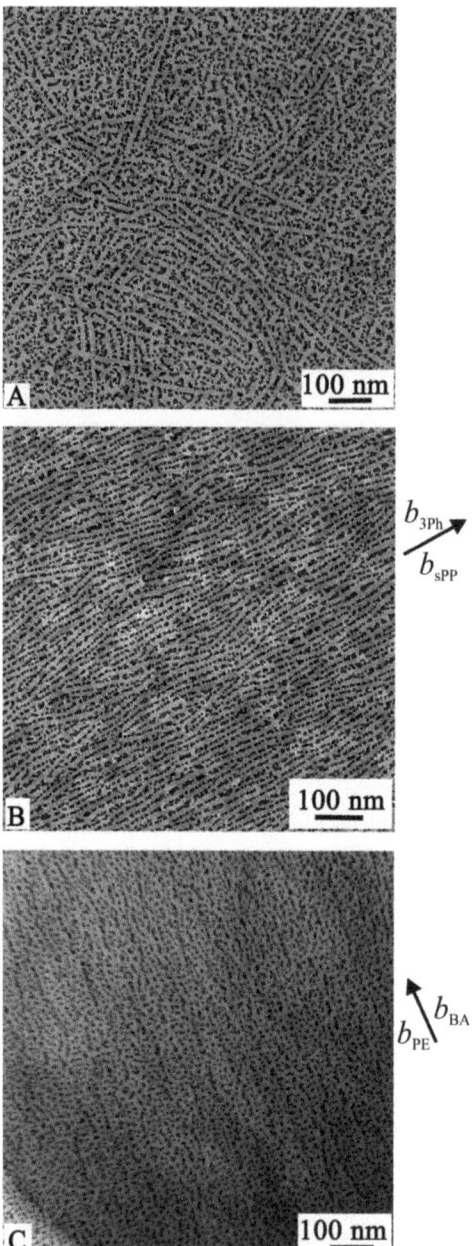

Figure 5. TEM bright-field images of thin films of the sample PE-*b*-sPP with f_{PE} = 13% crystallized by simple solution casting without epitaxy (**A**) and epitaxially crystallized on the (001) surface of crystals of 3Ph (**B**) and BA (**C**).

In all the images of Figure 5 the dark spots correspond to the gold particles that presumably are located in the amorphous intra-lamellar phases of PE and sPP, that is, in between the crystalline domains of PE or sPP, whereas the brighter regions correspond to PE and/or sPP crystalline lamellae. It is apparent that in the case of the films crystallized without epitaxy in Figure 5A, the PE and sPP crystalline lamellae (the light stripes) are randomly oriented and are not distinguishable. In the TEM images of the films epitaxially crystallized onto 3Ph (Figure 5B) and BA (Figure 5C), the crystalline lamellae of PE or sPP are in both cases highly aligned along one direction and oriented edge-on on the substrate surface. The epitaxy produces a highly aligned lamellar structure with long crystalline sPP and/or PE lamellae, with average thicknesses of 15 nm.

A single orientation of sPP lamellae on 3Ph and of PE lamellae on BA has been found for the two homopolymers [35,36] and also in epitaxially crystallized crystalline-amorphous BCPs, as in the case of sPP-b-EP [18] and PE-b-EP [15,19] (EP being an ethylene-propylene amorphous random copolymer). More complex morphology is instead expected for the crystallization of PE onto 3Ph, for which two different orientations of PE lamellae have been observed in the case of PE homopolymer crystallized onto 3Ph [33].

However, thanks to the use of different substrates, the images of Figure 5B,C, although very similar in term of induced single orientation of crystalline lamellae (sPP and PE), reveal the sequence of crystallization events during cooling from the melt and what is the dominant event that drives the final morphology. This information can be, indeed, revealed through interpretation of the images of Figure 5B,C and from the epitaxial relationships between polymer crystals and substrates crystals. The complex morphologies generated in the epitaxial crystallization of the sample PE-b-sPP result from interactions between all three components involved, sPP, PE and the crystalline substrate (3Ph or BA).

Both PE and sPP crystallize epitaxially onto crystals of 3Ph [28,33,34,36], and only PE crystallizes epitaxially onto BA [7,15,19,24,28,34,35], whereas no epitaxy exists for sPP onto BA. Epitaxial crystallization produces single crystal-like orientation of PE and sPP crystals onto the (001) exposed face of 3Ph [28,33,34,36] and of PE crystals onto the (001) face of BA crystals [34,35].

For sPP onto 3Ph, the (100) plane of crystals of form I of sPP is in contact with the (001) plane of 3Ph; therefore, the crystalline sPP lamellae stand edge-on on the substrate surface, oriented with the b and c axes of sPP parallel to the b and a axes of 3Ph, respectively (Figure 6A) [36]. The chain axis of the crystalline sPP lamellae lies flat on the substrate surface and oriented parallel to the a axis of 3Ph crystals (Figure 6A). This epitaxy is well explained in terms of the crystal structures of 3Ph (unit cell with $a = 8.05$ Å, $b = 5.55$ Å, $c = 13.59$ Å, $\beta = 91.9°$) [36] and form I of sPP (orthorhombic unit cell with axes $a = 14.5$ Å, $b = 5.6$ Å or 11.2 Å, $c = 7.4$ Å) [38,39] and matching of the $a_{3Ph} = 8.05$ Å and $b_{3Ph} = 5.55$ Å axes of 3Ph with the $c = 7.4$ Å and $b = 5.6$ Å axes, respectively, of form I of sPP [36]. The epitaxial relationships between sPP and 3Ph crystals are, therefore, (Figure 6A):

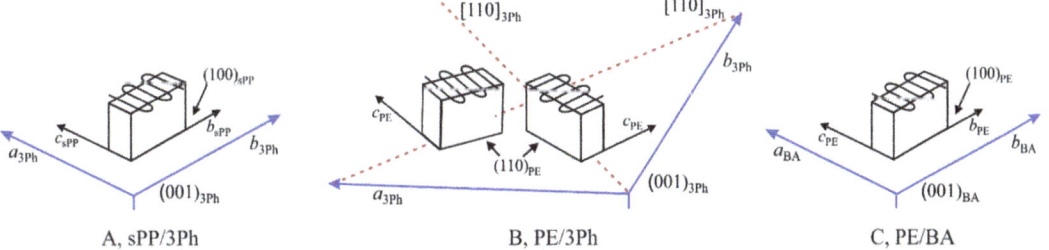

Figure 6. Schemes of the single orientation of crystalline lamellae of sPP (A) and double orientations of PE lamellae (B) onto the (001) face of 3Ph and of the single orientation PE lamellae onto the (001) face of BA (C), induced by epitaxial crystallization [33,35,36].

$(100)_{sPP} // (001)_{3Ph}$
$b_{sPP} // b_{3Ph}; c_{sPP} // a_{3Ph}$

In the case of PE/3Ph epitaxy, two different equivalent orientations of PE crystalline lamellae are generated by crystallization onto the (001) face of 3Ph (Figure 6B) [33]. The

(110) plane of PE is in contact with the (001) plane of 3Ph [33]. The PE lamellae stand edge-on with the chain axes oriented parallel to the [110] and [1$\bar{1}$0] directions of the 3Ph crystal about 74° apart, as shown in the scheme of Figure 6B. This epitaxy and the selection of the (110) plane as contact plane with the (001) plane of 3Ph is due to the matching between the 4.45 Å interchain distance in the (110) plane of PE and the 4.60 Å interplanar distance of the {110} planes of 3Ph [33]. The epitaxial relationships between sPP and 3Ph crystals are, therefore, (Figure 6B):

$(110)_{PE}//(001)_{3Ph}$

$c_{PE}//[110]_{3Ph}$ and $//[1\bar{1}0]_{3Ph}$

Therefore, the epitaxial crystallization of the sample PE-b-sPP onto 3Ph should give oriented overgrowth of both crystals of sPP and PE, with a single orientation of sPP lamellae (Figure 6A) and a double orientation of PE lamellae (Figure 6B) onto the (001) surface of the 3Ph substrate [23].

In the case of PE/BA epitaxy, a single orientation of PE lamellae is generated by crystallization of PE onto the (001) face of BA (Figure 6C). The chain axis of the crystalline PE lies flat on the substrate surface and oriented parallel to the a axis of BA crystals, as in the case of the PE homopolymer [35]. The (100) plane of PE is in contact with the (001) plane of BA [35]; therefore, the crystalline PE lamellae stand edge-on on the substrate surface, oriented with the b and c axes of PE parallel to the b and a axes of BA, respectively [35]. This epitaxy has been well explained in term of matching of the b = 4.93 Å and c = 2.53 Å axes of the unit cell of PE orthorhombic form (a = 7.40 Å, b = 4.93 Å, c = 2.53 Å) [40] with the b = 5.14 Å and a = 5.52 Å axes, respectively, of the BA unit cell (a = 5.52 Å, b = 5.14 Å, c = 21.9 Å, β = 97°) [35]. The epitaxial relationships between PE and BA crystals are, therefore, (Figure 6C):

$(100)_{PE}//(001)_{BA}$

$b_{PE}//b_{BA}; c_{PE}//a_{BA}$

Therefore, based on the epitaxial relationships found for sPP and PE homopolymers onto 3Ph and BA in Figure 6, a single orientation of sPP lamellae on 3Ph and of PE lamellae on BA and a double orientation of PE lamellae onto 3Ph would be expected in the epitaxial crystallization of the PE-b-sPP block copolymer. Moreover, no preferential orientation of sPP crystals onto BA is expected. The TEM images of films of the sample PE-b-sPP epitaxially crystallized onto 3Ph (Figure 5B) and BA (Figure 5C), instead, clearly show that a single orientation of crystalline lamellae (PE and/or sPP) is obtained onto both 3Ph and PE.

In the case of the crystallization onto 3Ph of Figure 5B, the obtained single orientation of the crystalline lamellae and the absence of double oriented lamellae of PE, as in Figure 6B, indicate that the observed parallel lamellae oriented along one direction are of the sPP blocks that, based on the Figure 6A, must have a single orientation with the c axis of sPP parallel to the a axis of 3Ph. The crystallization of the sPP block with the expected single lamellae orientation, therefore, defines the overall morphology of the whole epitaxially crystallized film with evident crystalline lamellae oriented along only one direction (Figure 5B). This means that sPP must have crystallized first. None of the expected PE lamellae with two different orientations 74° apart (Figure 6B) are visible. Therefore, PE crystallizes after sPP in the confined inter-lamellar regions prescribed by the oriented sPP lamellae. These trapped and thin PE lamellae are hardly visualized by the gold decoration. The final morphology (Figure 5B) is, therefore, driven by the crystallization of sPP, in agreement with the fact that the sPP block is longer than the PE block and according with the X-ray diffraction data of Figure 4B that have indicated that sPP crystallizes first upon cooling from the melt. A scheme of the final morphology representing the TEM image of Figure 5B is shown in Figure 7A. PE lamellae are confined between sPP lamellae and follow the orientation of the sPP lamellae that are aligned with the c and b axes of sPP parallel to the a and b axes of 3Ph, respectively. Since the growth of PE is confined between sPP lamellae and the epitaxy should produce different orientations of PE chain axes parallel to the [110] and [1$\bar{1}$0] directions of 3Ph (Figure 6B), it is most probable that PE lamellae are parallel to the sPP lamellae but are made of chains tilted with respect to their basal fold surface, as shown in the model of Figure 7A. The tilting of PE chains with tilt angle of 45° to the lamellar normal has already been described [47–49]. Thus, in these systems, the stem orientation is dictated by the epitaxy with 3Ph, but the fold surface orientation is dictated by the orientation of the lamellae of the block that crystallizes first (sPP). Therefore, in the confined sPP interlamellar regions, the trapped PE lamellae are parallel to the sPP lamellae and oriented along the direction dictated by the sPP crystallization, with the PE chains tilted at 74/2 = 37° to the lamellar normal (Figure 7A).

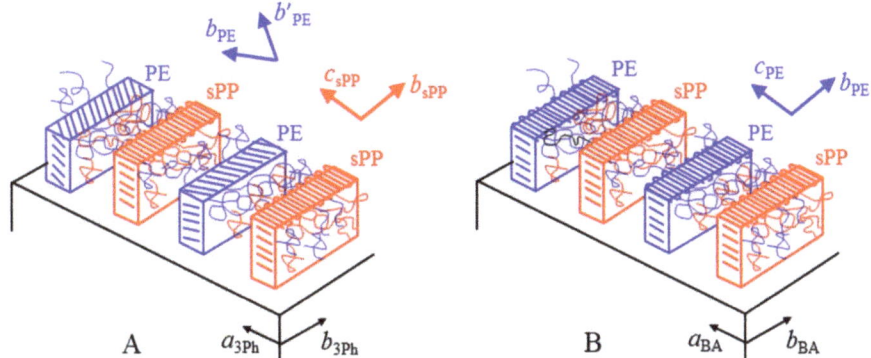

Figure 7. Models of the structures and morphologies that develop upon epitaxial crystallization of PE-b-sPP with f_{PE} = 13% onto the (001) surfaces of crystals of 3Ph (**A**) and BA (**B**). In A sPP crystallizes first onto 3Ph, forming lamellae aligned with the c and b axes of sPP parallel to the a and b axes of 3Ph. PE crystallizes after sPP in the confined inter-lamellar regions prescribed by the oriented sPP lamellae (**A**). In B PE crystallizes first onto BA forming lamellae aligned with the c and b axes of PE parallel to the a and b axes of BA. sPP crystallizes after PE in the confined inter-lamellar regions prescribed by the oriented PE lamellae (**B**).

The epitaxial crystallization of PE-b-sPP onto BA also produces single orientation of crystalline lamellae aligned along one direction, the b axis of BA (Figure 5C). Since no epitaxy exists for crystallization of sPP onto BA, random orientation of sPP lamellae is expected, as in Figure 5A for crystallization of PE-b-sPP without substrate. This random orientation is not observed in the morphology of Figure 5C. Therefore, the obtained single orientation of the crystalline lamellae and the absence of random orientation of sPP lamellae indicate that the observed parallel lamellae oriented along one direction are of the PE blocks that, based on Figure 6C, must have a single orientation, with the c and b axes of PE parallel to the a and b axes of BA, respectively. The crystallization of the PE block with the expected single lamellae orientation, therefore, defines the overall morphology of the whole epitaxially crystallized film with evident crystalline lamellae oriented along only one direction (Figure 5C). This may be explained considering that, even though the sPP block crystallizes first in the absence of substrates (Figure 4B) or onto 3Ph, the PE block must have crystallized first in the presence of BA, or nearly contemporarily to the sPP block. Therefore, sPP crystallizes after PE (or with PE) in the confined interlamellar regions prescribed by the oriented PE lamellae. However, since the epitaxial crystallization of the polymer blocks onto BA is preceded by the directional solidification of BA [7] that induces alignment of the BCP microdomains along the b axis of BA (the growth front direction) before and during the solidification and crystallization of the BCP, the process results in alignment of both PE and sPP crystalline lamellae parallel to the b axis of BA. Then, epitaxy of PE onto BA produces alignment of the c and b axes of PE parallel to the a and b axes of BA, respectively. A scheme of the final morphology representing the TEM image of Figure 5C is shown in Figure 7B. sPP lamellae are confined between PE lamellae and follow the orientation of the PE lamellae. Since there is no preferred orientation of the c axis of sPP onto BA, it is probable that sPP lamellae are parallel to the PE lamellae with the chains normal to their basal fold surface and parallel to the chain axis of PE, that is, the stem orientation dictated by the epitaxy of PE onto BA (Figure 7B).

4. Conclusions

A sample of crystalline–crystalline PE-*b*-sPP block copolymers with 13% volume fraction of the PE block was synthesized with a stereoselective living organometallic catalyst. The structure and morphology of the PE-*b*-sPP block copolymer have been studied in the bulk and in thin films epitaxially crystallized on crystals of 3Ph and BA substrates.

In both as-prepared and melt-crystallized samples of PE-*b*-sPP the sPP block crystallizes in the stable form I and the PE block crystallizes in the orthorhombic form. Crystals of PE and sPP melt at different temperatures, at 124 °C and 137 °C, respectively. The two blocks crystallize from the melt by cooling at a controlled rate (10 °C/min) almost simultaneously, and only one exothermic peak is observed in the DSC cooling curve. However, diffraction profiles recorded during cooling have demonstrated that the longer sPP block crystallizes first.

Thin films of the sample PE-*b*-sPP were epitaxially crystallized onto the (001) surfaces of crystals of *p*-terphenyl (3Ph) and benzoic acid (BA). The complex morphologies generated in the epitaxial crystallization result from interactions between all three components involved, sPP, PE and the crystalline substrate (3Ph or BA). The epitaxial crystallization produces oriented growth of both crystals of sPP and PE depending on the substrate, with a single orientation of sPP lamellae onto the (001) surface of 3Ph crystals and a single orientation of PE lamellae onto the (001) surface of BA, according to the epitaxy of sPP with 3Ph and PE with BA. Epitaxy of PE with 3Ph should instead produce a double orientation of PE lamellae onto the (001) surface of 3Ph. The process also produces development of ordered nanostructures composed of alternating lamellar domains of PE and sPP, guided by the orientation of the sPP or PE crystalline lamellae.

TEM bright-field images provide details of the resulting morphology and reveal the sequence of the crystallization events. In the case of the crystallization of PE-*b*-sPP onto 3Ph, highly oriented crystalline lamellae aligned along one direction are obtained. The expected double orientation of PE lamellae onto the (001) surface of 3Ph is not observed. This indicates that sPP crystallizes first and defines the overall morphology of the whole epitaxially crystallized film, forming, according to epitaxy onto 3Ph, long lamellae oriented along one direction with the *c* and *b* axis of sPP parallel to the *a* and *b* axes of 3Ph, respectively. PE crystallizes after sPP in the confined inter-lamellar regions prescribed by the oriented sPP lamellae.

The epitaxial crystallization of PE-*b*-sPP onto BA also produces single orientation of crystalline lamellae aligned along one direction, the *b* axis of BA. Since no epitaxy exists for crystallization of sPP onto BA, random orientation of sPP lamellae would be expected. Therefore, the obtained single orientation of the crystalline lamellae and the absence of random orientation of sPP lamellae indicate that the observed parallel lamellae oriented along one direction are of the PE blocks that, according to the epitaxy of PE with BA, must have a single orientation with the *c* and *b* axes of PE parallel to the *a* and *b* axes of BA, respectively. The crystallization of the PE block defines the overall morphology of the whole epitaxially crystallized film. This may be explained considering that the PE block must have crystallized first in the presence of BA, or nearly contemporarily to the sPP block. Therefore, sPP crystallizes after PE (or with PE) in the confined inter-lamellar regions prescribed by the oriented PE lamellae. However, since the epitaxial crystallization of the polymer blocks onto BA is preceded by the directional solidification of BA that induces alignment of the BCP microdomains along the *b* axis of BA (the growth front direction) before and during the solidification and crystallization of the BCP, the process results in alignment of both PE and sPP crystalline lamellae parallel to the *b* axis of BA. Then, epitaxy of PE onto BA produces alignment of the *c* and *b* axes of PE parallel to the *a* and *b* axes of BA, respectively.

These data show that the use of epitaxial crystallization and the choice of suitable substrate offer a means to produce oriented nanostructures and morphologies of BCP depending on the BCP composition and the substrates.

Author Contributions: C.D.R. conceived the experiments and wrote the manuscript. R.D.G. and G.T. synthesized the sample, R.D.G., A.C. and M.S. performed the experiments. All authors have read and agreed to the published version of the manuscript.

Funding: This research was funded by the Ministero dell'Istruzione, dell'Università e della Ricerca, Italy (PRIN 2017, project n. 20179FK77_003).

Informed Consent Statement: Informed consent was obtained from all subjects involved in the study.

Data Availability Statement: The data in this study are available on reasonable request from the corresponding author.

Acknowledgments: The task force "Polymers and biopolymers" of the University of Napoli Federico II is acknowledged.

Conflicts of Interest: The authors declare no conflict of interest.

References

1. De Rosa, C.; di Girolamo, R.; Malafronte, A.; Scoti, M.; Talarico, G.; Auriemma, F.; Ruiz de Ballesteros, O. Polyolefins based crystalline block copolymers: Ordered nanostructures from control of crystallization. *Polymer* **2020**, *196*, 122423. [CrossRef]
2. Hamley, I.W. Crystallization in Block Copolymers. *Adv. Polym. Sci.* **1999**, *148*, 113–137.
3. Loo, Y.L.; Register, R.A. Crystallization Within Block Copolymer Mesophases. In *Development in Block Copolymer Science and Technology*; Hamley, I.W., Ed.; John Wiley & Sons Ltd.: Chichester, UK, 2004; pp. 213–244.
4. Muller, A.J.; Balsamo, V.; Arnal, M.L. Nucleation and Crystallization in Diblock and Triblock Copolymers. *Adv. Polym. Sci.* **2005**, *190*, 1–63.
5. Muller, A.J.; Castillo, R.V. Crystallization and morphology of biodegradable or biostable single and double crystalline block copolymers. *Prog. Polym. Sci.* **2009**, *34*, 516–560.
6. He, W.N.; Xu, J.T. Crystallization assisted self-assembly of semicrystalline block copolymers. *Prog. Polym. Sci.* **2012**, *37*, 1350–1400. [CrossRef]
7. De Rosa, C.; Park, C.; Thomas, E.L.; Lotz, B. Microdomain patterns via directional eutectic solidification and epitaxy. *Nature* **2000**, *405*, 433–437. [CrossRef]
8. Loo, Y.L.; Register, R.A.; Ryan, A.J. Modes of Crystallization in Block Copolymer Microdomains: Breakout, Templated, and Confined. *Macromolecules* **2002**, *35*, 2365–2374. [CrossRef]
9. Rangarajan, P.; Register, R.A.; Adamson, D.H.; Fetters, L.J.; Bras, W.; Naylor, S.; Ryan, A.J. Dynamics of Structure Formation in Crystallizable Block Copolymers. *Macromolecules* **1995**, *28*, 1422–1428. [CrossRef]
10. Ryan, A.J.; Hamley, I.W.; Bras, W.; Bates, F.S. Structure Development in Semicrystalline Diblock Copolymers Crystallizing from the Ordered Melt. *Macromolecules* **1995**, *28*, 3860–3868. [CrossRef]
11. Quiram, D.J.; Register, R.A.; Marchand, G.R. Crystallization of Asymmetric Diblock Copolymers from Microphase-Separated Melts. *Macromolecules* **1997**, *30*, 4551–4558. [CrossRef]
12. Quiram, D.J.; Register, R.A.; Marchand, G.R.; Ryan, A.J. Dynamics of Structure Formation and Crystallization in Asymmetric Diblock Copolymers. *Macromolecules* **1997**, *30*, 8338–8343. [CrossRef]
13. Quiram, D.J.; Register, R.A.; Marchand, G.R.; Adamson, D.H. Chain Orientation in Block Copolymers Exhibiting Cylindrically Confined Crystallization. *Macromolecules* **1998**, *31*, 4891–4898. [CrossRef] [PubMed]
14. Park, C.; de Rosa, C.; Fetters, L.J.; Thomas, E.L. Influence of an Oriented Glassy Cylindrical Microdomain Structure on the Morphology of Crystallizing Lamellae in a Semicrystalline Block Terpolymer. *Macromolecules* **2000**, *33*, 7931–7938. [CrossRef]
15. De Rosa, C.; Park, C.; Lotz, B.; Wittmann, J.C.; Fetters, L.J.; Thomas, E.L. Control of Molecular and Microdomain Orientation in a Semicrystalline Block Copolymer Thin Film by Epitaxy. *Macromolecules* **2000**, *33*, 4871. [CrossRef]
16. Park, C.; de Rosa, C.; Fetters, L.J.; Lotz, B.; Thomas, E.L. Alteration of Classical Microdomain Patterns of Block Copolymers by Degenerate Epitaxy. *Adv. Mater.* **2001**, *13*, 724. [CrossRef]
17. Park, C.; de Rosa, C.; Lotz, B.; Fetters, L.J.; Thomas, E.L. Molecular and Microdomain Orientation in Semicrystalline Block Copolymer Thin Films by Directional Crystallization of the Solvent and Epitaxy. *Macromol. Chem. Phys.* **2003**, *204*, 1514. [CrossRef]
18. De Rosa, C.; Auriemma, F.; di Girolamo, R.; Aprea, R.; Thierry, A. Selective Gold Deposition on a Nanostructured Block Copolymer Film Crystallized by Epitaxy. *Nano Res.* **2011**, *4*, 241. [CrossRef]
19. De Rosa, C.; di Girolamo, R.; Auriemma, F.; Talarico, G.; Scarica, C.; Malafronte, A.; Scoti, M. Controlling Size and Orientation of Lamellar Microdomains in Crystalline Block Copolymers. *ACS Appl. Mater. Interfaces* **2017**, *9*, 31252–31259. [CrossRef]
20. Sun, L.; Liu, Y.; Zhu, L.; Hsiao, B.S.; Avila-Orta, C.A. Self-assembly and crystallization behavior of a double-crystalline polyethylene-block-poly(ethylene oxide) diblock copolymer. *Polymer* **2004**, *45*, 8181. [CrossRef]

21. Castillo, R.V.; Muller, A.J.; Lin, M.C.; Chen, H.L.; Jeng, U.S.; Hillmyer, M.A. Confined crystallization and morphology of melt segregated PLLA-b-PE and PLDA-b-PE diblock copolymers. *Macromolecules* **2008**, *41*, 6154. [CrossRef]
22. Myers, S.B.; Register, R.A. Crystalline-Crystalline Diblock Copolymers of Linear Polyethylene and Hydrogenated Polynorbornene. *Macromolecules* **2008**, *41*, 6773. [CrossRef]
23. De Rosa, C.; di Girolamo, R.; Auriemma, F.; D'Avino, M.; Talarico, G.; Cioce, C.; Scoti, M.; Coates, G.W.; Lotz, B. Oriented Microstructures of Crystalline-Crystalline Block Copolymers Induced by Epitaxy and Competitive and Confined Crystallization. *Macromolecules* **2016**, *49*, 5576–5586. [CrossRef]
24. De Rosa, C.; Malafronte, A.; di Girolamo, R.; Auriemma, F.; Scoti, M.; Ruiz de Ballesteros, O.; Coates, G.W. Morphology of Isotactic Polypropylene–Polyethylene Block Copolymers Driven by Controlled Crystallization. *Macromolecules* **2020**, *53*, 10234–10244. [CrossRef]
25. Coates, G.W.; Hustad, P.D.; Reinartz, S. Catalysts for the Living Insertion Polymerization of Alkenes: Access to New Polyolefin Architectures Using Ziegler—Natta Chemistry. *Angew. Chem. Int. Ed.* **2002**, *41*, 2236. [CrossRef]
26. Eagan, J.M.; Xu, J.; di Girolamo, R.; Thurber, C.M.; Macosko, C.W.; LaPointe, A.M.; Bates, F.S.; Coates, G.W. Combining polyethylene and polypropylene: Enhanced performance with PE/iPP multiblock polymers. *Science* **2017**, *355*, 814–816. [CrossRef]
27. Xu, J.; Eagan, J.M.; Kim, S.-S.; Pan, S.; Lee, B.; Klimovica, K.; Jin, K.; Lin, T.-W.; Howard, M.J.; Ellison, C.J.; et al. Compatibilization of Isotactic Polypropylene (iPP) and High-Density Polyethylene (HDPE) with iPP-PE Multiblock Copolymers. *Macromolecules* **2018**, *51*, 8585–8596. [CrossRef]
28. Wittmann, J.C.; Lotz, B. Epitaxial Crystallization of Polymers on Organic and Polymeric Substrates. *Prog. Polym. Sci.* **1990**, *15*, 909. [CrossRef]
29. De Rosa, C.; Scoti, M.; di Girolamo, R.; Ruiz de Ballesteros, O.; Auriemma, F.; Malafronte, A. Polymorphism in polymers: A tool to tailor material's properties. *Polym. Cryst.* **2020**, *3*, e10101. [CrossRef]
30. Tian, J.; Hustad, P.D.; Coates, G.W. A New Catalyst for Highly Syndiospecific Living Olefin Polymerization: Homopolymers and Block Copolymers from Ethylene and Propylene. *J. Am. Chem. Soc.* **2001**, *123*, 5134. [CrossRef]
31. Ruokolainen, J.; Mezzenga, R.; Fredrickson, G.H.; Kramer, E.J.; Hustad, P.D.; Coates, G.W. Morphology and Thermodynamic Behavior of Syndiotactic Polypropylene-Poly(ethylene-co-propylene) Block Polymers Prepared by Living Olefin Polymerization. *Macromolecules* **2005**, *38*, 851. [CrossRef]
32. Brandrup, J.; Immergut, E.H.; Grulke, E.A. (Eds.) *Polymer Handbook*, 4th ed.; John Wiley & Sons: Hoboken, NJ, USA, 2003.
33. Wittmann, J.C.; Lotz, B. Epitaxial crystallization of polyethylene on organic substrates: A reappraisal of the mode of action of selected nucleating agents. *J. Polym. Sci. Polym. Phys. Ed.* **1981**, *19*, 1837–1851. [CrossRef]
34. Wittmann, J.C.; Lotz, B. Epitaxial crystallization of monoclinic and orthorhombic polyethylene phases. *Polymer* **1989**, *30*, 27–34. [CrossRef]
35. Wittmann, J.C.; Hodge, A.M.; Lotz, B. Epitaxial crystallization of polymers onto benzoic acid: Polyethylene and paraffins, aliphatic polyesters, and polyamides. *J. Polym. Sci. Polym. Phys. Ed.* **1983**, *21*, 2495. [CrossRef]
36. Stocker, W.; Schumacher, M.; Graff, S.; Lang, J.; Wittmann, J.C.; Lovinger, A.J.; Lotz, B. Direct Observation of Right and Left Helical Hands of Syndiotactic Polypropylene by Atomic Force Microscopy. *Macromolecules* **1994**, *27*, 6948–6955. [CrossRef]
37. Park, C.; de Rosa, C.; Thomas, E.L. Large Area Orientation of Block Copolymer Microdomains in Thin Films via Directional Crystallization of a Solvent. *Macromolecules* **2001**, *34*, 2602–2606. [CrossRef]
38. De Rosa, C.; Corradini, P. Crystal Structure of Syndiotactic Polypropylene. *Macromolecules* **1993**, *26*, 5711–5718. [CrossRef]
39. Lotz, B.; Lovinger, A.J.; Cais, R.E. Crystal structure and morphology of syndiotactic polypropylene single crystals. *Macromolecules* **1988**, *21*, 2375. [CrossRef]
40. Bunn, C.W. The crystal structure of long-chain normal paraffin hydrocarbons. The "shape" of the CH2 group. *Trans. Faraday Soc.* **1939**, *35*, 482–490. [CrossRef]
41. De Rosa, C.; Auriemma, F.; Vinti, V. Disordered polymorphic modifications of form I of syndiotactic polypropylene. *Macromolecules* **1997**, *30*, 4137–4146. [CrossRef]
42. Zhang, J.; Yang, D.; Thierry, A.; Wittmann, J.C.; Lotz, B. Isochiral Form II of Syndiotactic Polypropylene Produced by Epitaxial Crystallization. *Macromolecules* **2001**, *34*, 6261–6267. [CrossRef]
43. Schumacher, M.; Lovinger, A.J.; Agarwal, P.; Wittmann, J.C.; Lotz, B. Heteroepitaxy of Syndiotactic Polypropylene with Polyethylene and Homoepitaxy. *Macromolecules* **1994**, *27*, 6956. [CrossRef]
44. Wittmann, J.C.; Lotz, B. Polymer Decoration: The Orientation of Polymer Folds as Revealed by The Crystallization of Polymer Vapors. *J. Polym. Sci. Polym. Phys. Ed.* **1985**, *23*, 205. [CrossRef]
45. Bassett, G.A. A New Technique for Decoration of Cleavage and Slip Steps on Ionic Crystal Surfaces. *Philos. Mag.* **1958**, *3*, 1042. [CrossRef]
46. Ayache, J.; Beaunier, L.; Boumendil, J.; Ehret, G.; Laub, D. *Sample Preparation Handbook for Transmission Electron Microscopy—Techniques*; Springer: Berlin/Heidelberg, Germany, 2010; Chapter 7; p. 279.
47. Khoury, F. General discussion. *Faraday Discuss. Chem. Soc.* **1979**, *68*, 404–405.
48. Bassett, D.C. *Principles of Polymer Morphology*; Cambridge University Press: Cambridge, UK, 1981.
49. Keith, H.D.; Padden, F.J.; Lotz, B.; Wittmann, J.C. Asymmetries of habit in polyethylene crystals grown from the melt. *Macromolecules* **1989**, *22*, 2230. [CrossRef]

Article

Morphology and Degradation of Multicompartment Microparticles Based on Semi-Crystalline Polystyrene-*block*-Polybutadiene-*block*-Poly(*L*-lactide) Triblock Terpolymers

Nicole Janoszka [1], Suna Azhdari [1], Christian Hils [2], Deniz Coban [1], Holger Schmalz [2,3,*] and André H. Gröschel [1,*]

1. Physical Chemistry, Center for Soft Nanoscience (SoN) and Center for Nanotechnology (CeNTech), University of Münster, Corrensstraße 28-30, 48149 Münster, Germany; nicole.janoszka@uni-muenster.de (N.J.); azhdari@exchange.wwu.de (S.A.); dcoban@exchange.wwu.de (D.C.)
2. Macromolecular Chemistry II, University of Bayreuth, Universitätsstraße 30, 95440 Bayreuth, Germany; christian.hils@basf.com
3. Bavarian Polymer Institute (BPI), University of Bayreuth, Universitätsstraße 30, 95440 Bayreuth, Germany
* Correspondence: holger.schmalz@uni-bayreuth.de (H.S.); andre.groeschel@uni-muenster.de (A.H.G.)

Citation: Janoszka, N.; Azhdari, S.; Hils, C.; Coban, D.; Schmalz, H.; Gröschel, A.H. Morphology and Degradation of Multicompartment Microparticles Based on Semi-Crystalline Polystyrene-*block*-Polybutadiene-*block*-Poly(*L*-lactide) Triblock Terpolymers. *Polymers* **2021**, *13*, 4358. https://doi.org/10.3390/polym13244358

Academic Editors: Holger Schmalz, Volker Abetz and Eamor M. Woo

Received: 24 November 2021
Accepted: 9 December 2021
Published: 13 December 2021

Publisher's Note: MDPI stays neutral with regard to jurisdictional claims in published maps and institutional affiliations.

Copyright: © 2021 by the authors. Licensee MDPI, Basel, Switzerland. This article is an open access article distributed under the terms and conditions of the Creative Commons Attribution (CC BY) license (https://creativecommons.org/licenses/by/4.0/).

Abstract: The confinement assembly of block copolymers shows great potential regarding the formation of functional microparticles with compartmentalized structure. Although a large variety of block chemistries have already been used, less is known about microdomain degradation, which could lead to mesoporous microparticles with particularly complex morphologies for ABC triblock terpolymers. Here, we report on the formation of triblock terpolymer-based, multicompartment microparticles (MMs) and the selective degradation of domains into mesoporous microparticles. A series of polystyrene-*block*-polybutadiene-*block*-poly(*L*-lactide) (PS-*b*-PB-*b*-PLLA, SBL) triblock terpolymers was synthesized by a combination of anionic vinyl and ring-opening polymerization, which were transformed into microparticles through evaporation-induced confinement assembly. Despite different block compositions and the presence of a crystallizable PLLA block, we mainly identified hexagonally packed cylinders with a PLLA core and PB shell embedded in a PS matrix. Emulsions were prepared with *Shirasu Porous Glass* (SPG) membranes leading to a narrow size distribution of the microparticles and control of the average particle diameter, $d \approx 0.4$ μm–1.8 μm. The core–shell cylinders lie parallel to the surface for particle diameters $d < 0.5$ μm and progressively more perpendicular for larger particles $d > 0.8$ μm as verified with scanning and transmission electron microscopy and particle cross-sections. Finally, the selective degradation of the PLLA cylinders under basic conditions resulted in mesoporous microparticles with a pronounced surface roughness.

Keywords: 3D confinement; ABC triblock terpolymers; degradation; emulsification; microparticles

1. Introduction

Block copolymers (BCPs) were demonstrated to be highly versatile materials to pattern surfaces and serve as templates for inorganic nanostructures [1,2]. They are also prime candidates for the self-assembly in solution, giving access to diverse micelle morphologies with tuneable dimensions [3,4], near-monodisperse fibres and platelets through crystallization-driven self-assembly [5–7], and complex nanostructures through hierarchical processes [8,9].

More recently, the microphase behaviour of BCPs in confinement has become another versatile method of forming BCP nanostructures with particular interest for the design of functional microparticles. With respect to AB diblock copolymers, a number of works clarified the effect of block composition [10], architecture [11,12], evaporation rate [13,14], type of surfactant [15–17] and emulsification method [18]. The focus progressively shifts towards more complex compositions (e.g., BCP nanoparticle hybrids) [19,20], as well as functionality, including response to magnetic fields, temperature, and light [21–23]. Another way to increase the functionality of the microparticles is the use of ABC triblock terpolymers [24–27] that feature three sequentially linked blocks with different chemistries that are known for their rich microphase behaviour [28–30]. Terpolymers in confinement have led to a variety of confinement-specific morphologies and nanoparticles, e.g., nanotoroids from lamella-ring morphology [31], nanocups from hemispherical lamella–lamellae [32] and perforated discs from axially stacked perforated lamellae [33]. Especially in combination with SPG membrane emulsification, the size of microparticles and nanoparticles can be controlled rather precisely.

Although a range of morphologies has been explored so far, individual microdomains or compartments have not been utilized much, regarding selective loading or removal, to create defined channel systems or porosity. The latter could, however, prove useful in the design of porous scaffolds for capture and release, catalysis, or the formation of mesoporous carbon. There are multiple, conceivable ways to hollow out individual compartments, e.g., the use of additives (homopolymer, hydrogen [34] or halogen [35,36] bonding) that can be washed out selectively or the use of degradable polymers. Among these, poly(ε-caprolactone) (PCL) and polylactide (PLA) are the most prominent examples, along with upcoming alternatives (e.g., polyphosphoesters) [37]. PLA and PCL and their copolymers are part of active research, where ecologically more viable catalysts are currently being identified [38–40], as well as degradable materials [41] and nanostructures with predetermined release profiles [42]. With respect to the confinement assembly of degradable BCPs, only a few studies exist that address polystyrene-*block*-poly(D-lactide) (PS-*b*-PDLA) BCPs [43–45]. There, the morphology was found to depend on the evaporation rate, solvent quality and temperature, leading to helical PDLA cylinders embedded in a PS matrix, axially stacked rings, or PDLA networks. Inspired by these works, we sought to add another polymer block/domain to the system, in order to introduce a functional surface layer after PDLA removal. In PS-*b*-PDLA microparticles, PDLA can be removed to form pores, while the purpose of PS is to retain the shape and structural integrity of the microparticles. Adding another block between PS and PDLA, e.g., polybutadiene (PB), makes it possible to further utilize the surface layer of the pore system of the microparticles for functionalization after PDLA removal (e.g., thiol-ene click chemistry). Therefore, the microparticles of PS-*b*-PB-*b*-PLLA (polystyrene-*block*-polybutadiene-*block*-poly(L-lactide), SBL) could provide a route toward surface-functional, porous microparticles with a versatile post-modification capability.

In this work, we lay the foundation for this direction by employing 3D confinement assembly as a structuring method to form MMs of SBL triblock terpolymers with a semi-crystalline and degradable PLLA microdomain. We studied the influence of block composition on inner morphology, as well as the effect of SPG membrane pore diameter on microparticle diameter and size distribution. Finally, microparticles were subjected to basic conditions to analyse the selective degradation of the PLLA microdomain, resulting in mesoporous microparticles.

2. Materials and Methods

All chemicals were used as received unless otherwise noted. Styrene ($\geq$99.9%, ReagentPlus®, Sigma-Aldrich, Taufkirchen, Germany) was purified over di-*n*-butylmagnesium ((n-Bu)$_2$Mg). 1,3-Butadiene (gaseous, Messer Industriegase GmbH, Bad Soden, Germany) was passed through columns filled with molecular sieve (4 Å) and basic aluminium oxide, condensed into a glass reactor, and stirred over (n-Bu)$_2$Mg at least two days prior to use [46]. Ethylene oxide (3.0, Linde GmbH, Pullach, Germany) was condensed onto calcium hydride (CaH$_2$) and stirred at 0 °C for 3 h before being transferred into a storage ampoule [47]. Additionally, ethylene oxide was purified over *n*-butyllithium at 0 °C directly before use, followed by condensation into a glass ampoule. L-lactide ($\geq$99.8%, PURASORB® L, Corbion, Amsterdam, The Netherlands) was recrystallized from toluene and stored under nitrogen until use. Dichloromethane (DCM, $\geq$99.5%, AnalaR NORMAPUR, VWR International

GmbH, Darmstadt, Germany) used for polymerization was dried by successive distillation over CaH$_2$. Triethylaluminium (Et$_3$Al, 1 M in hexane, Sigma-Aldrich, Taufkirchen, Germany), (n-Bu)$_2$Mg (0.5 M in heptane, Sigma-Aldrich), n-butyllithium (1.6 M in hexanes, AcroSeal®, Thermo Fisher Scientific, Geel, Belgium), sec-butyllithium (sec-BuLi, 1.3 M in cyclohexane/hexane (92/8), AcroSeal®, Thermo Fisher Scientific, Geel, Belgium), sodium dodecyl sulfate (SDS, >99%, Sigma-Aldrich, Taufkirchen, Germany), osmium tetroxide (OsO$_4$, 4 wt.% in H$_2$O, Science Services, Munich, Germany), trimethylsilyl chloride (TMSC, ≥99%, Sigma Aldrich, Taufkirchen, Germany), chloroform (CHCl$_3$, analytical grade, Merck Chemicals GmbH, Darmstadt, Germany), sodium hydroxide (NaOH, 99.1%, VWR International GmbH, Darmstadt, Germany), and methanol (MeOH, >99.9%, Thermo Fisher Scientific, Geel, Belgium) were used as received. All polymerizations were performed under a dry argon or nitrogen atmosphere using Schlenk techniques and glass reactors for anionic polymerization to exclude moisture and air during the polymerization. Ultrapure water from a Milli-Q Integral Water Purification System was used for the preparation of emulsions and for purification.

2.1. Synthesis of SB-OH Precursors

The OH-terminated polystyrene-block-poly(1,4-butadiene) (PS-b-PB-OH, SB-OH) diblock copolymers were synthesized by sequential anionic polymerization in toluene initiated with sec-BuLi. The end-functionalization of the polymer was completed with ethylene oxide. The polymer was precipitated from MeOH, filtered, and dried in vacuo.

2.2. Synthesis of SBL Triblock Terpolymers

The PLLA blocks were synthesized by anionic ring-opening polymerization according to established recipes [48,49]. In brief, the synthesis of polystyrene-block-poly(1,4-butadiene)-block-poly(L-lactide) (PS-b-PB-b-PLLA, SBL) was realized by polymerization of L-lactide with DBU as a catalyst and SB-OH as macroinitiator in DCM. In the terpolymer notation employed here (AxByCz), the subscripts denote the number average degree of polymerization of the respective block (Table 1).

Table 1. Composition of used SBL triblock terpolymers.

Code	P_n [a]	f_S [b]	f_B	f_L	M_n [c]	Đ [d]
SBL-10	S$_{118}$B$_{310}$LLA$_{27}$	0.49	0.41	0.10	40,100	1.12
SBL-37	S$_{118}$B$_{310}$LLA$_{150}$	0.34	0.29	0.37	57,900	1.14
SBL-52	S$_{118}$B$_{310}$LLA$_{280}$	0.26	0.22	0.52	76,000	1.17
SBL-56	S$_{295}$B$_{292}$LLA$_{408}$	0.29	0.15	0.56	105,300	1.08

[a] P_n is the number average degree of polymerization and [b] f_n the weight fraction of each block. [c] Number average molecular weight determined through a combination of SEC and ^{1}H-NMR. [d] Dispersity values determined with SEC (see Section 2.7).

2.3. Preparation of SBL Bulk Films

Before preparation of bulk films, each vial was hydrophobized with trimethylsilyl chloride prior to use. For each bulk film, 30 mg of SBL triblock terpolymer was dissolved in 1 mL DCM and stirred for 1 h, followed by slow evaporation of DCM over 5 days. After bulk film formation, the film was removed by cooling the vial with liquid nitrogen and breaking the glass vial.

2.4. Preparation of SBL Microparticles

SBL triblock terpolymers were dissolved in CHCl$_3$ at a concentration of $c = 10$ g L^{-1} and stirred overnight prior to emulsification. A stock solution of SDS with a concentration of $c = 5$ g L^{-1} was prepared as the continuous phase and 35 mL was used for each emulsification experiment. The emulsification process used a SPG membrane with different pore diameters of $d_{pore} = 0.3, 0.5, 0.8, 1.1$ and 2.0 µm. The 50 mL reservoir was stirred at 250 rpm while argon pressure continuously pressed the polymer solution through the membrane. After emulsification, the organic phase evaporated over three days at room temperature before the solid SBL microparticles were dialyzed against ultrapure water. The dialysis bath was exchanged 5 times to remove excess SDS.

2.5. Degradation of SBL Microparticles

The SBL microparticles were degraded in solution according to the literature [43]. The microparticles were dispersed in a NaOH solution (c = 20 g L^{-1}) of H$_2$O/MeOH (40/60 v/v) to yield a microparticle concentration of c = 1 g L^{-1}. The particle suspension was stirred for 5 days. After degradation, the remaining MMs were dialyzed against ultrapure water to remove NaOH and impurities.

2.6. Nuclear Magnetic Resonance (NMR) Spectroscopy

The SBL triblock terpolymers were characterized by ^{1}H-NMR spectroscopy (Bruker Ultrashield 300 spectrometer, Bruker Corporation, Billerica, MA, USA) using CDCl$_3$ as solvent. The signal assignment was supported by simulations with the NMR software MestReNova. Absolute number average block lengths were determined from the ^{1}H-NMR spectra, employing the number average molecular weight (M_n) of the PS precursor, determined by SEC with PS calibration for signal calibration.

2.7. Size-Exclusion Chromatography (SEC)

SEC measurements were performed on a SEC 1260 Infinity system (Agilent Technologies, Santa Clara, CA, USA) with two PSS-SDV gel columns (particle size = 5 μm) with porosity range from 10^2 to 10^5 Å (PSS, Mainz, Germany). CHCl$_3$ (HPLC grade) was used as eluent. All samples were dissolved and filtered through a 0.2 μm PTFE filter before analysis. The samples were measured at a flow rate of 0.5 mL min^{-1} at 23 °C, using a refractive index detector (Agilent Technologies, Santa Clara, CA, USA). The calibration was completed with narrowly distributed PS standards (PSS calibration kit), and toluene (HPLC grade) was used as internal standard.

2.8. Differential Scanning Calorimetry (DSC)

Calorimetric measurement of the SBL bulk films was performed on a Phoenix 204 F1 (Netzsch, Selb, Germany) instrument using aluminium crucibles (pierced lid) and nitrogen as carrier gas. The temperature range was selected from −150 °C to 200 °C (liquid N$_2$ cooling) with scanning rates of 10 K min^{-1}.

2.9. Transmission Electron Microscopy (TEM)

SBL microparticles were analysed on a JEOL JEM-1400 Plus TEM operating at an acceleration voltage of 120 kV. The diluted particle dispersion (c = 0.3 g L^{-1}) was dropped on a carbon-coated copper grid. The sample was blotted with a filter paper after 30 s and the grid was dried for at least 12 h. The samples were stained with OsO$_4$ for 1 h before analysis. The TEM images were processed with ImageJ (version 1.53c).

2.10. Scanning Electron Microscopy (SEM)

The surface and shape of the microparticles were recorded on a cryo-field emission Zeiss Crossbeam 540 FIB-SEM equipped with lens-, chamber-, and energy-selective detectors for 16 Bit image series acquisition with up to a 40,000 × 50,000-pixel resolution. Samples were prepared by dropping 20 μL of dispersion with a concentration of c = 0.3 g L^{-1} on a silicon wafer. The solution was blotted with a filter paper after 30 s and the wafer was dried for at least 12 h. Before recording images, a Pt-Cd layer of 6 nm was sputtered onto the sample using a Quorum PP3010T-Cryo chamber with integrated Q150T-Es high-end sputter coater and Pt–Cd target.

2.11. Ultra-Sectioning of SBL Microparticles

Freeze-dried SBL microparticles were embedded in 3D Rapid Resin CLEAR 3DR3582C, cured by UV light (λ = 365 nm), and sectioned with a Reichert/Leica Ultracut E microtome (Leica Microsystems, Wetzlar, Germany). A section velocity of around 1 mm s^{-1} with an inclination angle of 6 degrees was used to gain cross-sectional slices of about 70 nm thickness. For TEM imaging, cross-sections were transferred on a carbon-coated copper grid. The PB domains were selectively stained with OsO_4 for 3 h and followed by sputtering the samples with a carbon coating of 10 nm before imaging.

2.12. Raman Spectroscopy

The Raman spectra of the SBL-56 microparticles before and after hydrolytic degradation of the PLLA block were taken with a WITec alpha 300 RA+ imaging system (WITec GmbH, Ulm, Germany), equipped with an UHTS 300 spectrometer and a back-illuminated Andor Newton 970 EMCCD (electron multiplying charge-coupled device) camera. For the measurements an excitation wavelength of λ = 532 nm and a 50× objective (Zeiss LD EC Epiplan-Neofluar, NA = 0.55) together with the WITec suite FIVE 5.3 software package were used. Spectra were acquired using a laser power of 10 mW, an integration time of 0.5 s and 50 accumulations. All spectra were subjected to a cosmic ray removal routine and base line correction. Samples were prepared by placing some drops of the dispersions onto glass slides followed by drying.

2.13. Dynamic Light Scattering (DLS)

SBL microparticles were measured on a LSI spectrometer (Fribourg, Switzerland) operating with a DPSS Cobolt laser (max. 100 mW constant output at λ = 660 nm). Samples were prepared at a concentration of c = 0.06 g L^{-1} and purified three times from dust by passing the sample solution through a PTFE filter with a pore size of 5 µm directly into cylindrical quartz cuvettes (d = 10 mm), which were flushed with compressed air before measurement. Three intensity–time autocorrelation functions were measured at a scattering angle of 90° and an acquisition time of 60 s. Recorded data were analysed with the LSI software package (v.63).

3. Results and Discussion

We first synthesized SBL triblock terpolymers with different weight fractions of the poly(L-lactide) (PLLA) block (f_L = 0.1–0.56) to prepare MMs with different morphologies. For the SBL syntheses, a two-step procedure was employed as summarized in Scheme 1. First, a hydroxy-terminated polystyrene-*block*-poly(1,4-butadiene) (SB-OH) precursor was prepared by sequential living anionic polymerization of styrene and 1,3-butadiene in toluene, followed by quantitative end capping with ethylene oxide. Subsequently, SB-OH was used as macroinitiator for the anionic ring-opening polymerization of L-lactide in DCM in the presence of DBU as catalyst. This yielded a series of SBL triblock terpolymers (Table 1), SBL-10, SBL-37, and SBL-52, as well as SBL-56, synthesized from two different SB-OH precursors ($S_{118}B_{310}$-OH and $S_{295}B_{292}$-OH). In the used SBL-X notation, X stands for the weight fraction of the PLLA block.

According to SEC, all synthetic steps proceeded without significant side reactions, resulting in narrowly distributed SBLs with dispersities of $Đ \approx$ 1.1–1.2 (Figure 1A,B and Table 1). In the ^{1}H-NMR spectra, the signal at 5.2 ppm can be assigned to the methine protons of the PLLA repeating unit, showing an increasing peak area in the series SBL-10 to SBL-56 (Figure 1C). The final composition of the SBL triblock terpolymers was calculated by comparing the signals of PS, PB and PLLA, whereas the molecular weight of PS obtained from SEC served as a reference. In order to probe whether the PLLA block is able to crystallize, differential scanning calorimetry (DSC) was performed on SBL bulk films, which were cast from DCM (Figure 1D). Here, only the first heating traces, i.e., the thermal properties directly after film casting, were considered, as these came closest to the thermal properties of the SBL microparticles. For SBL-10, no melting endotherm could be observed for the PLLA block, most likely due to its low volume fraction. In addition, SBL-10 showed only one glass transition temperature (T_g), being located in between the expected T_g-values of PS (T_g = 100 °C) and PLLA (T_g = 50 °C), pointing to the formation of a mixed amorphous PS/PLLA phase. For SBL-37, SBL-52, and SBL-56 melting endotherms confirmed that a partial crystallization of PLLA occurs upon film casting with increasing PLLA content. This effect is most pronounced for SBL-56, which has the highest PLLA fraction as well as highest overall molecular weight.

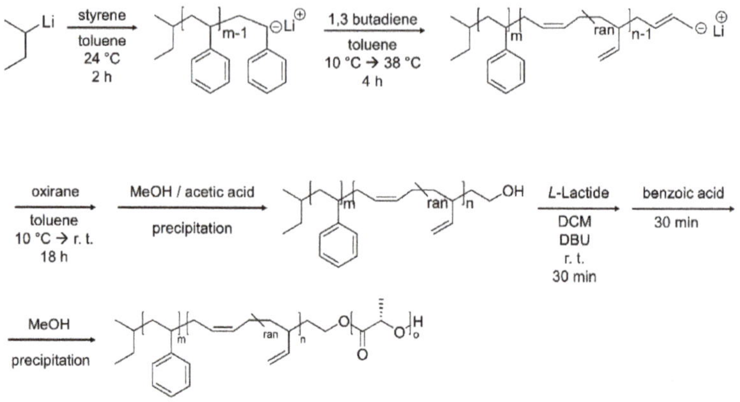

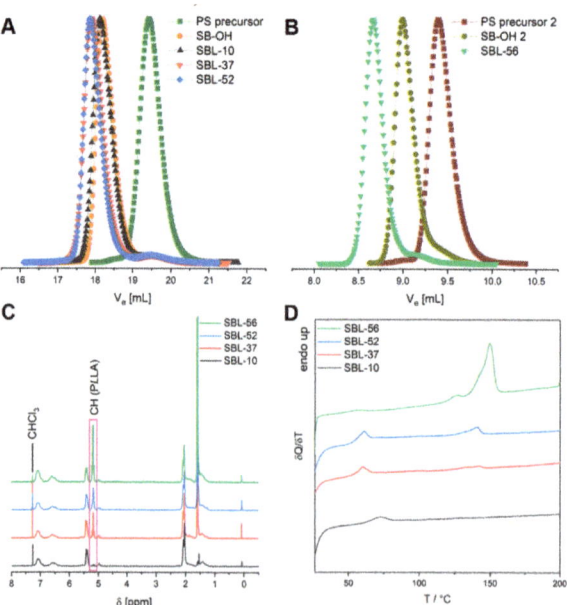

Scheme 1. Synthesis of SBL triblock terpolymers. Sequential anionic polymerization of styrene and 1,3-butadiene, and termination with oxirane leads to OH-modified polystyrene-*block*-poly(1,4-butadiene) (SB-OH). Anionic ring-opening polymerization of L-lactide afforded the SBL triblock terpolymers.

Figure 1. Characterization of the SBL triblock terpolymers. (**A,B**) SEC traces for the two series of terpolymers based on $S_{118}B_{310}$-OH and $S_{295}B_{292}$-OH macroinitiators and the respective PS precursors. (**C**) ^{1}H-NMR spectra of all SBLs with characteristic PLLA peak being highlighted and (**D**) DSC first heating traces of bulk films cast from DCM.

After SBL synthesis, we obtained MMs through SPG membrane emulsification followed by the evaporation of the organic solvent (evaporation-induced confinement assembly (EICA)). Scheme 2 outlines the process of obtaining SBL MMs. The SPG method utilizes porous glass membranes through which the polymer solution is pressed into the continuous surfactant phase to generate emulsion droplets of a homogeneous size. For that, SBL was first dissolved in CHCl$_3$ at a concentration of c_{SBL} = 10 g L^{-1} and stirred overnight prior to emulsification. To form the oil-in-water (O/W) emulsion, the SBL solution was pushed through the SPG membrane with argon pressure into the continuous phase consisting of an aqueous sodium dodecyl sulfate solution (SDS, c_{SDS} = 5 g L^{-1}). Depending on the pore size of the membrane, the SBL solution was completely emulsified within 2–3 h and CHCl$_3$ was evaporated over three days under an ambient atmosphere while gently stirring the emulsion. We utilized five different membrane pore sizes (d_{pore} = 0.3, 0.5, 0.8, 1.1, and 2.0 µm) to investigate the size distribution and morphology of each of the SBL MMs.

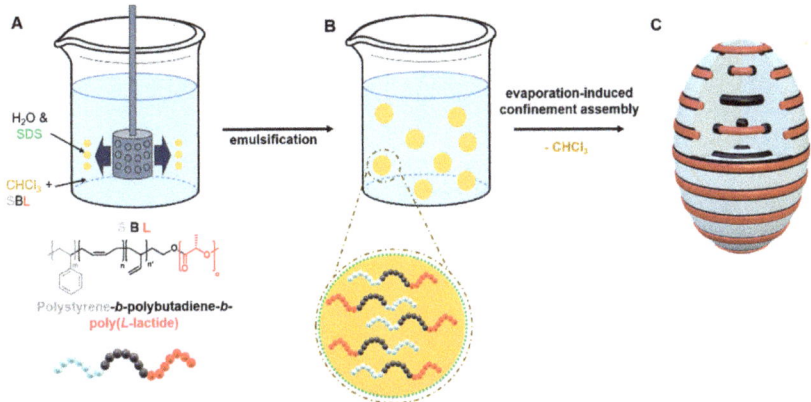

Scheme 2. Preparation of the SBL MMs through SPG membrane emulsification and evaporation-induced confinement assembly. (**A**) Emulsification of SBL with the SPG membrane. (**B**) Microemulsion droplets of SBL/CHCl$_3$ in aqueous solution stabilized by SDS. (**C**) Resulting MMs after solvent evaporation process.

We started our analytics by first verifying the suitability of the SPG method to form MMs from SBL triblock terpolymers. Figure 2 summarizes the results of SBL-37 MMs fabricated with a pore diameter of d_{pore} = 0.5 µm. DLS measurements suggest a narrow size distribution of the SBL-37 particles, as the CONTIN analysis shows a narrow fit with an average hydrodynamic diameter of D_h = 490 nm ($Đ$ = 0.15) (Figure 2A). According to SEM analysis (Figure 2B), the SBL-37 MMs exhibited an average particle diameter of d = 388 ± 48 nm and were frequently arranged in a hexagonal lattice upon drying on the silicon wafer corroborating the narrow size distribution realized with the SPG method. The D_h of the SBL-37 MMs is slightly larger than the diameter in SEM, which we attribute to shrinking due to e-beam damage (mostly degrading the PLLA block). Nevertheless, the D_h fits with the expected particle size produced with a 0.5 µm pore diameter of the SGP membrane.

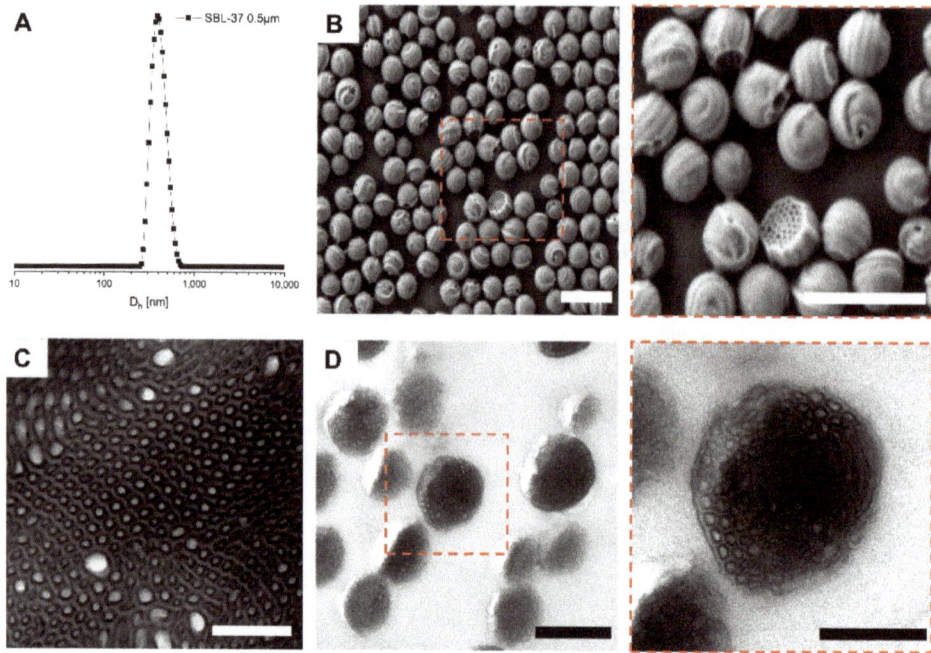

Figure 2. SBL-37 MMs prepared with the membrane d_{pore} = 0.5 µm. (**A**) Size distribution from DLS CONTIN plot. (**B**) SEM overview image and close-up (scale bars 1000 nm). (**C**) TEM image of the SBL-37 bulk morphology (scale bar 500 nm) and (**D**) TEM of cross-sections of embedded SBL-37 MMs (d_{pore} = 2.0 µm) visualizing the inner morphology (TEM samples stained with OsO$_4$ to enhance the contrast of PB; PLLA appears brightest, scale bars 1000 nm and 500 nm in inset).

SEM further revealed a striped surface of the SBL-37 MMs (Figure 2B) indicating a lamella or cylinder morphology. At weight fractions of $f_S = 0.34$, $f_B = 0.29$, and $f_L = 0.37$, a lamella–lamella morphology is more likely, but the close-up SEM image of Figure 2B clearly shows that some particles have a flat side revealing hexagonally packed cylinders. Although it is surprising to find a core–shell cylinder morphology with a PLLA core and PS shell at such a large f_L = 37, the corresponding TEM image of the bulk morphology corroborates this morphology (Figure 2C). Before TEM imaging, the sample was stained with OsO$_4$ to enhance the contrast of the PB microdomains, which appear dark (PS light grey, PLLA bright). As we will see later, the core–shell cylinder morphology is stable even at $f_L > 0.37$. Due to the anisotropy of the PLLA cylinder phase, the particles were not able to fully close into spheres in some cases, and therefore developed a flat side with a hexagonal perforation. The otherwise striped 'baseball' surface pattern is a typical defect pattern that cylinders develop in or on spherical surfaces [50–54]. In order to gain information about the internal structure of the microparticles, ultrathin cross-sections were generated from the largest d_{pore} = 2.0 µm (Figure 2D). The close-up TEM image of a single SBL-37 particle in Figure 2D verifies the core–shell cylinder morphology and their hexagonal packing. The cylinders do not appear to have a fully circular cross-sections, but instead show rectangular features. Such a deformation was observed before for core–shell cylinders of polystyrene-*block*-polybutadiene-*block*-poly(ε-caprolactone) (SBC) triblock terpolymers with a semi-crystalline PCL core block, but with a lower f_C = 0.16 [55,56]. We therefore attribute the deviation from the spherical cross-section of the cylinders to the semi-crystalline nature of PLLA. If the SBL core–shell cylinders undergo one complete revolution along the curved microparticle interface, they close into core–shell rings (or toroids), an idea that was also previously reported for PS-*b*-PB-*b*-P*t*BMA (P*t*BMA = poly(*tert*-butyl methacrylate) [26].

To investigate the effect of membrane pore diameter on the inner morphology, we emulsified SBL-37 with five different d_{pore} = 0.3, 0.5, 0.8, 1.1, and 2.0 µm, and analysed the size and inner structure with TEM, SEM and DLS (Figures 3 and S1). The SEM overview images exhibit a narrow size distribution for each sample implying a good control over the process (Figure 3A). The DLS measurements in Figure S1 confirm a monomodal and narrow size distribution for all SBL-37 MMs and an increasing D_h with pore size. As expected, the average diameter of the particles increases with the pore diameter from about 390 nm to 1800 nm. The cylinder morphology remains for all SBL-37 MMs, which is not that clear from the TEM images but more visible in SEM due to the characteristic surface pattern. For smaller particles (e.g., d_{pore} = 0.5 µm, Figure 3B middle row) the cylinders preferentially lie parallel to the particle surface (striped baseball pattern). In contrast, core–shell cylinders are able to orientate perpendicular to the particle surface for larger particles, leading to a hexagonal pattern and perforations (e.g., d_{pore} > 0.8 µm, Figure 3C–E, middle row). TEM images of all SBL-37 MMs were recorded to determine the internal structure and detect any size-dependent changes in the microphases (Figure 3B). The internal structure for the smallest SBL-37 MMs (d_{pore} = 0.3 and 0.5 µm) is visible in transmission (larger particles are rather dark). With decreasing particle diameter, the SBL-37 triblock terpolymer is more confined during emulsification. While the size of the microdomains remains unchanged, the smaller diameter means a stronger curvature, and thus higher degree of confinement, D/L_0. Each block is therefore forced to adapt to the curvature resulting in less-ordered structures. Conversely, microphase separation in larger particles approaches thermodynamic equilibrium and results in a more-ordered core–shell cylinder morphology (close to the bulk case). Nevertheless, all SBL-37 MMs exhibit a core–shell cylindrical morphology, suggesting that the pore diameter has little effect on the morphology itself, but rather on the orientation of the morphology.

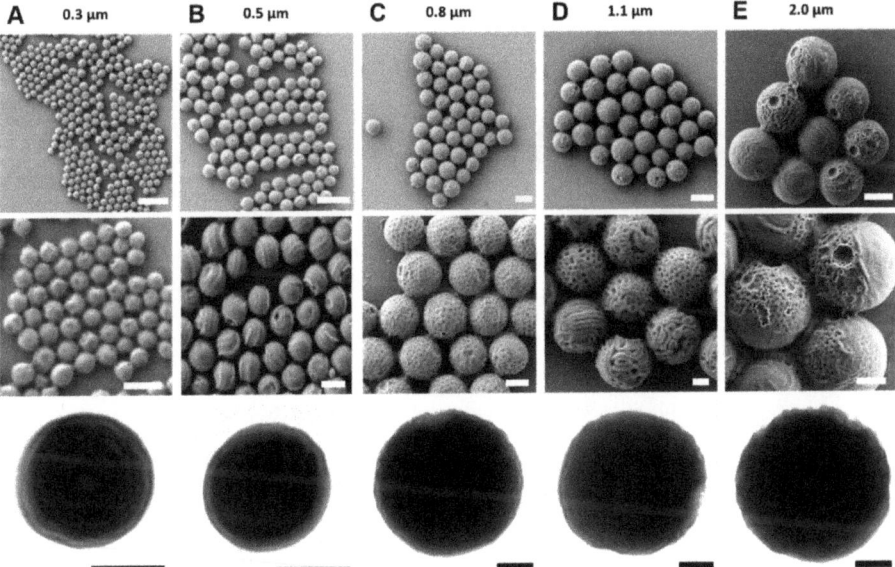

Figure 3. SBL-37 MMs prepared with different d_{pore}: (**A**) 0.3 µm, (**B**) 0.5 µm, (**C**) 0.8 µm, (**D**) 1.1 µm, and (**E**) 2.0 µm. Top row: SEM overview images (scale bars 1000 nm); middle row: close-up SEM images (scale bars 500 nm); bottom row: TEM images (scale bars 200 nm).

Next, we explored the shape and internal structure of all other SBLs. In this case, the f_L varied from f_L = 0.10 to f_L = 0.56. Since the inner order of the SBL MMs is best visible for smaller particle sizes, Figure 4 summarizes MMs fabricated with a d_{pore} = 0.5 µm. According to TEM, an onion-like morphology is obtained for SBL-10 (Figure 4A). SBL-10 exhibits the lowest f_L, and the PLLA microdomain does not seem to have any impact on the final morphology, most likely due to the mixing of PS and PLLA chains (supported by only one T_g in DSC, see Figure 1D). Since $f_S \approx f_B$, SBL-10 is arranged in an onion-like (concentric lamellar) structure of alternating PS and PB lamellae in which PLLA might even be mixed with PS. Both SBL-37 and SBL-52 MMs clearly form a cylinder morphology, although the f_L of both SBLs is quite different from each other (Figure 4B,C). Since SBL-37 with f_L = 0.37 already formed a core–shell cylinder morphology, as discussed in Figure 2, this morphology appears to have a rather large stability region (it is also in agreement with the bulk morphology). The main difference is the core diameter of the PLLA microphase, which increased from d_{cyl} = 13 nm for SBL-37 to d_{cyl} = 15 nm for SBL-52. Finally, SBL-56 has the largest f_L and exhibited a core–shell cylinder morphology, yet, was arranged as hexagonally packed and axially stacked hoops (Figure 4D). The PS still constitutes the matrix of the MMs, even though, with f_S = 0.29, it constitutes the minority phase.

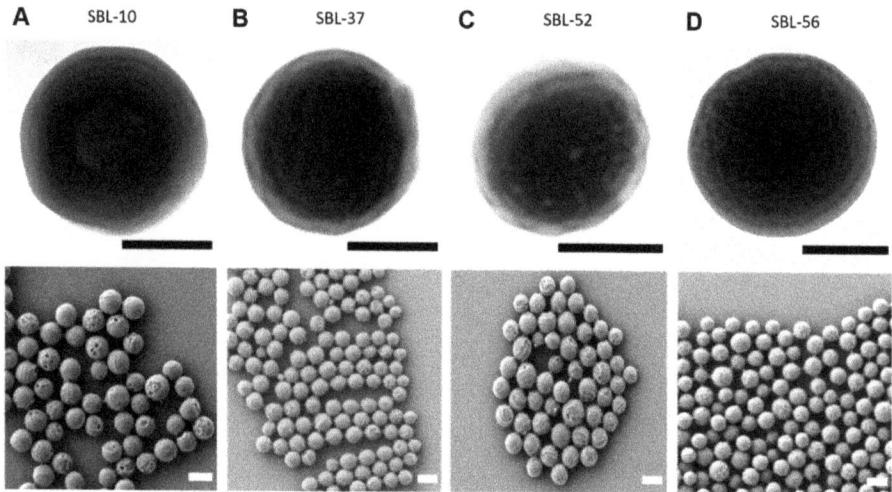

Figure 4. EICA of all SBLs (as indicated) employing a membrane with d_{pore} = 0.3 µm. (A) SBL-10, (B) SBL-37, (C) SBL-52, and (D) SBL-56. Top row: TEM images of individual MMs (scale bars 200 nm); bottom row: SEM overview images (scale bars 500 nm).

Finally, the hydrolysis of the SBL MMs prepared with d_{pore} = 0.3 µm was studied to selectively remove the PLLA domains, and thus fabricate porous microparticles. Starting with SBL-56, which had the highest PLLA fraction, hydrolysis was performed for 5 days under basic conditions by stirring the SBL-56 MMs in a H_2O/MeOH mixture (60/40 v/v) containing 20 g L^{-1} NaOH. We first analysed the morphology in TEM before (Figure 5A) and after degradation (Figure 5B,C). Before degradation, the MMs exhibited a core–shell cylinder structure looping into core–shell rings. As visible from the TEM images in Figure 5B,C, the ring structure remains after the hydrolysis of the PLLA domains. Note that the PB microphase is still present (dark rings in Figure 5C) and now covers the surface of the inner pores as well as the MM's outer surface. The PB provides double bonds for crosslinking, loading, and versatile post-modification possibilities (e.g., through thiol-ene click reactions). Furthermore, the microparticles now exhibit a rough surface pattern in SEM (Figure 5D). The surface structure is different than before degradation (see Figure 4D) indicating the successful removal of the PLLA microphase. To quantify PLLA hydrolysis, we performed ^{1}H-NMR and Raman measurements before and after degradation. The

^{1}H-NMR measurements corroborate a degradation of about 95% of the PLLA microdomain of SBL-56 (Figure 5E). We determined the degradation by comparing the resonance peaks of the methine group of PLLA at about 5.2 ppm relative to the aromatic signals of PS between 6.2–7.2 ppm before and after degradation. In addition, Raman measurements also confirm the removal of the PLLA phase after 5 days by the disappearance of the characteristic carbonyl stretching vibration at $\tilde{v}$ = 1765 cm^{-1} (Figure 5F). We furthermore tested the basic degradation of other SBL MMs under identical conditions (Figures S2–S4). For SBL-10 MMs, the concentric lamellar structure remains after 5 days of degradation. However, in some cases, the outermost lamella of the SBL-10 MMs detached from the particle, suggesting the degradation of a mixed PS and PLLA phase (Figure S2). According to the TEM images displayed in Figure S3, the initial core–shell cylinder morphology of SBL-37 MMs is retained upon hydrolysis, but the MMs appear porous after basic degradation of the PLLA microdomain. In particular, the close-up TEM image in Figure S3C and the SEM image in Figure S3D highlight the porous structure. A similar trend was also observed for SBL-52 MMs (Figure S4). The porosity of the hydrolysed SBL-37 and SBL-52 MMs further verified the formation of a PLLA cylinder core and agree with the morphology assignment of the SBLs after the confinement assembly discussed above.

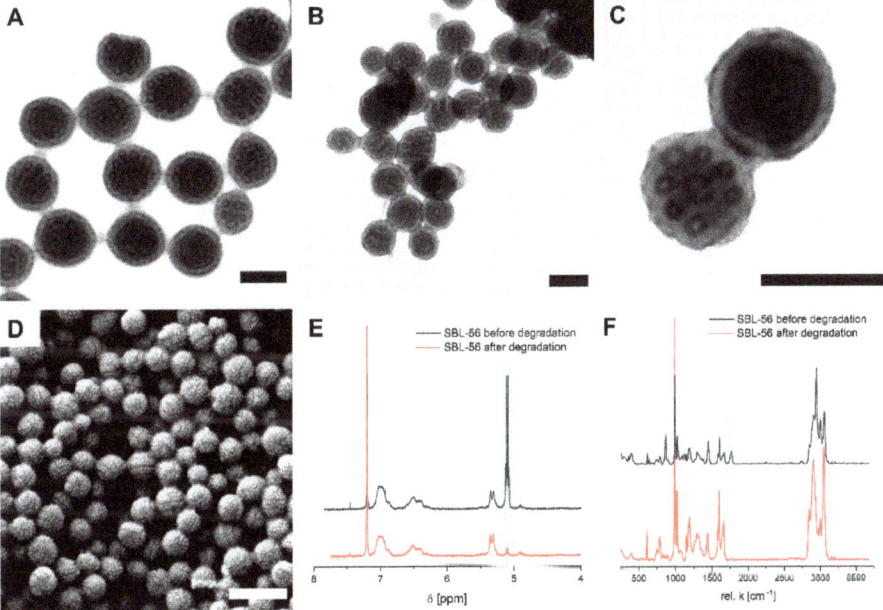

Figure 5. Degradation of SBL-56 MMs under basic conditions for 5 days (membrane with d_{pore} = 0.3 µm). (**A**) TEM overview image before and after degradation (**B**) (scale bars 200 nm). (**C**) TEM close-up and (**D**) SEM image after degradation (scale bars 500 nm). (**E**) ^{1}H-NMR and (**F**) Raman measurements before (black) and after degradation (red).

4. Conclusions

In summary, SBL triblock terpolymers with varying PLLA block length were synthesized and self-assembled in confinement, employing SPG membranes with different pore diameters to fabricate SBL MMs with a defined size and narrow size distribution. Despite different block compositions, this process mostly led to hexagonally packed core–shell cylinders consisting of a PLLA core, PB shell and a PS matrix. The increasing f_L did not have a noticeable effect on the inner structure of the MMs, which coincided with the bulk morphology. However, the average MM diameter did influence the morphology, i.e., for particle diameters $d < 0.3$ µm, the core–shell cylinders were less ordered; while for particles with 0.3 µm $< d < 0.8$ µm, the core–shell cylinders were oriented parallel to the particle surface, and preferred a perpendicular orientation of $d > 0.8$ µm. Selective degradation of the PLLA cylinders under basic conditions led to pronounced surface corrugations, as well as pores that were shaped according to the preceding PLLA cylinders. The degradation process did not affect MM structure or pore morphology. The PB microdomain became the surface coating of the pores and the overall surface. The confinement assembly of degradable triblock terpolymers, therefore, not only allows the formation of narrowly size-distributed microparticles with a controlled inner structure, but likewise provides functional microparticles after degradation.

Supplementary Materials: The following are available online at https://www.mdpi.com/article/10.3390/polym13244358/s1, Figure S1: CONTIN plots of SBL-37 MMs; Figures S2–S4: SBL-10, SBL-37, and SBL-52 MMs prepared with a membrane pore diameter of 0.3 µm before and after degradation.

Author Contributions: N.J. prepared the microparticles, performed TEM and SEM measurements, image analysis, DLS measurements, and wrote the manuscript. C.H. synthesized and analysed the SBL triblock terpolymers. S.A. performed the ultra microtome sectioning of the microparticles and recorded the cross-sections as well as some samples with TEM. D.C. performed and analysed the degradation experiments. The manuscript was read and approved by all authors before submission. H.S. and A.H.G. conceived and supervised the project. All authors have read and agreed to the published version of the manuscript.

Funding: C.H. and H.S. acknowledge support by the collaborative research centre SFB 840 (project A2) and by the Graduate School of the University of Bayreuth (C.H.). A.H.G. acknowledges funding from the DFG (GR 5075/6-1) and through the Emmy Noether Program (GR 5075/2-1).

Institutional Review Board Statement: Not applicable.

Informed Consent Statement: Not applicable.

Data Availability Statement: The data presented in this study are available on request from the corresponding author.

Acknowledgments: The authors would like to thank the Imaging Centre Essen (IMCES) at the University Clinic in Essen, where the SEM and most of the TEM data were recorded. We thank Emma Fuchs for help in the synthesis of some of the triblock terpolymers, Rika Schneider for SEC, Carmen Kunert for ultramicrotome cutting and TEM measurements of the SBL bulk film, the Keylab "Optical and Electron Microscopy" of the Bavarian Polymer Institute (University of Bayreuth) for support, and Corbion for the donation of L-lactide. Additionally, we would like to thank Fabian Herrmann for providing the Reichert/Leica Ultracut E microtome and the equipment for embedding the particles. We thank Jasper Berndt for providing the Carbon sputter device.

Conflicts of Interest: The authors declare no conflict of interest.

References

1. Stefik, M.; Guldin, S.; Vignolini, S.; Wiesner, U.; Steiner, U. Block copolymer self-assembly for nanophotonics. *Chem. Soc. Rev.* **2015**, *44*, 5076–5091. [CrossRef]
2. Bojer, C.; Ament, K.; Schmalz, H.; Breu, J.; Lunkenbein, T. Electrostatic attraction of nanoobjects-A versatile strategy towards mesostructured transition metal compounds. *CrystEngComm* **2019**, *21*, 4840–4850. [CrossRef]
3. Mai, Y.; Eisenberg, A. Self-assembly of block copolymers. *Chem. Soc. Rev.* **2012**, *41*, 5969–5985. [CrossRef]
4. Karayianni, M.; Pispas, S. Block copolymer solution self-assembly: Recent advances, emerging trends, and applications. *J. Polym. Sci.* **2021**, *59*, 1874–1898. [CrossRef]
5. Ganda, S.; Stenzel, M.H. Concepts, fabrication methods and applications of living crystallization-driven self-assembly of block copolymers. *Prog. Polym. Sci.* **2020**, *101*, 101195. [CrossRef]
6. MacFarlane, L.; Zhao, C.; Cai, J.; Qiu, H.; Manners, I. Emerging applications for living crystallization-driven self-assembly. *Chem. Sci.* **2021**, *12*, 4661–4682. [CrossRef]
7. Hils, C.; Manners, I.; Schöbel, J.; Schmalz, H. Patchy micelles with a crystalline core: Self-assembly concepts, properties, and applications. *Polymers* **2021**, *13*, 1481. [CrossRef]
8. Gröschel, A.H.; Walther, A.; Löbling, T.I.; Schacher, F.H.; Schmalz, H.; Müller, A.H.E. Guided hierarchical co-assembly of soft patchy nanoparticles. *Nature* **2013**, *503*, 247–251. [CrossRef]

9. Lunn, D.J.; Finnegan, J.R.; Manners, I. Self-assembly of "patchy" nanoparticles: A versatile approach to functional hierarchical materials. *Chem. Sci.* **2015**, *6*, 3663–3673. [CrossRef]
10. Jang, S.G.; Audus, D.J.; Klinger, D.; Krogstad, D.V.; Kim, B.J.; Cameron, A.; Kim, S.W.; Delaney, K.T.; Hur, S.M.; Killops, K.L.; et al. Striped, ellipsoidal particles by controlled assembly of diblock copolymers. *J. Am. Chem. Soc.* **2013**, *135*, 6649–6657. [CrossRef]
11. Steinhaus, A.; Pelras, T.; Chakroun, R.; Gröschel, A.H.; Müllner, M. Self-assembly of diblock molecular polymer brushes in the spherical confinement of nanoemulsion droplets. *Macromol. Rapid Commun.* **2018**, *39*, 1800177. [CrossRef] [PubMed]
12. He, Q.; Ku, K.H.; Vijayamohanan, H.; Kim, B.J.; Swager, T.M. Switchable full-color reflective photonic ellipsoidal particles. *J. Am. Chem. Soc.* **2020**, *142*, 10424–10430. [CrossRef] [PubMed]
13. Shin, J.M.; Kim, Y.; Yun, H.; Yi, G.R.; Kim, B.J. Morphological evolution of block copolymer particles: Effect of solvent evaporation rate on particle shape and morphology. *ACS Nano* **2017**, *11*, 2133–2142. [CrossRef]
14. Ku, K.H.; Lee, Y.J.; Kim, Y.; Kim, B.J. Shape-anisotropic diblock copolymer particles from evaporative emulsions: Experiment and theory. *Macromolecules* **2019**, *52*, 1150–1157. [CrossRef]
15. Jeon, S.J.; Yi, G.R.; Yang, S.M. Cooperative assembly of block copolymers with deformable interfaces: Toward nanostructured particles. *Adv. Mater.* **2008**, *20*, 4103–4108. [CrossRef]
16. Ku, K.H.; Yang, H.; Shin, J.M.; Kim, B.J. Aspect ratio effect of nanorod surfactants on the shape and internal morphology of block copolymer particles. *J. Polym. Sci. Part A Polym. Chem.* **2015**, *53*, 188–192. [CrossRef]
17. Ku, K.H.; Shin, J.M.; Kim, M.P.; Lee, C.H.; Seo, M.K.; Yi, G.R.; Jang, S.G.; Kim, B.J. Size-controlled nanoparticle-guided assembly of block copolymers for convex lens-shaped particles. *J. Am. Chem. Soc.* **2014**, *136*, 9982–9989. [CrossRef]
18. Shin, J.M.; Kim, M.P.; Yang, H.; Ku, K.H.; Jang, S.G.; Youm, K.H.; Yi, G.R.; Kim, B.J. Monodipserse nanostructured spheres of block copolymers and nanoparticles via cross-flow membrane emulsification. *Chem. Mater.* **2015**, *27*, 6314–6321. [CrossRef]
19. Ku, K.H.; Kim, M.P.; Paek, K.; Shin, J.M.; Chung, S.; Jang, S.G.; Chae, W.S.; Yi, G.R.; Kim, B.J. Multicolor emission of hybrid block copolymer-quantum dot microspheres by controlled spatial isolation of quantum dots. *Small* **2013**, *9*, 2667–2672. [CrossRef]
20. Yan, N.; Liu, X.; Zhu, J.; Zhu, Y.; Jiang, W. Well-ordered inorganic nanoparticle arrays directed by block copolymer nanosheets. *ACS Nano* **2019**, *13*, 6638–6646. [CrossRef] [PubMed]
21. Lee, J.; Ku, K.H.; Kim, J.; Lee, Y.J.; Jang, S.G.; Kim, B.J. Light-responsive, shape-switchable block copolymer particles. *J. Am. Chem. Soc.* **2019**, *141*, 15348–15355. [CrossRef]
22. Lee, J.; Ku, K.H.; Kim, M.; Shin, J.M.; Han, J.; Park, C.H.; Yi, G.-R.; Jang, S.G.; Kim, B.J. Stimuli-responsive, shape-transforming nanostructured particles. *Adv. Mater.* **2017**, *29*, 1700608. [CrossRef] [PubMed]
23. Hou, Z.; Ren, M.; Wang, K.; Yang, Y.; Xu, J.; Zhu, J. Deformable block copolymer microparticles by controllable localization of pH-responsive nanoparticles. *Macromolecules* **2020**, *53*, 473–481. [CrossRef]
24. Xu, J.; Yang, Y.; Wang, K.; Li, J.; Zhou, H.; Xie, X.; Zhu, J. Additives induced structural transformation of ABC triblock copolymer particles. *Langmuir* **2015**, *31*, 10975–10982. [CrossRef] [PubMed]
25. Xu, J.; Wang, K.; Li, J.; Zhou, H.; Xie, X.; Zhu, J. ABC triblock copolymer particles with tunable shape and internal structure through 3D confined assembly. *Macromolecules* **2015**, *48*, 2628–2636. [CrossRef]
26. Qiang, X.; Franzka, S.; Dai, X.; Gröschel, A.H. Multicompartment microparticles of SBT triblock terpolymers through 3D confinement assembly. *Macromolecules* **2020**, *53*, 4224–4233. [CrossRef]
27. Dai, X.; Qiang, X.; Hils, C.; Schmalz, H.; Gröschel, A.H. Frustrated microparticle morphologies of a semicrystalline triblock terpolymer in 3D soft confinement. *ACS Nano* **2021**, *15*, 1111–1120. [CrossRef]
28. Abetz, V.; Boschetti-de-Fierro, A. *Block Copolymers in the Condensed State*; Elsevier, B.V.: Amsterdam, The Netherlands, 2012; Volume 7, ISBN 9780080878621.
29. Qiang, X.; Chakroun, R.; Janoszka, N.; Gröschel, A.H. Self-assembly of multiblock copolymers. *Isr. J. Chem.* **2019**, *59*, 945–958. [CrossRef]
30. Wong, C.K.; Qiang, X.; Müller, A.H.E.; Gröschel, A.H. Self-assembly of block copolymers into internally ordered microparticles. *Prog. Polym. Sci.* **2020**, *102*, 101211. [CrossRef]
31. Steinhaus, A.; Chakroun, R.; Müllner, M.; Nghiem, T.L.; Hildebrandt, M.; Gröschel, A.H. Confinement assembly of ABC triblock terpolymers for the high-yield synthesis of Janus nanorings. *ACS Nano* **2019**, *13*, 6269–6278. [CrossRef] [PubMed]
32. Qiang, X.; Steinhaus, A.; Chen, C.; Chakroun, R.; Gröschel, A.H. Template-free synthesis and selective filling of Janus nanocups. *Angew. Chem. Int. Ed.* **2019**, *58*, 7122–7126. [CrossRef]
33. Steinhaus, A.; Srivastva, D.; Qiang, X.; Franzka, S.; Nikoubashman, A.; Gröschel, A.H. Controlling Janus nanodisc topology through ABC triblock terpolymer/homopolymer blending in 3D confinement. *Macromolecules* **2021**, *54*, 1224–1233. [CrossRef]
34. Deng, R.; Liu, S.; Li, J.; Liao, Y.; Tao, J.; Zhu, J. Mesoporous block copolymer nanoparticles with tailored structures by hydrogen-bonding-assisted self-assembly. *Adv. Mater.* **2012**, *24*, 1889–1893. [CrossRef]
35. Quintieri, G.; Saccone, M.; Spengler, M.; Giese, M.; Gröschel, A.H. Supramolecular modification of ABC triblock terpolymers in confinement assembly. *Nanomaterials* **2018**, *8*, 1029. [CrossRef]
36. Zheng, X.; Ren, M.; Wang, H.; Wang, H.; Geng, Z.; Xu, J.; Deng, R.; Chen, S.; Binder, W.H.; Zhu, J. Halogen-bond mediated 3D confined assembly of AB diblock copolymer and homopolymer blends. *Small* **2021**, *17*, 2007570. [CrossRef]
37. Haider, T.P.; Völker, C.; Kramm, J.; Landfester, K.; Wurm, F.R. Plastics of the future? The impact of biodegradable polymers on the environment and on society. *Angew. Chem. Int. Ed.* **2019**, *58*, 50–62. [CrossRef]

38. Ghosh, S.; Wölper, C.; Tjaberings, A.; Gröschel, A.H.; Schulz, S. Syntheses, structures and catalytic activity of tetranuclear Mg complexes in the ROP of cyclic esters under industrially relevant conditions. *Dalt. Trans.* **2020**, *49*, 375–387. [CrossRef]
39. Ghosh, S.; Huse, K.; Wölper, C.; Tjaberings, A.; Gröschel, A.H.; Schulz, S. Fluorinated β-ketoiminate zinc complexes: Synthesis, structure and catalytic activity in ring opening polymerization of Lactide. *Z. Für Anorg. Allg. Chem.* **2021**, *647*, 1744–1750. [CrossRef]
40. Ghosh, S.; Schäfer, P.M.; Dittrich, D.; Scheiper, C.; Steiniger, P.; Fink, G.; Ksiazkiewicz, A.N.; Tjaberings, A.; Wölper, C.; Gröschel, A.H.; et al. Heterolepic β-ketoiminate zinc phenoxide complexes as efficient catalysts for the ring opening polymerization of lactide. *ChemistryOpen* **2019**, *8*, 951–960. [CrossRef]
41. Middleton, J.C.; Tipton, A.J. Synthetic biodegradable polymers as orthopedic devices. *Biomaterials* **2000**, *21*, 2335–2346. [CrossRef]
42. Ahmed, F.; Discher, D.E. Self-porating polymersomes of PEG-PLA and PEG-PCL: Hydrolysis-triggered controlled release vesicles. *J. Control. Release* **2004**, *96*, 37–53. [CrossRef]
43. Li, H.; Mao, X.; Wang, H.; Geng, Z.; Xiong, B.; Zhang, L.; Liu, S.; Xu, J.; Zhu, J. Kinetically dependent self-assembly of chiral block copolymers under 3D confinement. *Macromolecules* **2020**, *53*, 4214–4223. [CrossRef]
44. Li, H.; Xiong, B.; Geng, Z.; Wang, H.; Gao, Y.; Gu, P.; Xie, H.; Xu, J.; Zhu, J. Temperature- and solvent-mediated confined assembly of semicrystalline chiral block copolymers in evaporative emulsion droplets. *Macromolecules* **2021**. [CrossRef]
45. Geng, Z.; Wang, H.; Jin, S.-M.; Yan, X.; Ren, M.; Xiong, B.; Wang, K.; Deng, R.; Chen, S.; Lee, E.; et al. Hierarchical microphase behaviors of chiral block copolymers under kinetic and thermodynamic control. *CCS Chem.* **2021**, *3*, 2561–2569. [CrossRef]
46. Löbling, T.I.; Hiekkataipale, P.; Hanisch, A.; Bennet, F.; Schmalz, H.; Ikkala, O.; Gröschel, A.H.; Müller, A.H.E. Bulk morphologies of polystyrene-*block*-polybutadiene-block-poly(*tert*-butyl methacrylate) triblock terpolymers. *Polymer* **2015**, *72*, 479–489. [CrossRef]
47. Bielawski, C.W.; Benitez, D.; Morita, T.; Grubbs, R.H. Synthesis of end-functionalized poly(norbornene)s via ring-opening methathesis polymerization. *Macromolecules* **2001**, *34*, 8610–8618. [CrossRef]
48. Touris, A.; Chanpuriya, S.; Hillmyer, M.A.; Bates, F.S. Synthetic strategies for the generation of ABCA' type asymmetric tetrablock terpolymers. *Polym. Chem.* **2014**, *5*, 5551–5559. [CrossRef]
49. He, Y.; Eloi, J.C.; Harniman, R.L.; Richardson, R.M.; Whittell, G.R.; Mathers, R.T.; Dove, A.P.; O'Reilly, R.K.; Manners, I. Uniform biodegradable fiber-like micelles and block comicelles via "living" crystallization-driven self-assembly of poly(L-lactide) block copolymers: The importance of reducing unimer self-nucleation via hydrogen bond disruption. *J. Am. Chem. Soc.* **2019**, *141*, 19088–19098. [CrossRef]
50. Wong, C.K.; Heidelmann, M.; Dulle, M.; Qiang, X.; Förster, S.; Stenzel, M.H.; Gröschel, A.H. Vesicular polymer hexosomes exhibit topological defects. *J. Am. Chem. Soc.* **2020**, *142*, 10989–10995. [CrossRef]
51. Gröschel, T.I.; Wong, C.K.; Haataja, J.S.; Dias, M.A.; Gröschel, A.H. Direct observation of topological defects in striped block copolymer discs and polymersomes. *ACS Nano* **2020**, *14*, 4829–4838. [CrossRef] [PubMed]
52. Khadilkar, M.R.; Nikoubashman, A. Self-assembly of semiflexible polymers confined to thin spherical shells. *Soft Matter* **2018**, *14*, 6903–6911. [CrossRef] [PubMed]
53. Zhu, X.; Guan, Z.; Lin, J.; Cai, C. Strip-pattern-spheres self-assembled from polypeptide-based polymer mixtures: Structure and defect features. *Sci. Rep.* **2016**, *6*, 29796. [CrossRef]
54. Vacogne, C.D.; Wei, C.; Tauer, K.; Schlaad, H. Self-assembly of α-helical polypeptides into microscopic and enantiomorphic spirals. *J. Am. Chem. Soc.* **2018**, *140*, 11387–11394. [CrossRef]
55. Balsamo, V.; von Gyldenfeldt, F.; Stadler, R. "Superductile" semicrystalline ABC triblock copolymers with the polystyrene block (A) as the matrix. *Macromolecules* **1999**, *32*, 1226–1232. [CrossRef]
56. Balsamo, V.; Gil, G.; Urbina de Navarro, C.; Hamley, I.W.; von Gyldenfeldt, F.; Abetz, V.; Cañizales, E. Morphological behavior of thermally treated polystyrene-*b*-polybutadiene-*b*-poly(ε-caprolactone) ABC triblock copolymers. *Macromolecules* **2003**, *36*, 4515–4525. [CrossRef]

MDPI
St. Alban-Anlage 66
4052 Basel
Switzerland
Tel. +41 61 683 77 34
Fax +41 61 302 89 18
www.mdpi.com

Polymers Editorial Office
E-mail: polymers@mdpi.com
www.mdpi.com/journal/polymers

www.ingramcontent.com/pod-product-compliance
Lightning Source LLC
LaVergne TN
LVHW070155120526
838202LV00013BA/1147